Noise
Mapping
in the EU

Models and Procedures

Noise Mapping in the EU

Models and Procedures

Edited by **Gaetano Licitra**

CRC Press
Taylor & Francis Group
Boca Raton London New York

CRC Press is an imprint of the
Taylor & Francis Group, an **informa** business

A SPON PRESS BOOK

CRC Press
Taylor & Francis Group
6000 Broken Sound Parkway NW, Suite 300
Boca Raton, FL 33487-2742

First issued in paperback 2019

© 2013 by Gaetano Licitra
CRC Press is an imprint of Taylor & Francis Group, an Informa business

No claim to original U.S. Government works

ISBN-13: 978-0-415-58509-5 (hbk)
ISBN-13: 978-0-367-86523-8 (pbk)

Library of Congress Cataloging-in-Publication Data

Noise mapping in the EU : models and procedures / editor, Gaetano Licitra.
    p. cm.
  Includes bibliographical references and index.
  ISBN 978-0-415-58509-5 (hardback)
    1. Noise pollution--European Union countries. 2. Noise control--European Union countries. I. Licitra, Gaetano.

TD893.5.E85N66 2012
363.74094′022--dc23   2012017946

Visit the Taylor & Francis Web site at
http://www.taylorandfrancis.com

and the CRC Press Web site at
http://www.crcpress.com

To my father

Contents

Preface

Nowadays, noise exposure monitoring and its reduction are among the main concerns for citizens, politicians, administrations, and technical-scientific bodies. Directive 2002/49/EC "relating to the assessment and management of environmental noise" (Environmental Noise Directive, END) is the last effort to harmonise European member states' policies and technical approaches concerning noise exposure reduction issues. Noise mapping of relevant environmental noise sources in agglomerations, following the directive, has presented the opportunity for assessing noise levels in main European agglomerations and testing different national or recommended calculation methods.

Noise mapping is a very complex and challenging issue: noise sources are widely diffuse, especially in urban areas; their characterisation is not simple and the quantification of a citizen's exposure to noise is a very difficult task. Moreover, the process requires different skills and experts: from geospatial analysis to uncertainty evaluation and psychoacoustic approaches to maximise cost benefits and action plan adherence.

The book is for students, researchers, acoustics consultants, environmental agencies, public administrations, and all stakeholders involved in protecting citizens from urban noise. This book tries to cover all the main issues about noise mapping in the framework of the END, collecting contributions from many experts from research bodies, consultancies, and environmental protection agencies.

After a brief introduction of some fundamental concepts in acoustics and a presentation of legal framework for noise mapping in Europe, numerical models are presented for roads, railways, airports, harbours, and industrial sites. Control and uncertainty in input data and output results, technical recommendations from working groups, and the *Good Practice Guide (GPG)* tool are discussed with a practical approach and worked examples. According to the aim of the book, a deep insight in geographic information system (GIS) techniques for noise management and the evaluation and management of noise exposure are covered. The last part of the book reviews noise mapping experiences in Europe, communication to the public, and future perspectives for mapping the effects of noise.

The authors intend to provide the readers with a full path in noise mapping issues: from the legislation to the critical topics in implementation of a map in a European agglomeration. I am convinced that only a complete picture of the problem can help all stakeholders in the noise mapping process.

I would like to acknowledge Tony Moore and Poole Siobhan for their support, encouragement, and cooperation. I thank all the contributors for their precious and invaluable work that made this book possible, in particular Diego Palazzuoli for his collaboration

Gaetano Licita
ARPAT
Firenze, Italy

Contributors

Elena Ascari
IDASC-CNR Institute of Acoustic
  "O.M. Corbino"
Rome, Italy
Elena.Ascari@gmail.com

Giovanni Brambilla
IDASC-CNR Institute of Acoustic
  "O.M. Corbino"
Rome, Italy
Giovanni.Brambilla@idasc.cnr.it

Rudolf Bütikofer
Laboratory of Acoustics/Noise
  Control
Empa, Swiss Federal Laboratories for
  Materials Science and Technology
Duebendorf, Switzerland
Rudolf.Buetikofer@empa.ch

Jose Luis Cueto
University of Cadiz
Cadiz, Spain
JoseLuis.Cueto@uca.es

Paul de Vos
DHV Amersfoort
Amersfoort, The Netherlands
Paul.Devos@dhv.com

Guillaume Dutilleux
ERA Acoustique Ifsttar
PCI Acoustique et Vibrations
Strasbourg, France
Guillaume.Dutilleux@
  developpement-durable.gouv.fr

Ricardo Hernandez
University of Cadiz
Cadiz, Spain
Ricardo.Hernandez@uca.es

Stylianos Kephalopoulos
European Commission, Joint
  Research Centre
Institute for Heath and Consumer
  Protection
Ispra, Italy
Stylianos.Kephalopoulos@jrc.it

Gaetano Licitra
ARPAT
Firenze, Italy
G.Licitra@arpat.toscana.it

Luigi Maffei
Seconda Università di Napoli
Avise, Italy
Luigi.Maffei@unina2.it

Douglas Manvell
Brüel & Kjær Sound & Vibration
  Measurement A/S
Nærum, Denmark
Douglas.Manvell@bksv.com

Paul McDonald
Sonitus Systems Ltd.
Dublin, Ireland
PMcdonal@tcd.ie

Gianluca Memoli
Memolix Environmental Consultants
Pisa, Italy
G.Memoli@memolix.eu
Currently at:
National Physical Laboratory
Teddington, United Kingdom

Diego Palazzuoli
ARPAT
Florence, Italy
D.Palazzuoli@libero.it

Marco Paviotti
(Private Consultant)
Udine, Italy
Marco@paviotti.it

Wolfgang Probst
DataKustik GmbH
Greifenberg, Germany
Wolfgang.Probst@datakustik.com

J. Rob Witte
Industry, Traffic, and Environment
dGm^R
Den Haag, The Netherlands
Wi@dgmr.nl

Chapter 1

Fundamentals

D. Palazzuoli and G. Licitra

CONTENTS

The whispering of wind in a wood and the roar of a traffic jam, a mountain waterfall and a road yard, noise and music: the same physical phenomenon but such different effects on human perception and well-being. This chapter introduces some fundamental concepts relating to the physics of noise, propagation, attenuation, and the main descriptors.

THE SOUND

The sound phenomenon can be described as a perturbation propagating in an elastic medium, causing a variation in pressure and particles displacement from their equilibrium positions. The term "perturbation" is used here because, if energy and information associated with the sound travel out from the source of the perturbation in the form of waves, single particles in the medium remain near their equilibrium positions.

In fluids, particles vibrate in the same direction of wave propagation: in this sense, perturbations like sound are defined longitudinal (or compressional) waves.

In a perfect gas the speed of sound, c [m/s], is:

$$c = \sqrt{\frac{kp_0}{\rho_0}}$$

where k is the adiabatic index (or heat capacity ratio) c_p/c_v, is the ratio between the specific heats at constant pressure (c_p) and constant volume (c_v), p_0 and ρ_0 are the equilibrium pressure [Pa] and density [kg m^{-3}].

For the speed of sound in air a useful, approximate formula gives

$$c = 331.6 + 0.6 \cdot T_C \text{ m/s}$$

with T_c in degrees Celsius.

In absence of attenuation, the speed of sound is the only parameter that enters the three-dimensional wave equation, which is given by

$$\frac{\partial^2 p}{\partial x^2} + \frac{\partial^2 p}{\partial y^2} + \frac{\partial^2 p}{\partial z^2} = \frac{1}{c^2} \frac{\partial^2 p}{\partial t^2} \tag{1.1}$$

where $p(x,y,z,t)$ is the instantaneous variation in pressure due to the acoustic phenomenon at the point expressed by the coordinates (x,y,z) at the time t.

The general solution[1] of the wave equation (Equation 1.1) in one dimension in terms of sound pressure, p, is

$$p(x,t) = f_1(ct - x) + f_2(ct + x)$$

in which f_1 and f_2 are arbitrary function (derivable in second order) representing a wave travelling in the positive x direction and in the negative one respectively with a speed c.

In the case of a source vibrating sinusoidally it can be showed p varies both in time, t, and space, x, in a sinusoidal manner:

$$p(x,t) = p_1 \sin(\omega t - kx + \phi_1) + p_2 \sin(\omega t + kx + \phi_2) \tag{1.2}$$

where k is the acoustic wavenumber ($k = \omega/c$), p_1 and p_2 are the amplitudes of the positive and negative direction travelling waves, ϕ_1 and ϕ_2 are phase angles, $\omega = 2\pi f$ is the angular frequency, and f is the frequency.

The wavelength λ [m] (the distance travelled by the wave in a complete oscillation) is related to the other wave parameters by the relation

$$\lambda = c \cdot T = \frac{c}{f}$$

T [s] the period for a complete oscillation and f [Hz] is the frequency $1/T$.

Having defined a fluid *particle* as the smaller element of volume that maintains the bulk properties of the fluid, Equation (1.2) remains valid also if pressure variation, p, is replaced with the particle displacement, ξ, or the particle velocity, u.

The (complex) ratio between the value of pressure and particle velocity defines the *specific acoustic impedance* Z_s [Pa s m^{-1}]. For one-dimensional propagation it can be shown that for any plane wave

$$\frac{p}{u} = \rho c$$

if the travelling direction is positive, or $-\rho c$ in the negative case.

The ratio is the *characteristic impedance*, Z_c, of the fluid. For air it is equal to 407 Pa s m^{-1} at 22°C and 10^5 Pa (density and speed of sound in a fluid are a function of temperature and pressure).

ACOUSTIC ENERGY, LEVELS, AND FREQUENCY SPECTRUM

Acoustic energy: Sound intensity and energy density

If we consider the energy flowing during sound propagation, it can be evaluated with regard to the *sound intensity*, I: the average rate of the energy that flows through an imaginary surface of a unit area in a direction perpendicular to the surface. The instantaneous acoustic intensity is:

$$I = p \cdot u \cdot cos(\theta)$$

where θ is the angle between the perpendicular to the unit surface and the propagation direction of the sound wave.

For a plane wave travelling in the positive x direction:

$$I = \frac{p^2}{\rho_0 c}$$

The effective value is simply the root mean square (rms) of the instantaneous one averaged over a time interval:

$$I = \frac{p_{rms}^2}{\rho_0 c} \tag{1.3}$$

For a sinusoidal signal with amplitude A and period T, the rms value is

$$rms = \sqrt{\frac{1}{T} \int_a^b (A \cdot \cos \omega t)^2 \, dt} = \frac{A}{\sqrt{2}}$$

In three dimensions I is represented by a three-dimensional vector, $\mathbf{I}$.[2]

It can be useful to define the *density of acoustic energy* (D) as the energy contained in a unit volume centred in a specified point in a space, by using Equation (1.2):

$$D = \frac{p_{rms}^2}{\rho_0 c^2}$$

Sound description

Audible sounds in air cover a very wide range of both pressure variations, from about 20 µPa to 10^4 Pa, and acoustic intensity so it is necessary to express these quantities in logarithmic rather than linear scale. Acoustic levels are generally expressed as 10 times the logarithm to the base 10 (decibel) of the ratio relative to a reference level:

$$L_p = 10 \log_{10} \left(\frac{p_{rms}^2}{p_{ref}^2} \right) \text{dB}$$

$$L_I = 10 \log_{10} \left(\frac{I}{I_{ref}} \right) \text{dB}$$

$$L_W = 10 \log_{10} \left(\frac{W}{W_{ref}} \right) \text{dB}$$

where p_{ref} = 20 µPa, I_{ref} = 10^{-12} W/m^2, and W_{ref} = 10^{-12} W.

If two or more sound sources are uncorrelated, the overall pressure (or power) level is determined by the *energetic* sum (the average squared sound pressure):

$$L_p = 10\log_{10}\left(\sum_{i=1}^{n} 10^{Lpi/10}\right) dB$$

Frequency and sound spectrum

Acoustics sources in an environment generally do not emit sounds characterised by only one frequency (pure tones), but they can be considered as a composition of different pure tones with specific frequencies, with a discrete or continuous distribution (spectrum). Fourier theorem[3] allows it to decompose any signal (under weak hypotheses) in a series of sine waves (harmonics) with suitable amplitudes and phases. The amplitudes associated to each frequency represent the *sound spectrum*.

Frequency analysis can be carried out by using different methods to define the energy content in bands at determined frequency intervals. Generally frequency analysis is carried out using filters with constant bandwidth or constant percentage filters. Each band is characterised by lower (f_1) and upper (f_2) cutoff frequencies, a band centre frequency (f_c), and a bandwidth $\Delta f = f_2 - f_1$. In constant percentage filters the upper and lower frequencies are in a geometric progression:

$$f_2 = 2^n \cdot f_1$$

with

$$f_c = \sqrt{f_1 \cdot f_2}$$

For $n = 1$ we obtain the *one-octave bands*, and the relevant parameters are defined as

$$f_2 = 2 \cdot f_1$$

$$\Delta f = f_c(\sqrt{2} - 1/\sqrt{2}) = f_c / \sqrt{2}$$

For one-third-octave bands $(n = 1/3)$:

$$f_1 = \frac{f_c}{\sqrt[6]{2}} \quad f_2 = \sqrt[6]{2} f_c$$

$$\Delta f = f_c\left(2^{1/6} - 2^{-1/6}\right)$$

Preferred frequencies for the analysis in bands are defined by the technical norm ISO 266-1997 Acoustics.[4]

LOUDNESS, FREQUENCY WEIGHTING, AND THE EQUIVALENT PRESSURE LEVEL

Loudness and frequency weighting

The concept of loudness is related to the perceived intensity of sound. The human ear, for instance, has a different sensibility to sounds depending on their frequencies. It is more sensitive in the range of 1000–4000 Hz (a range of frequencies corresponding to speech), with a poor response at lower and higher frequencies. Even before being interpreted by the brain, two sounds with the same energetic level (in decibels [dB]), but different spectral content, are perceived differently by the human ear.

Loudness is therefore a "subjective" property, unlike pressure levels (in decibels) or frequency spectra, which are objectively measurable physical quantities. For a given individual, sound loudness can vary not only with pressure level of the sound itself but also with frequency and sound. It is, however, possible to define the correction due to a "standard ear," that can be applied to more objective measurements (e.g., those taken by a microphone) to simulate what a human would "perceive."

By a procedure of comparison of a reference 1000 Hz pure tone sound with one varying both in frequency and pressure level, standardised equal-loudness contours are designed. A pure tone level in a defined contour line has a loudness equal to the pressure level of the reference 1000 Hz: if the level of a 1000 Hz is 50 dB then all tones on the same line have a loudness of 50 phons.

The 40 phons iso-loudness contour is chosen as the reference of A-weighting sound pressure: when a sound is weighted by using a transfer function inverse of that contour, an A-weighted pressure level is obtained.

Other commonly used filters are B-, C-, and D-weighting designed for specific applications. For instance, a D-weighting curve was used for the evaluation of disturbance from aircraft noise.

After the pioneering research done by Fletcher and Munson (1933, Bell Laboratories) the ISO 226:2003 adopted the curves from Robinson and Dadson.

Equivalent sound pressure level

In evaluating exposure to noise sources, the most commonly used single index quantity is the *equivalent sound pressure level* (L_{eq}). It is useful to rating sounds with levels varying in time. L_{eq} is the sound pressure level averaged over a suitable period, T:

$$L_{eq} = 10\log\left(\frac{1}{T}\int_0^T 10^{L(t)/10}dt\right) = 10\log\left(\frac{1}{T}\int_0^T p_{eff}^2(t)\Big/p_{ref}^2\; dt\right)dB$$

If pressure levels are weighted by the A-weighting curve it is obtained the *equivalent continuous sound pressure level* (L_{Aeq}):

$$L_{Aeq} = 10\log\left(\frac{1}{T}\int_0^T 10^{LA(t)/10}dt\right) = 10\log\left(\frac{1}{T}\int_0^T p_{Aeff}^2(t)\Big/p_{ref}^2\; dt\right)$$

where $LA(t)$ is the instantaneous sound level A-weighted and $p_{Aeff}(t)$ is the sound pressure measured A-weighting frequency filter.

L_{Aeq} is one of the most widely used descriptors in evaluating environmental noise from roads, railways, and industry. In community noise evaluation, the evaluation time, T, is a period representing day, night, or evening.

In the framework of noise management, the Directive 2002/49/EC,[5] prescribed the indicator L_{den}, which represents the day–evening–night level in dB:

$$L_{den} = 10\log\frac{1}{24}\left(12\cdot 10^{\frac{L_{day}}{10}} + 4\cdot 10^{\frac{L_{evening}+5}{10}} + 8\cdot 10^{\frac{L_{night}+10}{10}}\right)$$

where

- L_{day} is the A-weighted long-term average sound level as defined in ISO 1996-2:1987, determined over all the day period of a year
- $L_{evening}$ is the A-weighted long-term average sound level as defined in ISO 1996-2:1987, determined over all the evening period of a year
- L_{night} is the A-weighted long-term average sound level as defined in ISO 1996-2:1987, determined over all the night period of a year"

(Appendix I, Directive 2002/49/EC)

The day is 12 hours, the evening 4 hours, and the night 8 hours. "The Member States may shorten the evening period by one or two hours and lengthen the day and/or the night period accordingly" and "the start of the day (and consequently the start of the evening and the start of the night) shall be chosen by the Member State (that choice shall be the same for noise from all sources); the default values are 07.00 to 19.00, 19.00 to 23.00 and 23.00 to 07.00 local time" (Appendix I, Directive 2002/49/EC).

SOURCES AND PROPAGATION

Directivity

Most sources are not isotropic, that is, their pattern of emission is not constant in all the direction. Directivity is usually a function of frequency; many sources are omnidirectional at low frequency when their dimensions are lower than wavelength of the emitted sound but become directive with increasing frequency.

Directivity factor $D(\theta,\phi)$ is defined as the ratio of mean square sound pressure $p^2_{rms}(\theta,\phi)$ (at angles θ,ϕ and distance r from the source, and the value of p^2_{rms} at the same distance r due to an omnidirectional source of the same emitting power:

$$D(\theta,\phi) = \frac{p^2_{rms}(\theta,\phi)}{p^2_{rms}}$$

The *directivity index* $DI(\theta,\phi)$ is then defined as

$$DI(\theta,\phi) = 10\log(D(\theta,\phi)) = 10\log\left(\frac{p^2_{rms}(\theta,\phi)}{p^2_{rms}}\right)$$

Absorption, reflection, and refraction

When an acoustic wave hits a surface, its energy is partly absorbed (E_a) and partly reflected (E_r) and transmitted (E_t) (see Figure 1.1):

$$E_i = E_a + E_t + E_r$$

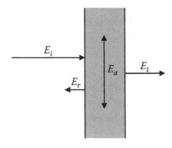

Figure 1.1 Schematic picture of absorption, reflection, and transmission.

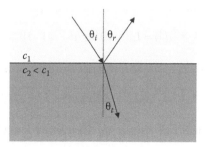

Figure 1.2 Reflection and refraction at different media boundary.

The adsorption coefficient, α, is defined as

$$\alpha = \frac{E_a + E_t}{E_i}$$

Similarly, the transmission coefficient, τ, is

$$\tau = \frac{E_t}{E_i}$$

The phenomena of reflection and refraction of a sound wave are analogous to optical rays. If we consider two different ones, media 1 and 2, with speed sound c_1 and c_2, respectively (Figure 1.2), the incident angle (θ_i) is equal to the reflection angle (θ_r), and for the transmission angle (θ_t) Snell's law holds:

$$\frac{c_1}{c_2} = \frac{\sin(\theta_i)}{\sin(\theta_t)}$$

If c_2 is smaller than c_1, the wave is refracted toward the normal; θ_t will be smaller than θ_i.

Outdoor sound propagation

A sound propagating without any obstacle or absorption phenomena is defined as the free field propagation condition.

For a point source of power, W, sound intensity, I, at distance, d, in a free field condition is

$$I = \frac{W}{4\pi d^2}$$

and if $\rho_0 c = 407$ Pa s m^{-1}:

$$L_p = L_W - 10\log(4 \cdot \pi \cdot d^2) = L_w - 20\log(d) - 11 \text{ dB}$$

In the case of a directive source, the pressure level in a point (r, θ, ϕ) is

$$L_p = L_W - 10\log(4 \cdot \pi \cdot d^2) = L_w - 20\log(d) - 11 + DI(\theta, \phi)$$

Similarly, for an ideal line source on an infinite length pressure level can be calculated as

$$L_p = L_w - 10\log(d) - 8 + DI(\theta, \phi)$$

where L_w is the sound power level per unit of length.

Sound attenuation from sources to the receivers depends on the following:

- Air absorption
- Ground or vegetation absorption
- Meteorological conditions
- Obstacles along the propagation path

A detailed description of attenuation in atmosphere and a calculation procedure is shown in the technical norms ISO 9613 parts 1[6] and 2,[7] but a brief description of the different terms will be reported here. First, sound energy is dissipated as heat during its propagation in air, but the effect becomes significant only at high frequencies and at a long distance from the source.

For a plane wave the attenuation of sound in air can be evaluated by

$$Att_{air} = \alpha \cdot r \text{ dB}$$

where α depends on frequency, humidity, pressure, and temperature; and r is the distance source-receiver.

For low frequencies, air attenuation is smaller than 1 dB/km, whereas for frequencies higher than 12 to 13 kHz the attenuation is very high.

The attenuation due to foliage is generally small. The excess attenuation, when a dense forest is present between the source and the receiver, may be estimated using[8]

$$A_f = 0.01 \cdot r_f \cdot f^{\frac{1}{3}}$$

where f is the sound frequency (Hz) and r_f is the length (in metres) of the path through the forest. Values of foliage attenuation are generally reported in decibels when sound travel distances r_f between 10 and 20 m; and reported in dB/m when r_f is between 20 and 200 m.

The distance of sound in foliage depends on the relative height of the source and receiver, and on the curvature sound path due to meteorological conditions.

The interference between the direct sound from the source and the reflected sound from the ground modify the overall level to the receiver. The ground effect depends on the geometry of source–receiver position and the properties of the ground surface.[9] ISO 9613-2:1996 presents a calculation procedure that takes into account the ground effect on sound attenuation. It considers the worst case of sound propagation downwind from the source to the receiver. Considering the height of the source and the receiver and their distance, the norm considers three zones: near the source, middle, and near the receiver. The acoustic characteristic of each zone is defined by the parameter, G, varying from 0 (hard ground) to 1 (soft ground). The ground excess attenuation (A_g) is then the sum of the attenuation due to the three zones (source, middle, receiver):

$$A_g = A_s + A_m + A_r$$

Meteorological conditions modify sound propagation mainly by the effects of wind gradient and vertical temperature gradient. With temperature inversion phenomena (positive temperature gradient near the ground surface) sound rays are diffracted downward resulting in an increasing sound level. On the other hand a temperature lapse (negative temperature gradient) reduces the sound level on the ground.[9]

Noise barriers or obstacles, large in comparison with the wavelength of the incident sound, reduce noise level at the receiver (R) in their shadow zone (Figure 1.3) where acoustic rays reach the receiver only for diffraction phenomena.

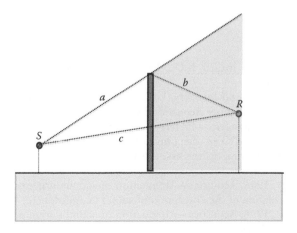

Figure 1.3 Sound barrier.

The sound level in the shadow zone is determined by the fraction of acoustic energy passing through the screen and diffracted by its edge. In order to reduce the fraction of the energy transmitted by the barrier, the surface density has to be >20 kg/m^2.

The attenuation [dB], or insertion loss (IL), is used to define the performance of a barrier:

$$Att = IL = L_{p,withoutbarrier} - L_{p,withbarrier}$$

Barrier maximum insertion loss is about 20 dB. For a long barrier, in order to ignore the contribution of the diffraction from the lateral edges, the attenuation can be evaluated by using the empirical Maekawa relation[10]:

$$Att = 10log(3 + 20N) \text{ dB}$$

where N represents the Fresnel number

$$N = \frac{2\delta}{\lambda}$$

and δ is the difference between the diffracted path and the direct one, $\delta = a + b - c$.

In order to evaluate the noise level reduction from road, the semiempirical formula of Kurze and Anderson[11] can also be used:

$$Att = 20log_{10}\left[\frac{\sqrt{2\pi N}}{\tanh\sqrt{2\pi N}}\right] \text{ dB } + 5 \text{ dB}$$

REFERENCES

1. Morse P.M., and Ingard K.U., 1968, *Theoretical Acoustics* (New York: McGraw-Hill).
2. Fahy F.J., 1995, *Sound Intensity* (London: E&FN Spon, Chapman & Hall).
3. William E., 1999, *Fourier Acoustics* (London: Academic Press).
4. ISO 266: 1997, Acoustics—Preferred frequencies.
5. Directive 2002/49/EC of the European Parliament and of the Council of 25 June 2002 relating to the assessment and management of environmental noise.
6. ISO 9613-1, 1996, Acoustics—Attenuation of sound during propagation outdoors. Calculation of the absorption of sound in atmosphere.
7. ISO 9613-2, 1996, Acoustics—Attenuation of sound during propagation outdoors. General methods of calculation.

8. Hoover R.M., 1961, Tree zones as barriers for the control of noise due to aircraft operations. Bolt, Beranek, and Newman, Report 844.
9. Attenborough K., Li K.M., and Horoshenkov K., 2007, *Predicting Outdoor Sound* (London: Taylor & Francis).
10. Maekawa Z., 1968, Noise reduction by screens. *Applied Acoustics* 1, 157–173.
11. Kurze U.J., and Anderson G.S., 1971, Sound attenuation by barriers. *Applied Acoustics* 4, 35.

Part I

Noise evaluation and mapping

Chapter 2

Legal basis on noise mapping in the European Union and the Directive 69/2002/EC

P. de Vos and G. Licitra

CONTENTS

INTRODUCTION

Environmental policy in general and noise policy in particular are shared matters and responsibilities between the European Council and its member states. Local actions and regulations have to be supported by a global approach to the problem for an effective reduction of noise exposure.

The European normative interventions started in the early 1990s and often regulatory requirements become part of already existing national legislation.

After an overview of the main community measures relating to the mitigation of environmental noise, Directive 2002/40/EC (also known as Environmental Noise Directive [END]) is presented with reference to the main implementation issues.

NOISE POLICY IN THE EUROPEAN UNION

If we could fix a date for noise policy action in Europe, 1996 would be a reference year. On 1 February 1993 the European Council and the representatives of the member states approved the "European Community programme of Policy and action in relation to the environment and sustainable development Towards Sustainability" better known as the "Fifth Environmental Action Programme." The programme recognises that the urban environment has "difficulties in reconciling the need to meet the demands of modern commerce and transport with the desire to provide a good quality living environment" with "resulting congestion, pollution, noise, deterioration of streets, public places and architectural heritage and general loss of amenity."

After 3 years, on 10 January 1996, the "Progress Report on implementation of the European Community Programme of Policy and Action in relation to the environment and sustainable development Towards Sustainability" enlightened progresses made and more critical issues where more efforts were needed toward sustainability. The Progress Report underlined that data on noise exposure levels were not complete and homogeneous in the European Union (EU) with different calculus methods and noise descriptors. On the other hand, since the adoption of the Fifth Programme either the council or the commission had adopted or proposed various acts and norms to reduce noise emissions from "motor vehicles, two and three wheelers, aircraft and construction machinery."[1]

At the member-states level, the report revealed a diverse situation. Austria, Denmark, Finland, Greece, the Netherlands, the United Kingdom, and Germany have started noise reduction programmes, and Italy with the issuing of the decree D.P.C.M. 1 March 1991 "Limiti massimi di esposizione al rumore negli ambienti abitativi e nell'ambiente esterno"[2] had set limit values for noise exposure in the environment following a land zoning. Whereas the Netherlands with the "First and Second National Environmental Policy Plan" tried to reduce noise pollution to the level of 1985, Greece focussed on the development of a noise monitoring network and noise survey and the United Kingdom stated that "no new residential development, where necessary with noise mitigation measures, should be exposed to noise levels from transport or mixed transport and industrial sources above 66 dBLAeq at night."[1] Some first efforts in noise mapping had also been made, revealing: 25% of Finland's municipalities had completed a noise inventory, and Austria had completed the inventory of rail and road traffic noise exposure levels also setting limit values at 65 dB during the daytime and 55 dB during the nighttime by 2003 on federal Austrian roads and highways. It is worthwhile considering that Austria had enhanced its legislation on admittance of civil aircraft and legislation on admission of rail wagons requiring a reduction of 5 dB(A) from 1995 for goods wagons, Finland (1992) had provided a "mechanisms for funding abatement measures for those who fall within certain zones coupled with a reinforcement in control measures." During the same period many member states had launched information programmes on noise and its effects on health.

Conclusions from the Progress Report could be summarized as

- With the support measures at EU level MS [member states] would have to improve the information on noise exposure also producing "accurate inventories of noise situation,"[1] to "work on the comparability of measurement and calculation systems and the definition of common/equivalent rating units."[1]
- It had recognised the need for a "collective effort and a comprehensive approach at EU level to take action on noise abatement,"[1] introducing general noise quality standards, developing information to the public and an accurate land-use planning taking into account noise issues.
- The need for a more integrated systematic and simplified EU legislative approach in relation to noise emissions from point sources.

"Future Noise Policy—European Commission Green Paper" (Brussels, 04/11/1996 COM(96) 540) marks the first step toward the development of the Fifth Environmental Action Programme following the proposal on the review of it (COM(95)647). After a review of the overall noise situation in Europe and "Community and national action taken to date followed by the outline of a framework for action covering the improvement of information and its comparability and future options for the reduction of noise from different sources," the Green Paper stated the need for a new approach in noise abatement policy focussed on "shared responsibility involving target setting, monitoring of progress and measures to improve the accuracy and standardisation of data to help improve the coherency of different actions."[3] Moreover, the Green Paper sought a proposal for a new directive "providing for the harmonization of methods of assessment of noise exposure and the mutual exchange of information" including "recommendations on noise mapping and the provision of information." The commission also specified actions to undertake in order to reduce noise from major transport systems: road, rail, and air. In particular for

- Road traffic noise—The need for addressing tyre–road interaction and promoting low noise surface.
- Rail noise—"Investigate the feasibility of introducing legislation setting emission limit values, negotiated agreements with the rail industry on targets for emission values and economic instruments such as a variable track charge."
- Air transport noise—The definition of more stringent emission values and the implementation of economic instruments to encourage "the development and use of lower noise aircraft."

The Green Paper, furthermore, sought the definition of target values at the EU level and the need for member states to take action to reach the targets.

"Directive 2002/30/EC of the European Parliament and of the Council of 26 March 2002 on the establishment of rules and procedures with regard to the introduction of noise-related operating restrictions at Community airports,"[4] "Directive 2000/14/EC of the European Parliament and of the Council of 8 May 2000 on the approximation of the laws of the Member States relating to the noise emission in the environment by equipment for use outdoors," and Directive 2002/49/EC are implementations of the needs enlightened in the Green Paper.

Few years later, with the "Decision No. 2179/98/EC of the European Parliament and of the Council of 24 September 1998 on the review of the European Community programme of policy and action in relation to the environment and sustainable development towards sustainability" the European Community recognized that in relation to transport, the priority objectives were (Article 2) "to tighten provisions on emissions and noise from road and off-road." With regard to the environmental themes, Article 11 stated that "particular consideration will be given to the development of a noise abatement programme which could address comprehensively the provision of information to the public, common noise exposure indices, and targets for noise quality and noise emissions from products." Moreover, in Article 1 it had been requested that "at the end of the Programme, the Commission will submit to the European Parliament and the Council a global assessment of the implementation of the Programme, giving special attention to any revision and updating of objectives and priorities which may be required, and accompanied, where appropriate, by proposals for the priority objectives and measures that will be necessary beyond the year 2000."

In response to the request from the council and the European Parliament (Article 1 of Decision No. 2179/98/EC) the commission presented a global assessment on the implementation of the Fifth Programme (Europe's environment: what directions for the future? COM [1999] 543[5]) identifying some suggestions for the debate. With regard to the noise problem, the assessment showed that 32% of the population is exposed to a high level of noise (3.5 in an urban environment) and that in spite of progress made the environment status "continues to affect public health and the quality of life of citizens" recognising that noise exposure "disturbs sleep, affects children's cognitive development and may lead to psychosomatic illnesses." One of the main outcomes of the assessment remains the need for addressing the environment "together with the economic and social dimensions."

With Decision 1600/2002/EC the Sixth Community Environment Action Programme[6] was established, covering the period from 22 July 2002 to 21 July 2012. One of the "Objectives and priority areas for action on environment and health and quality of life" (Article 7) is the "reduction of the number of people regularly affected by long-term average levels of noise, in particular from traffic which, according to scientific studies, cause detrimental effects on human health and preparing the next step in the work with the noise directive." The programme recognizes research and scientific expertise as a key issue for enforcement of the environment protection strategy. Besides the environment and health issues (the definition and development of new indicators, the definition of priority areas for research and the updating of current health standards and limit values with particular attention to the potentially more vulnerable groups) the programme about noise defines the priorities of (Article 7)

- "Supplementing and further improving measures, including appropriate type-approval procedures, on noise emissions from services and products, in particular motor vehicles including measures to reduce noise from the interaction between tyre and road surface that do not compromise road safety, from railway vehicles, aircraft and stationary machinery."
- "Developing and implementing instruments to mitigate traffic noise where appropriate, for example by means of transport demand reduction, shifts to less noisy modes of transport, the promotion of technical measures and of sustainable transport planning."

The issuing of the END in 2002 with the aim of defining a common EU approach to noise from transport infrastructures in order to avoid, prevent, or reduce harmful effects (including annoyance) due to exposure to environmental noise supplemented the noise legislation already articulated. In Table 2.1 a general view of European noise legislation is shown.

DIRECTIVE 2002/49/EC

In the following paragraphs the main issues of Directive 2002/49/EC (END) will be described. Member states' tasks and obligations will also be discussed with reference to the implementation of the noise mapping in the European Union, picturing critical issues and opportunities.

Article 4: Implementation and responsibilities

Article 4 of END addresses the member states of the European Union:

1. Member States shall designate at the appropriate levels the competent authorities and bodies responsible for implementing this Directive, including the authorities responsible for:
   (a) making and, where relevant, approving noise maps and action plans for agglomerations, major roads, major railways and major airports;
   (b) collecting noise maps and action plans.

Table 2.1 European Political Instruments Addressing Noise Issues

Directive/Decision/Report	Year
Council Directive 70/157/EEC of 6 February 1970 on the approximation of the laws of the Member States relating to the permissible sound level and the exhaust system of motor vehicles	1970
Council Directive 80/51/EEC of 20 December 1979 on the limitation of noise emissions from subsonic aircraft	1979
Council Directive 89/629/EEC of 4 December 1989 on the limitation of noise emission from civil subsonic jet aeroplanes	1989
Council Directive 92/14/EEC of 2 March 1992 on the limitation of the operation of aeroplanes covered by Part II, Chapter 2, Volume 1 of Annex 16 to the Convention on International Civil Aviation, second edition (1988)	1992
Council Directive 96/48/EC of 23 July 1996 on the interoperability of the trans-European high-speed rail system	1996
Directive 97/24/EC of the European Parliament and of the Council of 17 June 1997 on certain components and characteristics of two or *three-wheel motor vehicles*	1997
Communication from the Commission to the Council, the European Parliament, the Economic and Social Committee and the Committee of the Regions: Air transport and the environment—Towards meeting the challenges of sustainable development (COM/99/0640 final) of 1 December 1999	1999
Directive 2001/43/EC of the European Parliament and of the Council of 27 June 2001 amending Council Directive 92/23/EEC relating to tyres for motor vehicles and their trailers and to their fitting	2001
Directive 2001/16/EC of the European Parliament and of the Council of 19 March 2001 on the interoperability of the trans-European conventional rail system	2001
Directive 2001/16/EC of the European Parliament and of the Council of 19 March 2001 on the interoperability of the trans-European conventional rail system	2001
Directive 2000/14/EC of the European Parliament and of the Council of 8 May 2000 on the approximation of the laws of the Member States relating to the noise emission in the environment by equipment for use outdoors	2001
Directive 2002/30/EC of the European Parliament and of the Council of 26 March 2002 on the establishment of rules and procedures with regard to the introduction of noise-related operating restrictions at Community airports	2002
Commission Decision of 30 May 2002 2002/735/EC concerning the technical specification for interoperability relating to the rolling stock subsystem of the trans-European high-speed rail system referred to in Article 6(1) of Directive 96/48/EC	2002
Technical specification for interoperability (TSI) relating to high-speed railway infrastructures—Commission Decision 2002/732/EC	2002
Commission Decision of 29 April 2004 specifying the basic parameters of the 'Noise', 'Freight Wagons' and 'Telematic applications for freight' Technical Specifications for Interoperability referred to in Directive 2001/16/EC	2004

Table 2.1 European Political Instruments Addressing Noise Issues (Continued)

Directive/Decision/Report	Year
Corrigendum to Directive 2004/50/EC of the European Parliament and of the Council of 29 April 2004 amending Council Directive 96/48/EC on the interoperability of the trans-European high-speed rail system and Directive 2001/16/EC of the European Parliament and of the Council on the interoperability of the trans-European conventional rail system (OJ L 164, 30.4.2004)	2004
Directive 2005/88/EC of the European Parliament and of the Council of 14 December 2005 amending Directive 2000/14/EC on the approximation of the laws of the Member States relating to the noise emission in the environment by equipment for use outdoors	2005
Report from the Commission to the Council and the European Parliament Noise Operation Restrictions at EU Airports (Report on the application of Directive 2002/30/EC) of 15 February 2008	2008
Regulation (EC) No 219/2009 of the European Parliament and of the Council of 11 March 2009 adapting a number of instruments subject to the procedure referred to in Article 251 of the Treaty to Council Decision 1999/468/EC with regard to the regulatory procedure with scrutiny Adaptation to the regulatory procedure with scrutiny—Part Two	2009

2. The Member States shall make the information referred to in paragraph 1 available to the Commission and to the public no later than 18 July 2005.

This article defines, in a concise way, the obligations of the 27 member states of the European Union, namely,

- To designate competent authorities
- To collect the information
- To make the information available to the commission and the public

In addition, Article 1(a) defines the four types of noise maps to be produced:

- Maps for agglomerations, which, according to Article 3, Definition k in the directive, are defined as areas "delimited by the member state"
- Maps for major roads, which are defined in the text of the directive
- Maps for major railways (similar)
- Maps for major airports (similar)

The legal requirements, including the requirement to produce strategic noise maps, following the European Directive on the Assessment and Management of Environmental Noise, refer to all 27 member states of the European Union. These are in alphabetical order: Austria, Belgium, Bulgaria,

Cyprus, Czech Republic, Denmark, Estonia, Finland, France, Germany, Greece, Hungary, Ireland, Italy, Latvia, Lithuania, Luxemburg, Malta, the Netherlands, Poland, Portugal, Romania, Slovakia, Slovenia, Spain, Sweden, and the United Kingdom. In addition to these obligatory countries, Norway has supplied data on a voluntary basis.

Among the member states, there was a big difference with respect to their previous experience in dealing with noise issues, particularly in having noise related data available in a structured way. Some countries may have had a noise legislation in place for decades, Italy for instance, and it is likely that road and rail administrations and city authorities would have a variety of data available that would allow them to operate according to the legal rules of such legislation. On the other hand, there would be countries that had hardly any experience in the field of environmental noise and would have at their disposal no relevant data at all.

In the framework of the directive, the member states were to report to the European Commission in various so-called data flows. The data supplied by the member states included the names and contact details of national contact persons, indicated as "competent authorities"; and the data on major roads, major railways and major airports as required by directive.

After having produced the noise maps, the member states were required to supply data on the number of people exposed to environmental noise from one of the four particular noise sources and divided in classes of noise exposure. For the average day–evening–night level (L_{den}), the number of exposed citizens should be supplied, rounded to the nearest hundred and expressed in hundreds, in exposure classes of 5 dB from 55 dB L_{den} up to over 75 dB L_{den}. For the nighttime noise level (L_{night}) the data should cover noise exposure classes from 50 dB L_{night} upward, again in 5 dB classes. This data should be supplied to the commission in another data flow.

In addition, the information from the noise maps was supposed to be communicated to the general public. The directive does not require evidence of this communication actually happening.

Following the noise mapping, noise action plans should be drawn. Summaries of the noise action plans should be sent to the commission in another data flow (Table 2.2).

For the first round of noise mapping, the member states were obliged to submit these data at predefined dates, as indicated in the following table. The first round of noise mapping was supposed to be concluded 30 June 2007 (Article 7) and action planning should have concluded 18 July 2008 (Article 8). In practice, very few member states succeeded to finalise this work on schedule. Some were very late, even by a few years, resulting in infringement procedures by the Commission.

Table 2.2 Data Flows as Defined by EEA/DG ENV

Data Flow Number	Description	Deadline	Update
DF1	Major Roads, major railways, major airports, and agglomerations designated by member states for first round mapping	30 June 2005	Possible at all times
DF2	Competent bodies for strategic noise maps, action plans, and data collection	18 July 2005	Possible at all times
DF3	Noise limit values in force or planned and associated information	18 July 2005	Possible at all times
DF4	Strategic noise mapping related data (i.e., the results) as listed in Annex VI of END for major roads, railways, and agglomerations mapped during the first round	31 December 2007	Obligatory every 5 years
DF5	Major roads, major railways, major airports, and agglomerations designated by member states for second round mapping	31 December 2008	Possible at all times
DF6	Noise control programmes that have been carried out in the past and noise measures in place	31 December 2008	No update
DF7	Action plan related data as listed in Annex VI of END for major roads, railways, and agglomerations mapped during the first round, together with any criteria used in drawing up action plans	18 January 2009	Obligatory every 5 years
DF8	Strategic noise mapping related data (i.e., the results) as listed in Annex VI of END for major roads, railways, and agglomerations mapped during the second round	31 December 2012	Obligatory every 5 years
DF9	Noise control programmes that have been carried out in the past and noise measures in place	18 January 2014	No update
DF10	Action plan related data as listed in Annex VI of END for major roads, railways, and agglomerations mapped during the second round, together with any criteria used in drawing up action plans	18 January 2014	Obligatory every 5 years

Organisation of responsibility: What has happened?

During the first round of noise mapping there have been large variances in the way member states organised the responsibility over the various governance levels. For instance:

- In the United Kingdom, the Department for Environment, Food and Rural Affairs took the full responsibility and carried out the mapping for main roads, railways, airports, and agglomerations in England. Specialised acoustic consulting companies were hired to carry out this work.
- In Germany, the responsibility for rail noise maps was delegated to the national rail authority (DB Netz) at national level and for road noise to the federal states, covering the noise maps of national roads and agglomerations within their territory.
- In France, agglomerations represent well-defined areas around the big cities. These agglomerations constitute legal entities with certain governmental tasks, and these were made responsible for the agglomeration noise maps.
- In the Netherlands, several cities with populations ranging from a few thousand up to 800,000 constituted the six agglomerations defined by the national government. The responsibility for noise mapping was delegated to the city councils. In this approach, city councils of municipalities with merely 20,000 inhabitants were made responsible for the tasks of the environmental noise directive, simply because they were part of an agglomeration, and there was no central governmental entity available at the agglomeration level.
- In Italy, Directive 4972002/EC was adopted with the Legislative Decree No. 194/1995 (Decreto Legislativo 194/2005). Italian policy defines agglomeration as an urban area, defined by regional or provincial administration, formed by one or more contiguous built-up areas (according to the Legislative Decree n. 285/1992) with an overall population higher than 100,000 inhabitants. In this way the Italian government defines compliance authority the regional or provincial administration.

The lack of a clear definition of an agglomeration was signalled as a point of improvement for the directive's technical annexes in the future revision.

The organisation of competance levels coincides with advantages and disadvantages in the implementation of the directive's obligation. If the level of competance is very central, the project management, progress control, and monitoring will be more straightforward than with a more fragmented competance. On the other hand it is important to realize that action plans should be drafted based on public consultation, which is likely to be more efficient if organised at a local, more decentralized level.

The organisation alternatives have been accompanied by different ways of funding or financing the mapping operation. In the majority of cases, commercial consulting companies carried out the actual technical work. For the first round, with many new tasks, this seemed a sensible choice. In some cases, regional environmental agencies took up the task to carry out the noise mapping for participating cities. In view of the structural character of the mapping this is sensible as well. The funding that allowed hiring a consultant or involving additional manpower from an environmental agency was sometimes provided by the national government (e.g., in the Netherlands), in other cases it had to be allocated from the federal state, agglomeration, or city budget.

Fragmented responsibilities would likely lead to fragmented input data collection, with differing conclusions of the mapping operation as a result. The more parties involved in the mapping, the more synchronization and coordination required at the interfaces. It is likely that this coordination has not always been adequate.

Article 3: Definitions—What effects of differences?

Article 3 of END defines the sources for noise mapping, terms, and indicators. Definitions that with regard to the identification of sources are clear but have showed uncertainty in the application, include:

(n) 'major road' shall mean a regional, national or international road, designated by the Member State, which has more than three million vehicle passages a year;

(o) 'major railway' shall mean a railway, designated by the Member State, which has more than 30,000 train passages per year;

(p) 'major airport' shall mean a civil airport, designated by the Member State, which has more than 50,000 movements per year (a movement being a take-off or a landing), excluding those purely for training purposes on light aircraft.

Member states interpreted these definitions differently. In the directive, the cut-on values had been clearly defined in terms of annual number of vehicles on that road. In spite of this clear definition, some member states interpreted this as an absolute minimum, leading to stretches of road with intensities just below the limit value being excluded from the mapping operation. In other countries the cut-on value was interpreted as an efficiency indicator, dividing between entire road links being either busy (where the traffic intensity was more or less at the indicated level) or quieter (where the traffic intensity was clearly below the indicated level). Again, the different interpretations may have led to differences in the outcomes, and there is a clear need for a tighter definition when the directive gets revised. Things were even more complex in countries where "major road" is used as a legal designation of a certain type of road, irrespective of the traffic intensity on that road. In these countries, the criterion for selecting road stretches to be mapped becomes a two-step procedure: First, select the major roads from the national road network, then decide for which stretches of major road within that selection the traffic intensity threshold is exceeded.

Interface problems were recognized, particularly for major roads on a certain agglomeration's territory. Many countries had set up a clear division of tasks, with the national road authority being responsible for the noise maps of the major national roads, and the agglomerations being responsible for the noise maps of all roads on their territory. Clearly, there is an overlap in these definitions. The compliance for these national roads on agglomeration territory needed to be clearly defined, and in any case it should be ensured that the public would be presented with only one or at least two identical results and that the counting of exposed citizens, the end result of the mapping, would be straightforward and would avoid any doubling.

REFERENCES

1. Commission of the European Communities Com(95) 624 Brussels, 10.1.1996 Communication from the Commission "Progress Report on implementation of the European Community Programme of Policy and Action in relation to the environment and sustainable development towards sustainability."
2. Decreto del Presidente del Consiglio dei Ministri del 01/03/1991 "Limiti massimi di esposizione al rumore negli ambienti abitativi e nell'ambiente esterno, Gazzetta Ufficiale Italiana," no. 57 del 08/03/1991 (D.P.C.M. 1st March 1991 "Limiti massimi di esposizione al rumore negli ambienti abitativi e nell'ambiente esterno").
3. "Future Noise Policy—European Commission Green Paper" (Brussels, 04/11/1996 COM(96) 540).
4. Directive 2002/30/EC of the European Parliament and of the Council of 26 March 2002 on the establishment of rules and procedures with regard to the introduction of noise-related operating restrictions at Community airports.
5. Communication from the Commission "Europe's Environment: What directions for the future? The global assessment of the European Community programme of policy and action in relation to the environment and sustainable development, 'Towards Sustainability'" Brussels, 24/11/1999 COM (1999) 543 final.
6. Decision No. 1600/2002/EC of the European Parliament and of the Council of 22 July 2002 laying down the Sixth Community Environment Action Programme.

Chapter 3

Measurements

J.L. Cueto and R. Hernandez

CONTENTS

This chapter will try to review the state of the art in environmental noise measurements and monitoring, and finally provide a wide perspective on the role of measurements in Noise Directive 2002/49/EC.[1] This also includes the estimation of the indicators L_{den} and L_{night}, as defined by this directive. It is well known that to develop noise maps from noise measurements is not a good idea. There are many inconveniences in making maps in that way. (See Figure 3.1.)

Basically, what we try to capture during the measuring interval is not the total amount of noise detected at the microphone. What we really want to know is the specific sound,[2] in other words, the noise generated by the target noise sources. The rest is the residual noise. Probably one of the most important problems that concern technicians is how to manage to avoid this residual noise, and sometimes it is quite complicated, even impossible.

But what is a real disadvantage of measurements and made noise experts prefer noise prediction software is the capability of the latter to develop

Disadvantages of Prediction with Regard to Measurement in the Development of Noise Maps	Advantages of Prediction
▪ Prediction software needs a great quantity of data at the input ▪ The data output accuracy depends on the data input accuracy ▪ To use the software correctly, the users have to be acoustic experts ▪ Communities assign more credibility to measurements	✓ Provide detailed information about: 　✓ The contribution from each noise source in the overall noise level 　✓ Not influenced by residual noise ✓ Not influenced by meteorological conditions ✓ Calculation assesses not only the actual environmental situation but the future one as well (what if). This is the reason it can be used to evaluate the effectiveness of the corrective and preventive measures against noise ✓ The map can be easily updated ✓ Measuring implies a trade-off between uncertainty and measurement effort to obtain levels representative of a year (in time) and of large areas (in space) ✓ The last means the map is less expensive to produce using prediction

Figure 3.1 Noise measurements versus computer predictions: pros and cons in the development of noise maps.

action plans. A noise model is a powerful tool that evaluates multiple scenarios, so what is the role of noise measurements in noise mapping and action plans?

THE AIM AND THE SCOPE OF NOISE MEASUREMENT SURVEYS

Nevertheless, outdoor noise measurements have an important position inside the overall process covered by the Noise Directive:

- Diagnosis of the current acoustic pollution in Europe
- Action plans designed to prevent the excess of noise
- Control and management of environmental noise

There are many types of environmental noise sources caused by human activities. At this point, we are going to focus on the sources proposed as a target by the Noise Directive: large transport infrastructures, harbours, big industrial facilities and built-up urban areas, and leave out noise from leisure, construction, military and other small activities located inside and close to residential areas. One could think that we are acousticians, and sooner or later acousticians find a reason to carry out measurements, but actually, there are good practical reasons to measure environmental noise:

1. Estimating the sound power of noise sources when we have no previous knowledge about their noise emission behaviour. A good example is the application to industrial noise, as an application for large industrial facilities.[3]

2. Assessing noise environment by using an appropriate rating equivalent continuous level when necessary. The measurement surveys are carried out under a set of technical procedures defined by the relevant administration:

   • Scheduling inspections of actual environmental noise situations in areas where people complain or even wherever noise level is suspected to exceed limits assigned to noise zoning. Irregular noise activity patterns can be captured in the same way as noise adjustment as tonalities, impulsiveness, and low frequency sound. The relevant administration is in charge of judging the environmental situation and establishing who is responsible.

   • Acquiring data to support planning developments. Verifying the quality and correctness of conclusions predicted in environmental noise impact assessment studies of new transport infrastructures, new industrial facilities, and, on the other hand, new sensitive areas.

   • Checking the consequences of potential or adopted decisions by comparing "before and after" noise situations. Analysing the global effectiveness of measures against noise.

   • Controlling the environmental situation. Control implies a feedback process, but the response cannot always be in real time. Sometimes local residents request monitoring their residential areas and then noise monitoring is consequently used as a way to provide communities information about noise pollution, for example, by Internet. Then the noise producer can check the public response (annoyance) to different scenarios and make decisions. Controlling environmental noise can also be used to understand the variables surrounding the noise problem and to verify future noise trends in order to anticipate preventive measures against noise.

3. Making descriptive works in which monitoring and measuring provide information about the noise climate at a particular location. These could be the basis for the estimation of noise indicators covered by the Noise Directive 2002/49/EC with various purposes:

   • Noise monitoring and measurements can be used to confirm the validity of strategic noise maps developed by modelling.

- Defining hot spots accurately. In certain circumstances a deeper approximation to the noise problem than the one provided by strategic noise maps is advisable. For instance, when the noise exhibits special characteristics it is recommended that it be taken into account in the noise assessment. Another special situation arises in areas where the local noise is considerably different from expectations. This is usually caused by anomalies in noise sources that are usually beyond the scope of strategic noise maps. A good example of the former is the measurement of tonalities, impulsive behaviour, low frequency energy, peak levels, and other noise magnitudes. Some possible examples of the latter are the evaluation of noise produced by road humps, rumble strips, and parking lots. These characteristics of environmental noise levels help noise engineers to objectively analyse the annoyance caused by noise that usually forms an important part of action plan designs.

Now the reader is aware of the benefits of scheduling a noise measurement as a support for noise mapping and action plans processes, although the estimation of the sound power of industrial facilities and other outdoor noise sources are beyond this chapter's scope. Then, this review will focus essentially on four methodology steps that cover the total task, irrespectively of the previous objectives we chose:

- How to design noise surveys[4-6]
- How to conduct the noise tests properly[7]
- How to estimate the uncertainty of every test[4-6,8]
- How to post process the raw data until it is transformed into relevant information[4,6]

BASIS FOR THE METHODOLOGY

Unfortunately, outdoor noise measurement methodology has nothing to do with a controlled laboratory test and systematic approaches are sometimes inapplicable. The acoustic measurement is a process influenced by a complex of variables, some static and others dynamic, which means noise constantly changes in the time and space domain. Some of these variables can be controlled and measured during tests and others cannot. Some of them can be predicted, or at least statistically predicted, and others cannot. (See Figure 3.2.) Anyway, the variables affecting the noise measurements results include:

- The time and spatial behaviour of the specific noise
- The time and spatial behaviour of the residual noise

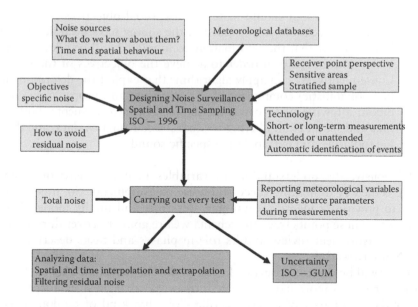

Figure 3.2 Phases in the environmental noise measurement process and information managed.

- The propagation-related variables and the evolution of some of them in time and space inside the path between emitter and receiver
- The standard or procedure used for testing
- The instrumentation and human resources

All decisions about designing noise surveillance depend directly on the complexity of the situation and, at the same time, they also depend on the level of approximation demanded and the availability of personnel and technology. The uncertainty of the whole measurement process is hidden behind all these decisions. So, the sampling criterion states the number and distribution of the measuring points, how many measurements are considered for each point, how long the short-term measurements are going to be, where to monitor for the long term, and when it is possible to leave short-term measurements unattended. No matter what you decide, two aspects of the measurement process should be guaranteed to consider the methodology a success:

- The quality assurance of the test should be fulfilled. Tracing the variables involved in the noise measurement process guarantees the uncertainty can be estimated and, finally, the magnitude measured and indicators calculated guarantee the comparability (reliability) of the results of the study.

- The survey design should match the expected objectives. The techniques and technologies included in the survey have been developed in such a way that the information accumulated will be relevant and usable (meaningful) in order to achieve the objectives of the study. Consequently, we can apply algorithms that exploit the deterministic (or probabilistic) component of the operating and propagation condition, allowing us to overcome the lack of measurements, both in time and space. First, filtering residual sound from raw data, and then interpolating and extrapolating specific sound.

Obviously, the understanding of variables related to environmental noise is the key to making successful noise measurements. So, it is necessary to previously identify, catalogue, describe, and finally model these variables. These points trace the schema we are going to cover during this noise measurement review and its role in phases and tasks described by the Noise Directive.

What will be described during the rest of the chapter is not a set of closed procedures. It is more like a flexible framework[6] that must be adapted to cover the major part of cases, regardless of what kind of outdoor noise sources and receiver areas we manage. The size of the project (the order of magnitude) is determined by the level of approximation demanded, which is usually directly proportional to the measurement effort and indirectly related to the uncertainty.

TIME PROFILE OF ENVIRONMENTAL NOISE

The first task is to identify and catalogue the total (or at least the most representative) number of noise sources presented in the study area; no matter whether they are specific or residual noise sources. Next, these major noise contributors have to be studied in relation to their sound emission characteristics with time. To do so, this catalogue has to include a set of parameters describing their source operating conditions. These parameters sometimes are practically the same as those gathered for noise mapping purposes but with different time resolution. This means that these dynamic parameters have to be considered in short time variations during the measurement time interval, not averaged over the whole year. How accurately we need to register and describe these variations depends upon the sensitivity of noise levels to the variation of the parameters.

Previous knowledge about the time behaviour of the noise source tells us about the statistical representativeness of short-term measurements in the long-term levels. This information extracted from long-time operative conditions of the source could reveal, for example, intraday significant events, probability of occurrence of noisy and quiet episodes, long-term periodicities, and trends. So the catalogue of noise sources must be extended until it includes the significant noise emission classes of each source. The major part of this environmental noise has to do with human activities and their normal periodicities like traffic, but others do not, like wind turbines.

As the target noise sources are the same as in the Noise Directive—road traffic, railway traffic, aircraft flights and industrial areas—we can summarize the time behaviour in three types:

- Areas affected by transport infrastructures where the flow is low (railways, aircrafts, and secondary roads). The noise caused by vehicle passing-bys can be considered as repetitive single-events or fluctuating noise.
- However, those places affected by major roads have a perception of steady noise (slowly varying really).
- Industrial areas could generate repetitive events, steady noise (or maybe slowly time varying or step-type noise), or a combination of both.

Finally, technicians have to determine the percentage of time covered by each one of the emission classes per each noise source during periods of day, evening, and night. These emission conditions, during which measurements can be performed with limited and known variations in measurement results due to emission variations, is called the emission window.[4]

The second task is to identify and catalogue the total (or at least the most representative) number of meteorological situations presented in the study area, unless the noise measurement equipment stays close to the target source. These weather conditions, during which measurements will be performed, is called the meteorological window.[4]

Noise source operation and propagation conditions (we can add the receiver's habits, so space becomes an important information, too) can give some of the clues that help technicians to simplify the measurements survey in such a manner that a number of short measurements can be combined until they provide annual average levels. The total time interval in which a series of measurements is conducted is known as the observation time interval, and the measurement time interval[6] refers to the duration of every single measurement.

DENSE ROAD TRAFFIC NOISE

Road traffic noise is considered the largest source of noise pollution in modern nations and is especially disturbing during rush hours. No matter if it is talking about ring highways or crowded main streets inside cities, all of them have one thing in common: the emission can be described as slow varying and its noise time profile describes clear periodicities in the long term.

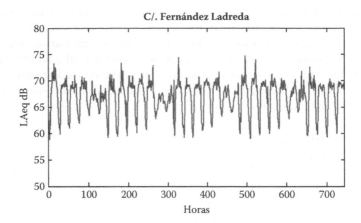

Figure 3.3 L_{Aeq} noise profile during one month.

Figure 3.3 represents one month of noise registered in one of the permanent monitoring stations placed in the streets of Madrid.[9] What we can see is the $L_{Aeq,T}$ time profile, taken every hour ($T = 1$ hour). Urban noise looks like a damped signal repeated every week. A great percentage of the variability is explained by intraday, daily, and weekly components. But part of the components is of random nature. Apparently, every day passes without notable changes regarding the traffic flow, without roadwork, accidents, and so forth. If they occur, these random events tend to be overcome in the long term. With enough numbers of independent measurements during a long period of observation, it is possible to give a good estimation of L_{den}. But for short time series, filtering is needed prior to an estimation of L_{den}.

When the measuring test of traffic noise is carried out, there are some parameters to be registered. One of them consists on counting the total number of vehicles passing by during the measurement time interval. Normally the official statistics from authorities provide the traffic data flow per hour in roads, making distinctions between days of the week. (See Figure 3.4.)

When it is not possible to measure continuously over the observation time interval, this information database establishes which hours and days are best to measure that road; and with the samples (measurement time interval) reconstruct (estimate) the long-term noise figures. As far as we know from the IMAGINE project WP2, the sensitivity of noise regarding the AADT shows that duplicating the traffic flow (while the other traffic parameters remain constant implies an equivalent noise level increase of 3 dB).[10] In these road traffic official statistics, at least two categories of vehicles have to be taken into account: light and heavy vehicles. Although during measurement intervals, the distinction could be increased until it covers all vehicle classes per acoustic characteristics: motorcycles, mopeds, passenger cars, medium heavy

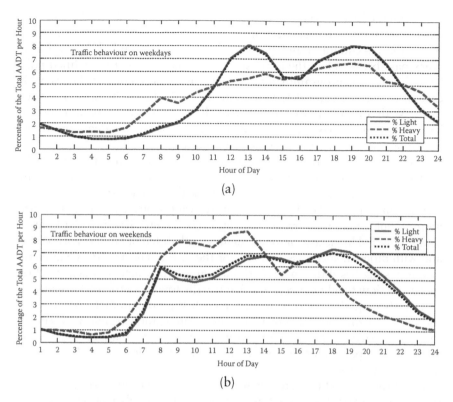

Figure 3.4 Percentage of the total average annual daily traffic (AADT) per hour during a 24-hour period. (a) Statistics for working days are presented, and (b) only for weekends. Distinctions between percentages of heavy and light vehicles are considered, too. **(See colour insert.)**

vehicles (two axles), heavy trucks (more than two axles), and so on. To guarantee the representativeness of the noise measurements, other parameters that represent the condition of the noise source during the period of measurement surveillance (observation time interval) should be traced (and then reported):

- Average traffic speed shall be measured.
- Type and maintenance condition of road surface shall be noted.
- Acceleration and deceleration over the section shall be reported.

LOW DENSITY ROADS, RAILWAYS, AND AIR TRAFFIC NOISE

The noise test method is similar to the previous case, but it is not necessary to measure continuously, only during pass-bys and flyovers. A unique identification of every event by a reliable clock is fundamental for the correct recognition of every event. The rest of the time there is simple background noise. Normally, what is determined and reported are the exposure level (L_{AE}), the maximum sound pressure level (L_{AFmax}), and the duration of every pass-by.[4-6] When the noise station is unattended, a predefined noise level threshold is programmed to trigger the sound record of the event. In case of railways a train detection system could trigger the noise measurement. With a representative number of these single-event sound exposure levels for vehicles passing by, we can calculate the equivalent-continuous sound pressure level with the required accuracy.

Low density roads require a most extended distinction between vehicles for the designing of measurements. What is relevant is the previous classification of vehicles according to their acoustical characteristics.

Regarding trains, the classification in relevant acoustical classes is a little more complex than road traffic. Usually, every country has different kinds of trains: powered by electric or diesel engines, with different numbers of coaches, and with different speed assignations for the sector. The acoustic characteristics of passing-bys differ in strength and duration, which is evident for high speed and slow freight diesel trains that usually tow more than 50 wagons.

It is also equally relevant to know each type of aircraft operating on the airport examined. There is another important variable in airports: the traffic pattern. The noise measurements have to manage the type of airplane in a particular fly path to understand the long-term L_{Aeq} at that measurement spot.

When the variability increases per vehicle class, because we cannot control the total variables implied in passing-bys, the number of the minimum vehicles per category passing-bys need to be increased in order to get a good estimator for L_{max}. At least 30 passing-bys of every vehicle category is recommended, although, occasionally, fewer events are acceptable, especially in railway and airport surveys.

INDUSTRIAL PLANT NOISE

Industrial noise can be generated by multiple sound sources placed inside the factory perimeter, and that is the reason for distinct noise behaviour from every source. Fans and cooling towers are good examples of steady sources; compressors are a good example of how machinery produces a step-type noise profile according to its different working modes. If the noise steps are steady and the distance is long enough to introduce propagation factors in measurements, this noise must be at the receiver point, at least, stationary during the time measuring interval. In this case the measurement should include every step in which the noise remains steady. The probability of occurrence of each one of these stages allows estimating the noise level in the long term. Other sources are unpredictable and composed by short-term events like cutters and truck pass-bys. Even industrial sources can produce noise with tonality, impulsivity, and low frequency. The combination of all of these defines the noise from industrial plants.

The catalogue of noise sources has covered a set of parameters that describes the total noise major contributors in the area to be analysed, regardless of whether they are specific or residual. Some of the parameters are dynamic in nature; they need to be monitored in real time, synchronized with noise, and finally reported with the rest of parameters:

- Georeferred situation of every noise source and height over surrounding ground.
- In a simplified approximation to the problem, it is important to understand if the radiating source's shape can be described as a point, a line, or a surface from the measure point's perspective. The dimension of the radiating source.
- The direction and face of the shape and the position with regard to reflective surfaces and buildings.
- Directivity due to design or installation.
- Power level emitted for each working mode and machinery. A third octave sound power spectrum.
- Timetable of the industrial plan. Time behaviour of the sources of noise during each period (day, evening, and night), which is dependent upon the working modes.
- Tonality, impulsivity, and low frequency characteristics.
- Number of mobile sources over a track.

SPATIAL SAMPLING

The election of the outdoor receiver point's locations should be chosen looking for an appropriate acoustic spatial description of the phenomenon under consideration. And this is not always a straightforward task. From the noise study objectives we define the alternatives:

- Studying the role of a noise source in total noise. The measurements take place in the proximity of that specific noise source in order to minimise the influence of the rest of residual sources.
- Adopting the receivers' perspective in sensitive areas or inside exposed buildings. If the measurement survey's purpose is to study the contribution of one specific noise source, some measurement strategies can be exploited to avoid residual noise, and some of these strategies are related to the measurement point's spatial election. When a long period of measurement must be programmed, the selection of points

most exposed to specific noise should be carefully chosen. Sometimes it is impossible to install the instrumentation in the best spot. As we know, some situations of environmental noise can only be evaluated in the long term, not only regarding aspects related with the Noise Directive, but for other legal requirements, such as noise zoning, which usually demand environmental noise testing in relation to the quality objectives over the entire year.

- Describing the total environmental noise in the agglomeration. The stratified sampling is performed when selected points are designed with a prior knowledge of the urban soundscapes. It is possible to categorise different areas in relation to their predominant sources of environmental noise and their characteristics in time behaviour.

The density of receiver points and their distribution over the territory depend on the purposes of the noise survey and the spatial resolution needed to comply with these requirements. With the purpose of validation of noise maps, 5 dB is a normal figure for this resolution. Strong gradients of noise levels can be located near the noise sources and in the proximity of noise barriers. The directivity and height of the noise source has to be taken into account. The height of the microphone should be 4 m high, unless this is impossible to achieve. The lower the microphone is situated, the higher the influence of impedance of ground, topography, and barriers.

What appears to be the best acoustic position is not always available. It is recommended that the measurements carried out to characterise the exposition of buildings be executed[4,6]:

- Far enough from the façade to minimise the influence of reflections. This is difficult to achieve when the façade and the noise source are very close. Otherwise, some corrections must be included in the estimation of noise at that façade.
- Fitted to the façade by a reflective surface at the considered height. This position needs a correction of −6 dB.
- When it is not physically possible to follow these two previous ways, it is preferable to place the sound level meter inside a dwelling positioned in the most exposed façade of the building.
- A great variety of noise measurement guides recommend not placing the microphone close to the façade of buildings. A distance of 0.5 to 2 m from the walls need a correction of 3 dBs, but highly increase the uncertainty.

Often the measurement results have to be combined with calculations with different intentions: sometimes to interpolate and extrapolate spatial data, and in other occasions to include in the long-term analysis of the receiver points operating or propagation conditions not really measured.

METEOROLOGICAL SITUATIONS

There are no concerns about long-term measurements because these include all possible combinations of meteorological situations and source conditions. Changes in the ground impedance through the path during a year are not considered here. The corrections over the year noise time series usually include the deletion of some episodes like unusually noisy events (road work, transitory diversion of traffic, etc.) and when rain and strong winds affect the microphone. But for short-term noise measurements intended for inspection purposes, it is essential to carry out measurements under favourable propagation conditions, especially when the specific noise source is far away from the receiver. Doing so, we guarantee the reproducibility of the measure under favourable propagation situations. But, when we desire to estimate the indicators L_{den} and L_{night} in the long-term using short-term measurements, we need to select different stable propagation conditions. Here, stable means with a limited and controlled variability of the propagation conditions in every measurement over the total observation time interval. Favourable conditions of propagation are independent of weather conditions when the relation between the distance, r, between receiver and source, and the respective heights of the source, h_s, and receiver, h_r, are related by the inequality

$$\frac{h_s + h_r}{r} \geq 0.1 \tag{3.1}$$

In case of reflecting grounds, longer distances can be acceptable.

A favourable situation in outdoor sound propagation arises when sound paths are refracted downward. The meteorological variables that play a great role in that sound propagation are the wind and the temperature gradient near ground. The direction and speed of the wind have to be measured at least at 10 m height or reported by a meteorological station near the area. *Downwind* implies predominant winds within an angle of no more than 60° during day and 90° during night, from the imaginary line that links the noise source to the receiver. The level of insulation influences the temperature inversion near the ground. This is why the elevation angle of the sun over the ecliptic and the coverage of cloud are important factors to understand the propagation condition with calm winds. With practical implementation purposes, some classifications of specific situations regarding the propagation have been detailed. We extract one of them easily employable for *in situ* tests (see Table 3.1).[4,6]

The statistical data has to be obtained from local meteorological stations and must be obtained from climate time series with a resolution of no less than an hour. In common situations these stations can provide a good approximation to the predominant winds, percentage of hours with clouds

Table 3.1 Meteorological Classes

Propagation Condition	Insulation Level	Wind Speed and Direction
Very favourable	Night	0 m/s or downwind
Very favourable	Day	>6 m/s or downwind
Favourable	Totally cloudy	0 m/s or downwind
Favourable	Day	3–6 m/s downwind
Neutral	Day	1–3 m/s downwind
Unstable*	Sunrise and sunset	—
Unfavourable	Day	0–1 m/s downwind
Unfavourable	Day	Upwind
Unfavourable	Night	Upwind
Unfavourable	Totally cloudy	Upwind

*Measurements are not recommended in these particular parts of day.

and, possibly, daily and seasonal temperature gradient data. In climate studies, 30 years is the minimum to establish good average values, but with shorter series good approximations can be achieved for our purposes.

METEOROLOGICAL WINDOWS COMBINED WITH SOURCE EMISSION WINDOWS

To estimate the long-term noise, the short-term measurement survey should be conducted in such a way that the long-term equivalent sound pressure level can be calculated by taking into account the probability of occurrence of all possible operating and propagation conditions. For each of these conditions the sound pressure level is measured several times and then the results are combined until obtaining an average year extrapolation. In other words, the number and time length of measurements per combination of each noise source class with each distinctive propagation condition has to be representative and then able to be extrapolated for the long term. All of this could be done for every receiver point and every one of the evaluation periods: day, evening, and night.

Following the example of Figure 3.5, the long-term L_{day} estimation from short-term measurements is given by

$$L_{day} = 10\log\left(\sum_{k,n} p_{k,n} 10^{0,1L_{k,n}}\right) \tag{3.2}$$

where $p_{k,n}$ is the probability of occurrence of the combination between every emission and meteorological class. In each of these combined

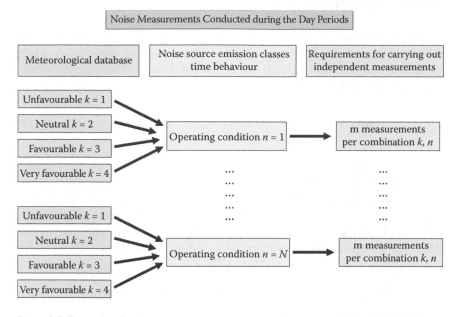

Figure 3.5 Example of a short-term measurement program, used in an industrial area, for a long-term noise description during the day period.

windows k,n, a set of measurements must be carried out. If these measurements time interval is fixed but the number of measurements, $M_{k,n}$, per combined window is variable, the L_{eq} level for every one of these windows, $L_{k,n}$, is given by

$$L_{k,n} = 10 \log \left(\sum_m \frac{1}{M_{k,n}} 10^{0,1L_m} \right) \tag{3.3}$$

where L_m is every single measurement in the overall test. When it was impossible to measure in different combined windows, these levels can be estimated by calculations. Seasonal differences and other anomalies of emission and meteorological classes can also be included. The techniques used for corrections comprise noise prediction software, traffic models, and a variety of interpolation–extrapolation analysis tools. Simultaneously, the source operating conditions have to be monitored using all the parameters required for the input of the prediction software. Separately, it is necessary to measure during evening and nighttime. In case one or few measurements have been executed per window, the uncertainty has to be estimated for every separate measurement. On the contrary, if we have enough measurements per window, we can estimate the uncertainty from this set of registers.

MANAGING RESIDUAL NOISE

We know that we can consider total noise as specific when the residual noise is below 10 dB. We have to correct the measured noise levels when residual sound is 3 dB to 10 dB below the total noise. The correction is a simple energy subtraction:

$$L_{spe} = 10\log(10^{0,1L_{tot}} - 10^{0,1L_{res}}) \tag{3.4}$$

where L_{spe} is the estimation of specific sound, L_{tot} the measurement sound pressure level, and L_{res} the residual noise pressure level.

Unfortunately, one of the facts in noise measurement practice is the impossibility to take a separate register from two noise components presented at the same time in the same place. But we can estimate one or the other exploiting different strategies for avoiding residual noise:

- Exploiting the time variability of every kind of noise. Previous knowledge about the time evolution of the characteristics of both residual and specific sounds enable the recognition and filtering of (for example):
  - A specific steady noise from fluctuating residual noise.
  - A tonality from a specific source in the broadband residual noise.
- Exploiting the space variability of every kind of noise.
  - When the requirements for placing the microphone are not constrained to a defined point or with monitoring purposes, it is better to be far away from the residual sources and guarantee that differences between total and residual levels remains in 10 dB.
  - Using correlation techniques of time–frequency data from different sound level metres measuring synchronized. The analysis could establish a causal relationship between the noise records from points located near principal noise sources (residual and specific) and the receiver point record. The microphones have to be placed near sources to minimise the effects of other sources and meteorological conditions but not too close to get influenced by the near field. In long distances, the delay in propagation of sound and the divergence and atmospheric absorption has been taken into account.
- Exploiting control over the noise emissions (on/off) from specific and residual noise sources.
  - Usually, carrying out a total noise test and then a residual noise test, with the exigency that all factors underlying residual noise remain the same.

- Exploiting the capacities of instruments.
  - Using directive microphones.
  - Programming the sound level metre with specific time resolution or to record special variables like frequency statistics.
- Unattended instruments triggering sound records to test and classify noise events as specific or residual.

MONITORING INSTRUMENTATION

The noise measurement chain that compounds a sound level metre is composed of the microphone, the preamplifier, cables, windshields, an extendable 4 m tripod, and the sound analyser. A windshield is compulsory for outdoor microphone use. A noise monitoring station is equipped with special features, allowing long-term outdoor work and the registration of great quantities of data referred to environmental noise. Both measurement systems must comply with the requirements of IEC-61672-1 for sound level metres type I and IEC-61670 concerning the bank of filters characteristics.

The instrumentation for long-term monitoring purposes exhibits some special characteristics (see Figure 3.6):

- Usually the analyser unit remains inside a roughed box that protects the circuitry from the environment and extreme weather.
- The microphone is the most exposed part of the measurement chain; that is why it needs to be designed to resist fauna, wide variations of temperature, heavy rain, lightning bolts, and corrosive environment.

Figure 3.6 Image of a noise monitoring system comprising: a metal housing for the sound level metre, mobile phone, and other electronics; a meteorological station; a weatherproof microphone; and an aerial for transferring data over cellular network. The equipment is mounted on a lamppost by the easy availability of an uninterrupted power supply.

Many airports have installed permanent stations to monitor noise. In these cases directional microphones show great advantages, maximising the relationship between the levels of aircraft passing by from the rest of surrounding noise.

- As a sound level metre the monitor measurement chain has to be automatically calibrated with regularity. The microphone has the possibility to autocalibrate.
- The associated weather station assures the record of temperature, humidity, wind speed, and direction, although many factories, airports, and wind farms already have meteorological stations that can provide this info in real time. Besides, when certain weather conditions are presented at the microphone, the noise data files have to be discarded, for example, presence of rain, extreme temperature conditions and wind speed above 5 m/s.
- Power supply guarantees continuous operation.
- Sound recording, video recording, road traffic automatic counts and speed, and even automatic recognition of events (for example, airplanes passing by).
- Data and control signals transmission. There are great differences between installations regarding the data transfer and alarm systems. Sophisticated central operation stations use a centralized server that makes the acquisition of data from several monitoring terminals via broadband wireless communication like WIFI or current cellular protocol HSDPA, which support data transfer speeds up to 7.2 Mbit/s. Time profile graphs can be plotted, and summarized reports can be associated to maps to present the information to the public.

Deciding the time resolution during the measurement time interval depends on various circumstances:

- The instrument's storage capacity, the frequency of downloading data, and also the number of noise variables to be recorded.
- The minimum duration of a significant event that it is aimed to define precisely (which in turn depends on the type of noise source). For instance, urban noise and crowded motorways have relevant noise periods of hourly, daily, weekly, and yearly. Resolutions of minutes and hours could be a good idea. Railways, empty roads, and aircraft flights are well defined by its pass-by events of seconds or minutes. Defining the measurement time resolution of these events second by second is the right choice. Industrial noise could be generated by the composition of different sources. Some industrial processes have short and long periodicities, and then the resolution must be adjusted to include short-term ones.

- The nature of residual noise. In case of continuous noise affected by residual noise provoked by intermittent noise sources (including railways, empty roads, and aircraft flights) an accurate time resolution could be the key to filter this class of noise from the specific one.

The environmental noise data parameters to record usually include $L_{Aeq,T}$; L_{AE}; L_{AFmax}, L_{AFmin}, $L_{AFn,T}$; and the start–end time and date. When the resolution of every measurement is $T = 1$ s, profile graphs can be created with all previous data. When industrial sites or other kinds of noise activities are analysed, the spectrum $L_{Leq,T}$ and the impulsiveness are also added. Finally both indexes, L_{den} and L_{night}, can be calculated or estimated.

CALCULATING UNCERTAINTY

The final value recorded during the measurement process by itself has a lack of real meaning. The reason is that the true value of the measurement can only be achieved through a procedure of exact and perfect measurement. Measurement yields only estimations of the real magnitude. To complete the knowledge of the measurement and guarantee the comparability with other measurements it is necessary to provide these test results accompanied by uncertainty. Uncertainty indicates the probability that the real value is located inside a certain interval.

This part of the text introduces a description of uncertainty in accordance with the standard ISO "GUM,"[8] which defines measurement uncertainty as "a parameter, associated with the result of a measurement which characterises the dispersion of the values that could reasonably be attributed to the measurand." Uncertainty about a magnitude's estimation does not imply doubt about the validity of the estimation; on the contrary, knowledge of the uncertainty implies increased confidence in the validity of a model result.

The challenge is to identify and model all the variables influencing the measurement process, to finally combine them until the total uncertainty is obtained. These variables were already explained, and we realised that uncertainty is hidden behind all decisions and processes surrounding the test.

The real values of noise pressure levels must be estimated from the *in situ* measurements L_m, which are a function of independent variables x_i ($i = 1$, 2, ...), in accordance with the following relation

$$L_m = f(x_1, x_2, ..., x_i) \tag{3.5}$$

The model function identifies the sources of uncertainty affecting the measurements and represents the influence of those variables over the final quantity shown in the display of the instrument L_m. It is necessary to take into account these input variables, x_i, are also an estimation of real variables.

$$u(L_m) = \sqrt{\sum_i (c_i \cdot u(x_i))^2} \qquad (3.6)$$

where c_i is the sensitivity coefficient,

$$c_i = \frac{\partial f}{\partial x_i} \qquad (3.7)$$

c_i will be equal to one, while the input variables remain independent. The sensitivity coefficient is the partial derivate of model function $f(\bullet)$ with respect to X_i evaluated from the estimations x_i.

Basically, there are two types of uncertainty: type A and B.

- Type A uncertainties are caused by complex sources difficult to model. Noise engineers could decide not to evaluate part of the components of uncertainty in an analytical way. The key here is to determine a certain part of the total uncertainty on the basis of a set of measurements and their statistical treatment.
- Type B uncertainties are caused by systematic effects and can be predicted and modelled.

In general the uncertainty is expressed through the combined standard uncertainty, u_c, which is a composition of type B standard deviation, u_i, and type A standard deviation, s_i. It is usual to report the measurement with a coverage in which the measurand is within a given probability. The probability density function is Gaussian, so a confidence level of 95% corresponds to a coverage factor $k = 2$. The output of measurement is expressed by the interval $L_{real} = L_m \pm k \bullet u_c(L_m)$, around the estimator L_m. So, what we have now is the expanded uncertainty $U = k \bullet u_c(L_m)$, and the true value becomes $L_{real} = L_m \pm U$. This means that we have a 95% level of confidence in that the real value is in between U dB with regards to the estimator.

$$U = k \bullet u_c = 2 \bullet \sqrt{s_A^2 + u_B^2} \qquad (3.8)$$

In Figure 3.7 it is shown how to manage this uncertainty in situations in which the measurements need to be compared with legislation limits for the area. There are no ambiguities for measurements 1 and 4; the first one overcomes the threshold and the fourth complies with the noise requirements for the area. But for measurements 2 and 3, an indetermination impossible

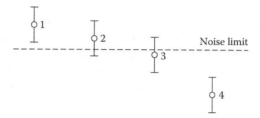

Figure 3.7 Four different cases of measurements with the expanded uncertainty in comparison with legislative noise limits.

to solve exists. The laboratory must abstain from giving conformity with the measurements.

In the References, a lot of sources of uncertainty and their influence in the final measurement are identified. We are going to extract some of them, probably the most important one. The combined uncertainty can be expressed in a simplified way as

$$u_B = \sqrt{u_{slm}^2 + u_{sou}^2 + u_{met}^2 + u_{pos}^2 + u_{res}^2} \tag{3.9}$$

where

slm = measurement chain
sou = deviation from the expected operation condition of source
met = deviation from the ideal condition in meteorological conditions, bearing in mind the distance between receiver and source and their relative height over the ground
pos = indetermination in the influence of the microphone's position over the measurements
res = indetermination in the influence of residual noise in the total noise measured

A contribution of measuring instrumentation could be decomposed again as a function of several other sources of uncertainties. Although, some guides and standards recommend to directly apply a standard uncertainty for a sound level meter type 1, of 0.5 dB.[6] In the case of ISO-1996, the recommendation of 1 dB quantifies the reproducibility of an acoustic measurement.[4] Following the performance specifications for maximum tolerances of instruments conforming to IEC-EN-61672-1 yield a combined standard uncertainty of the sound level metre of 0.7 dB.[11] It is interesting to calculate the standard uncertainty of the instrumentation chain on the basis of its technical specification, declared by the manufacturer and its calibration sheet, where the instrumentation results for periodic tests are expressed. A comprehensive study about the standard uncertainty of 22 different sound level metres class 1 has been estimated to be 0.4 dB.[12] The main contributions for that uncertainty are expected from the deviation associated with time window, especially the fast window, the accuracy response of the root mean square (RMS) detector, and the deviation of the true measurement due to imperfection in the linearity in a specific dynamic range. However, the uncertainty of sound level metre depends on other components, as we can see in the following list:

- The resolution of the equipment display
- The influence of environmental variables such as temperature, humidity, and pressure
- The influence of nonperfect A-weighting filter
- Directivity of the microphone
- Windshield, etc.

It is clear that in many outdoor practical situations, the contribution to the combined uncertainty is due to time and space varying environmental factors (including propagation factors, source conditions, and residual noise behaviour) and sound reflections from nearby surfaces. The microphone location introduces close to zero uncertainty in an ideal free-field state. But this microphone location is difficult to achieve in practical outdoor situations, especially when the surveillance is planned to estimate noise exposure over buildings in residential and sensitive areas. The installation and typical positioning of microphones was previously listed; the uncertainty is related to the impossibility to apply the true reflections correction. There are few variables implicated: the angle of sound incidence over the microphone, the impedance of the walls, the geometry of the receiver area, and so on. (See Figure 3.2.) Basically, with line sound sources with a broad vision angle over the microphone the uncertainty for a microphone fitted to a reflective surface is shown in Table 3.2.[6,13]

Table 3.2 Microphones in Different Installation Situations and Its Standard Uncertainty Figures

Microphone Assembly	Angle of Incidence of Sound from Line Noise Source	Standard Uncertainty
Flux-mounted over a reflective surface	Broad vision angle over extended source	0.25 dB
Flux-mounted over a reflective surface	Grazing incidence or point source	0.5 dB
0.5 to 2 m from walls	Broad vision angle over extended source	3.5 dB
0.5 to 2 m from walls	Grazing incidence or point source	1.8 dB

Source: IMAGINE project.

Residual noise is assumed to be exactly the same during the total noise measurement and during the measurement of residual used for the correction. When L_{res} is between 3 dB and 10 dB below the total noise level, L_{total}, in the receiver point, the estimation of the uncertainty of that residual noise follows the same steps and takes the same standard uncertainty, except for the source, and assuming the sensitivity coefficient is no longer 1.

$$c_{res} = \frac{10^{0,1L_{res}}}{10^{0,1L_{total}} - 10^{0,1L_{res}}} = 10^{-0,1(L_m - L_{res})} \tag{3.10}$$

The standard uncertainty due to meteorological conditions[6,13] is estimated $u_{met} = 2$ dB when the propagation conditions are favourable and very favourable, and the distance between receiver and source, r, is less than 400 m (u_{met} can be 1.5 dB or less in distances source–receiver up to 50 m when the microphone is positioned 4 m high and the terrain surface on the path is hard). For larger distances we can employ the following formula:

$$u_{met}(fav, vfav) = 1 + \frac{r}{400} \tag{3.11}$$

To reach a better estimation it is necessary to carry out n independent measurements until obtaining a decrease in uncertainty proportional to $n^{-1/2}$.[13]

The traffic flow's standard uncertainty depends on the number of vehicle classes, the number of vehicle pass-bys per class and time test interval, and the distribution of speed inside the flow. Roughly speaking, we can expect noise variations of 1 dB per variations in speed of 10 Km/h. The IMAGINE project gives some values when no other information is available.

$$u_{sou} \approx \frac{C}{\sqrt{v}} \tag{3.12}$$

where v represents the vehicles during pass-bys and C is the vehicle class according to Table 3.3.[6,13]

This estimation is obtained by taking into account deviation from real mean speed between ±10 Km/h. A greater deviation in speed should introduce 1 dB/10 Km/h.

However, the evaluation of type A uncertainty of every combination of source emission window and meteorological window could be calculated from measurements using the predefined set of measurements $L_{k,n}$. This can be done for the steady periods of emissions in industrial areas and for

Table 3.3 Standard Uncertainty Figures for Different
Types of Vehicle Classes

Vehicle Category	Standard Uncertainty within Category (C dB)
1. Light vehicles	2.5 dB
2. Heavy trucks	5 dB
3. Mixture fleet	10 dB
4. Same train and number of cars	3 dB
5. Freight trains	5 dB
6. Mixture fleet	10 dB

Source: IMAGINE project.

the period of emissions from transport infrastructures in which the noise
is stationary. During these periods the meteorological conditions have to
remain stable during the measurement interval.

$$S_{met,sou}(k,n) = \sqrt{\frac{1}{(W-1)} \circ \sum_{k,n=1}^{W} (L_{k,n} - \overline{L})^2} \qquad (3.13)$$

where W is the total number of measurements per each combined win-
dow k,n and $\overline{L}$ is the set of measurements $L_{k,n}$ arithmetic average. Another
option is achieved when the measurement is carried out so close to the
source that the measurement results become independent of meteorological
variables. In that case, L_n represents the values recorded during every phase
of noise source emission.

GLOSSARY OF TERMS AND ACRONYMS

AADT (annual average daily traffic)   It is the total volume of vehicle traffic in a highway or road for a year divided by 365 days and expressed
in vehicles per day. To measure AADT on individual road segments,
traffic data is collected by an automated traffic counter.

Combined window   The time period in which both the emission window
and the meteorological window remain stable simultaneously. Thus,
noise surveys can be conducted with limited and controlled variability
in measurements results.

Emission window   The time period in which a noise source (or a set
of noise sources) produces noise steadily in relation to their sound
emission characteristics with time. So, a survey near the noise source
can be conducted with limited and controlled variability in measurements results.

$L_{Aeq,T}$ A noise level index that illustrates the equivalent continuous noise level over the time period, T.

$$L_{Aeq,T} = 10\log\left[\frac{1}{t_2 - t_1} \int_{t_1}^{t_2} \frac{p_A^2(t)}{p_0^2} \, dt\right]$$

$t_2 - t_1$ = a certain time interval, broad enough to encompass all significant sounds of an event

p_0 = the reference sound pressure level (20 µPa)

$p_A(t)$ = the instantaneous A-weighted sound pressure level

L_{AE}, **exposure level**   SEL (sound exposure level in decibels) is computed by converting the total noise energy measured during a noise event to an equivalent decibel level for a single event that would only be one second in duration. The frequency weighting should be specified; otherwise, A weighting will be understood. The expression is the following:

$$L_{AE} = 10\log\frac{1}{t_0} \int_{t_1}^{t_2} \frac{p_A^2(t)}{p_0^2} \, dt$$

where

$t_2 - t_1$ = a certain time interval, broad enough to encompass all significant sounds of an event

p_0 = the reference sound pressure level (20 µPa)

$p_A(t)$ = the instantaneous A-weighted sound pressure level

t_0 = the reference time (1 s)

SEL accounts for both the noise event magnitude and duration.

L_{AFmax}, **maximum sound pressure level**   A noise level index defined as the maximum root mean square (RMS) noise level during the period T. Sometimes it is used to assess occasional loud noises, which may have little effect on the overall L_{eq} noise level but will still affect the noise environment.

L_{res}, **residual sound**   The noise remaining when the specific noise source is turned off.

L_{spe}, **specific sound**   The sound component received from a particular noise source. In most occasions, a specific sound has to be filtered from the total sound using the appropriate measurement procedures.

L_{tot}, **total sound**   The total noise present in an area, usually generated by noise sources of different types.

Long-term noise measurements   It is a period of time long enough to cover all possible combinations of meteorological and source emission windows. Results estimated from these long-term measurements should be representative of the annual average noise.

Measurement time interval   The duration of each single measurement.

Meteorological window   The time period in which sound propagation conditions remain stable. Accordingly, a stable noise source survey can be conducted with limited and controlled variability in measurement results.

Observation time interval   The total time interval in which a series of measurements were carried out.

Short-term noise measurements   It is a period of time long enough to cover one combination of meteorological and source emission window.

Step or staircase function   A function that increases or decreases abruptly from one constant value to another.

REFERENCES

1. Directive 2002/49/EC of the European Parliament and of the Council of 25 June 2002 relating to the assessment and management of environmental noise, European Commission, 2002.
2. ISO 1996-1:2003: Acoustics—Description, measurement and assessment of environmental noise—Part 1: Basic quantities and assessment procedures.
3. ISO 8297:1994: Acoustics—Determination of sound power levels of multi-source industrial plants for evaluation of sound pressure levels in the environment—Engineering method.
4. ISO 1996-2:2007: Acoustics—Description, measurement and assessment of environmental noise—Part 2: Determination of environmental noise levels.
5. ISO 20906:2009: Acoustics—Unattended monitoring of aircraft sound in the vicinity of airports.
6. IMAGINE (2006), "Determination of L_{den} and L_{night} using measurements." IMAGINE report IMA32TR-040510-SP08, 11 January 2006. http://www.imagine-project.org/.
7. ISO/IEC 17025:2005: General requirements for the competence of testing and calibration laboratories.
8. ISO/IEC Guide 98-3:2008: Uncertainty of measurement—Part 3: Guide to the expression of uncertainty in measurement.
9. Environment Department of Madrid City Council, http://www.mambiente.munimadrid.es.
10. IMAGINE (2004), "Review of data needs for road noise source modelling." Document IMA2TR-040615-M+P10, June 2004, http://www.imagine-project.org/.
11. IEC 61672-1:2002: Electroacoustics—Sound level meters—Part 1: Specifications.
12. Richard Payne (2004), "Uncertainties associated with the use of a sound level meter." NPL REPORT DQL-AC 002. National Physics Laboratory. Middlesex.
13. Hans G. Jonasson (2005). "Uncertainties in measurements of environmental noise." INCE Congress on Managing Uncertainties in Noise Measurement and Prediction. Le Mans.

Chapter 4

Road traffic noise

G. Dutilleux

CONTENTS

INTRODUCTION

During the first round of noise mapping in the Environmental Noise Directive 2002/49/EC (END) framework,[1] the French method NMPB 96 (NMPB96, 1997) in combination with emission data from *Guide du Bruit*[2] was selected as the interim method for road traffic noise prediction. *Interim* means recommended before a reference method would become mandatory for the next rounds as foreseen by the END. Interim does not mean mandatory, since the member states are left free to use their national method instead of the interim one for a given noise source, provided that they would demonstrate the equivalence of their national method with respect to the interim one.

For the 2007 deadline, 13 member states have chosen the interim method for road traffic noise. Besides the implementation of END, NMPB 96 has been in use for several years in many countries for noise impact studies of road infrastructures.

Instead of defining a common mandatory method, an intermediate solution for Europe could have been to enforce a quantitative comparison framework with respect to the interim methods. Sticking to the initial objectives of the END, Europe has decided to keep on with the development of a common method to all member states. The transition to a common reference method is scheduled for the third round of noise mapping in 2017.

At the time of publication of the END, the emission data in the *Guide du Bruit* was considered obsolete in the more demanding context of noise impact studies. The experimental basis it was built upon dated back to the 1970s and was poorly documented. During the last decades the vehicle fleet had changed a lot. The same applied to road surfaces, with the introduction of porous pavements, and more generally the diversification of pavement formulations. SETRA* coordinated a working group to draft an updated emission model for road sources.

In parallel, NMPB has also undergone a thorough revision process steered by SETRA. The essential motivation for the revision comes from the experimental validation carried out shortly after NMPB 96 was released. Although the agreement between measurement and prediction was good, these campaigns have shown a trend of overestimating noise levels in downward-refraction conditions. Overestimation of noise levels leads to noise abatement solutions more costly than necessary. Moreover, two improvements of the method were also requested. The first one is the possibility to simulate noise barriers smaller than 2 m. The second comes from noise consultants who carry out sound pressure level measurements at the top of embankments to characterise the sound power level of the infrastructure of interest. A major concern during the revision process steered by SETRA was to keep the method simple to use and to find a good trade-off between speed and accuracy.

---

* Technical Department for Transport, Roads and Bridges Engineering and Road Safety of the French Ministry of Ecology, Sustainable Development, Transport and Housing (http://www.setra.equipement.gouv.fr/English-presentation.html).

Besides SETRA, the most significant contributions to the revision process were provided by INRETS, and LRPC Strasbourg for the emission part; and CSTB, LCPC,* and LRPC Strasbourg for the propagation part.

The revision of the French road traffic noise prediction model was completed in 2008. The revision of the related AFNOR NF S 31-133 French standard was achieved in late 2010. Albeit NMPB 96 (or XP S 31-133:1997) and *Guide du Bruit* actually appeared in the text of the END, this framework is now superseded by the so-called NMPB 2008. Here NMPB 2008 stands both for the emission of road sources and a general-purpose model for sound propagation. So it seems more relevant to present the updated framework than the old one.

The aim of this section on road sources is to describe the main aspects of the NMPB 2008 framework, first for road traffic noise emission and second for propagation. Although NMPB 2008 is fully specified, this section is not intended to be a complete description of the method. For an exhaustive specification the reader shall refer to NMPB 2008 guides for emission[3] and propagation.[3,4,7] The last part of this section also provides a comparison of NMPB 2008 with Harmonoise/IMAGINE, which is another candidate to become the common method enforced by the revised END, although its current design makes it more suitable for high-end expertise calculations on comprehensive input data in small-scale projects than on large-scale computations on limited input data. The latter are however the characteristics of noise maps and action plans in the END framework. The comparison addresses emission and propagation separately.

The whole method is expressed in one-third octave bands from 100 Hz to 5 kHz median nominal frequencies. The method works also in octave bands. This would be a harmless simplification for traffic noise sources.

EMISSION

The emission part of NMPB 2008 is fully specified in NMPB 2008.[3] For more details on the design of this emission model, the reader shall refer to Hamet et al.[6]

The model addresses two categories of vehicles: light vehicles (LV) and heavy vehicles (HV; more than 3.5 tons). Whatever the category, a vehicle boils down to a single point source located 0.05 m above the ground. This rather low height has been derived from two experiments based on different principles: microphone array processing[7] and an interference-based method.[8] This low value is consistent with the fact that tyre–road noise is

---

* LCPC and development and network INRETS merged into IFSTTAR, the French Institute of Science and Technology for Transport (http://www.ifsttar.fr).

dominant over engine noise, even at low speeds and can also be explained by the fact the engine cannot be seen as a point source. The source is supposed to be omnidirectional.

The range of application is 20–130 km/h for LV (20–120 km/h for HV). The model considers three different paces: steady speed, deceleration, and acceleration. In deceleration and acceleration, the lower speed limit is 25 km/h. A qualitative description of the traffic flow type has been preferred to a numerical value of acceleration. The latter is currently not available in France and probably in many other countries as well.

NMPB 2008 covers the road surface types that are used in France. The modelling of road surfaces is based on a large database of ISO 11819-1-like statistical pass-by measurements.

The model is expressed in $L_{A\max}$. The link to a sound power level per unit length per vehicle is immediate by the relationship

$$L_{w/m/veh}(v) = L_{A\max}(v) - 10\log_{10} v - 4.4$$

where v is the speed in kilometres per hour (km/h).

The model distinguishes two contributions:

- Rolling noise, L_r
- Power unit noise, L_p

The overall level is the energetic sum of the two contributions (see Figure 4.1):

$$L_{A\max} = L_p \oplus L_r = 10\log_{10}[10^{0.1L_p} + 10^{0.1L_r}]$$

Here the two terms are conventional. For instance, there is no claim that L strictly corresponds to rolling noise.

Rolling noise component

The rolling noise component depends on vehicle speed and on the road surface. On the road surface side the method takes into account the type of formulation and age as influence factors. In NMPB 2008, the pavement formulations available in France are split in three categories named R1, R2, and R3, going from the least noisy pavements to the noisiest ones (see Figure 4.2). Cobblestones are considered to have a higher emission than R3. This classification must be understood statistically. A pavement whose surface type is R1 might behave like an R2 pavement. This classification

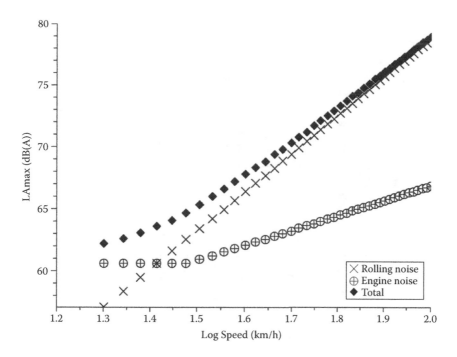

Figure 4.1 NMPB 2008 emission model. Rolling noise, engine noise component, and energetic sum for a LV on an RI pavement.

is based on 380 SPB measurement campaigns on LV where each site is described by the following speed dependence:

$$L_{A\max}(v) = L_{A\max}(v_{ref}) + b\log_{10}\frac{v}{v_{ref}}$$

where v_{ref} is the reference speed (90 km/h for LV, 80 km/h for HV).

In the database used, the age of pavement ranged from a few months to 18 years, but the majority of surfaces were less than 3 years old.

The rolling noise component is the subtraction of the power unit noise component from the global noise.

The sound power level by unit length of source line $L_{rW/m}$ for a pavement has the following expression:

$$L_{rW/m} = a_i + b_i\log_{10}\frac{v}{v_{ref}} + \Delta L_{rW/m}$$

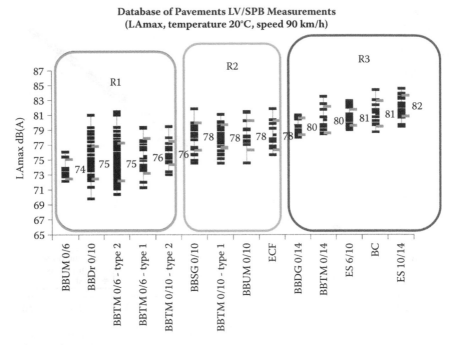

Figure 4.2 Definition of pavement categories. BBTM means very thin asphalt concrete; BBUM, ultra-thin asphalt concrete; BBDR, drainage asphalt concrete; BBSG, dense asphalt concrete; ECF, cold mix; BC, cement concrete, and ES, surface dressing.

where i is the index of the category of pavement, v_{ref} equals 90 km/h for LV or 80 km/h for HV, and $\Delta L_{rW/m}$ is a corrective term to take ageing into account.

The model provides scattering with 95% confidence for the three pavement categories. They range from 2.5 to 3.4 dB(A) depending on the vehicle (LV, HV) and pavement categories (R1, R2, R3).

The evolution of $L_{rW/m}$ over time depends on the vehicle category, the category of road surface, and the age class (see Table 4.1).

Table 4.1 Corrections for HV Depending on Gradient and Traffic Flow Type

	Gradient		
	$0\% \leq p \leq 2\%$	Upward $2\% \leq p \leq 6\%$	Downward $2\% \leq p \leq 6\%$
Stabilised pace	0 dB(A)	$2 \cdot (p-2)$	$1 \cdot (p-2)$
Acceleration	5 dB(A)	$5 + \max[2 \cdot (p-4.5); 0]$	$5 \cdot$ dB(A)
Deceleration	0 dB(A)	0 dB(A)	$1 \cdot (p-2)$

Power unit component

The propulsion noise component depends on speed, traffic flow type, and road gradient. As mentioned before, pace is either steady speed or accelerating or decelerating, in a qualitative way. This component depends both on the vehicle and on the driving style. It was obtained essentially from controlled pass-by measurements on a representative sample of the French vehicle fleet.

The propulsion component, $L_{p,W/m}$, of LV is as follows:

$$L_{p,W/m} = a_{t,i} + b_{t,i} \log_{10} \frac{v}{90}$$

where tuples $(a_{t,i}, b_{t,i})$ depending on traffic flow type t allow for covering the range 20 to 130 km/h in two or three intervals.

The propulsion component of HV is slightly more elaborate:

$$L_{p,W/m} = a_i + b_i \log_{10} \frac{v}{80} + \Delta L_{p,W/m}, i \in \{1,2\}$$

where $\Delta L_{p,W/m}$ depends on traffic flow type and road gradient as shown in Table 4.2.

For the lowest speed interval, $b_{t,i}$ or b_i is typically negative, whereas it is obviously positive for higher speeds.

Starting and stopping segments

For speeds below 20 or 25 km/h, NMPB 2008 considers so-called starting and stopping segments. For LV the sound power level per unit length is a constant. For HV, $L_{W/m}$ in some cases depends on road gradient and ageing (see Table 4.3).

Table 4.2 Age Effect $\Delta L_{rW/m}$ in dB(A) as a Function of Age a in Years and Category of Pavement

Age of Surface		LV		HV	
		≤2 Years	2 to 10 Years	≤2 Years	2 to 10 Years
Pavement category	R1	−4	$0.5 \cdot (a - 10)$	−2.4	$0.3 \cdot (a - 10)$
	R2	−2	$0.25 \cdot (a - 10)$	−1.2	$0.15 \cdot (a - 10)$
	R3	−1.6	$0.2 \cdot (a - 10)$	−1	$0.12 \cdot (a - 10)$

Table 4.3 Sound Power Level $\Delta L_{W/m/veh}$ in dB(A) for Starting and Stopping Section Taking Age of Pavement into Account

	LV		PL	
	All Gradients	*Horizontal Road (0% ≤ p ≤ 2%)*	*Upward Gradient (2% ≤ p ≤ 6%)*	*Downward Gradient (2% ≤ p ≤ 6%)*
Starting section	51.1	62.4	62.4 + max[2.(p − 4.5); 0]	62.4
Stopping section	44.5	58.0	58.0	58.0 + (p − 2)

Spectrum

NMPB 2008 considers the pavement type in the spectrum. It distinguishes

- Drainage asphalt (BBDr)
- Nondrainage asphalt

As expected, the drainage spectrum exhibits its maximum at a lower frequency than the nondrainage one (see Figure 4.3). If one assumes two pavements with the same noise level, one drainage and one nondrainage, then at low frequencies, the drainage asphalt is noisier than the nondrainage one.

The emission spectrum does not depend on speed or on traffic flow type. This spectrum assumes a traffic with 15% HV. So it does not distinguish between LV and HV. Comparisons with experiments have shown that this simplification generates discrepancies below 500 Hz, depending on the percentage of HV, but the incidence on the level in A-weighted decibels is negligible.

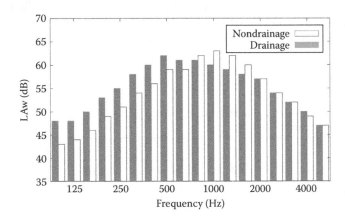

Figure 4.3 NMPB 2008 traffic noise spectrum on drainage or nondrainage asphalt.

The sound power level per unit length for third octave j is obtained by introducing the traffic:

$$L_{w/m}(j) = (L_{w/m/LV} + 10\log_{10} Q_{LV}) \oplus (L_{w/m/HV} + 10\log_{10} Q_{HV}) + R(j)$$

where $R(j)$ is the A-weighted 0-dB-standardised traffic spectrum for third octave j, and Q_{LV} stands for the hourly flow rate of LV (Q_{HV} *for* HV).

Comments

In this model, the rolling noise component starts to overcome the engine noise from 40 km/h on. The pavements of the 1970s can be classified as R3. The rolling noise component above 70 km/h for R3 is almost equal to the one of *Guide du Bruit,* whereas at lower speeds, the rolling noise component of NMPB 2008 is systematically lower.

The emission model for road sources in NMPB 2008 is defined on a pragmatic basis. Not easily available parameters have been avoided. It is built on more than a decade of standard pass-by measurements.

Of course one may regret that the model does not cover two-wheelers and medium-sized vehicles. For LDEN calculations two-wheelers almost never contribute significantly in France. A significant basis of experimental data on these vehicles is now available from a recent French research project.[9,10] So an additional category could be added in the near future.

The problem with medium-sized vehicles is that the traffic counting devices used in France do not provide enough information to split more than 3.5 tons in two categories. The situation is likely to be the same in many other member states.

PROPAGATION

For specification of the propagation part see NMPB 2008.[4] For further details on the physical aspects, the reader shall refer to Dutilleux et al.[11] As in any engineering model for outdoor sound propagation, the general approach of NMPB 2008 is to break down the physical noise sources into elementary sources and to do a point-to-point calculation between one source and one receiver. Meteorology is taken into account by a set of meteorological classes. Each class corresponds to a certain vertical sound speed gradient. This gradient is supposed to be constant throughout the range. NMPB 2008 assumes a linear sound speed gradient and two meteorological classes.

For a given source, S, whose sound power is L_w and a given receiver, R, the engineering methods use ray-tracing or image-source methods to identify the set of propagation paths between S and R. A particular path may include reflections and diffractions, whether on the ground or obstacles like noise barriers or buildings.

Sound levels

Elementary contribution of a propagation path

The A-weighted sound pressure level generated by S at a distance d of R in propagation condition C is defined by

$$L_{A,C} = L_w - (A_{div} + A_{atm} + A_{bnd,C})$$

where

$A_{div} = 20\log_{10} d + 11$ is the geometrical spreading.

A_{atm} is the atmospheric absorption. A_{atm} is computed like in ISO 9613-1 for a reference atmosphere whose temperature is 15°C, relative humidity is 70%, and atmospheric pressure is 101325 Pa.[12]

$A_{bnd,C}$ is the attenuation relating to the boundary characteristics and the only attenuation term that is propagation-condition dependent. In $A_{bnd,C}$, C stands for either "homogeneous" or "downward refraction."

The boundary is composed of the ground and the occasional man-made obstacles like buildings or noise barriers. In NMPB 2008, the ground attenuation is not taken into account by reflected path but by an attenuation term representing the ground effect. This term is included in $A_{bnd,C}$. More details on $A_{bnd,C}$ are provided in the next sections.

Long-term sound level

Several paths between a source and a receiver exist *a priori*, depending on topography and constructions. A long-term sound level, $L_{Ai,LT}$, is associated to each path, i. This long-term sound level is derived from two computations on each path, one for homogeneous conditions, and one for downward-refraction conditions. The sound level in homogeneous conditions, $L_{Ai,H}$, is a safe-side estimate of the level in upward-refraction conditions, because it is well-known that homogeneous conditions are only a transient state of the atmosphere over the day–night cycle. The sound level in downward-refraction conditions, $L_{Ai,F}$, is obtained assuming a standard atmosphere with a range-independent sound-speed gradient. Thus, the site- and orientation-dependent probability, p_i, of occurrence of downward-refraction conditions allows one to compute

$$L_{Ai,LT} = (p_i 10^{0.1 L_{Ai,F}}) \oplus ((1 - p_i) 10^{0.1 L_{Ai,H}})$$

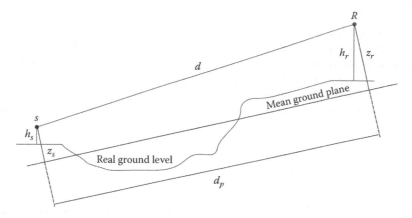

Figure 4.4 Concept of mean ground plane.

Ground description

Mean ground plane

Like in ISO 9613-2, along a propagation path, the description of the ground in NMPB 2008 is based on the concept of a mean ground plane as shown in Figure 4.4. A specific procedure based on least squares regression describes how to calculate in a robust way the mean ground plane from the available topographic data. Heights and distances are recomputed with respect to the mean ground plane. The ground effect is computed under this simplified geometrical framework. When a diffraction occurs, the same approach applies except that two mean ground planes are considered on each side of the diffraction point (see Figure 4.5).

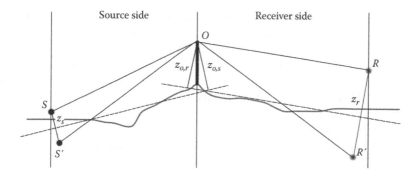

Figure 4.5 Two mean ground planes are computed when diffraction occurs.

Absorption

Whereas advanced noise prediction models use impedance, the local absorption is represented in NMPB 2008 by a frequency-independent adimensional parameter, G. $G = 0$ corresponds to a reflecting ground like a road pavement and $G = 1$ for an absorbing one like a grass-covered ground. Even if G varies between 0 and 1, G must be seen more as a normalized airflow resistivity than as an absorption coefficient. Indeed G is linked to the effective airflow resistivity, σ, by the empirical formula[12]

$$G = \min\left[\left(\frac{300}{\sigma}\right)^{0.57}, 1\right]$$

where σ is expressed in cgs rayls units. The average of G along the mean ground plane between source and receiver stands for the absorption of the propagation path, G_{path}. When source and receiver are close to each other, it is assumed that the reflection on the ground takes place on the road. So G'_{path} is introduced as

$$G'_{path} = \begin{cases} \dfrac{d_p}{30(z_s + z_r)} G_{path} & 30(z_s + z_r)/d_p < 1 \\ G_{path} & otherwise \end{cases}$$

where d_p is the orthogonal projection on the mean ground plane, z_s and z_r are the source and receiver heights (see Figure 4.4) over the mean ground plane.

This correction is necessary for the common case of propagation over an absorbing ground besides the reflecting road. Otherwise, the reflection on the ground would be underestimated.

Ground effect

In NMPB 96, there were two totally different expressions for the ground effect A_{ground}: one for homogeneous conditions and one for downward-refraction conditions.[13] The latter was the same as in ISO 9613-2. Since the equivalent source for road traffic appears to be rather close to the road surface, the empirical ISO 9613-2 expression is not well suited because it has been obtained from measurements on industrial sources sufficiently high above the ground.[14]

NMPB 2008 introduces a new formulation of $A_{ground,F}$ that is based on the expression of $A_{ground,H}$ and valid for any height[15] and already used in NMPB 96:

$$A_{ground,H} = \max[A_{ground,H}^{unbounded}, -3(1 - G'_{path})]$$

where

$$A_{ground,H}^{unbounded} = -10\log_{10}\left[4\frac{k^2}{d_p^2}\left(z_s^2 - \sqrt{\frac{2C_f}{k}}z_s + \frac{C_f}{k}\right)\left(z_r^2 - \sqrt{\frac{2C_f}{k}}z_r + \frac{C_f}{k}\right)\right]$$

where k is the wavenumber. See Figure 4.4 for z_s, z_r, d_p. G_{path}' has been defined earlier and

$$C_f = d_p \frac{1 + 3w(f)d_p e^{-\sqrt{w(f)d_p}}}{1 + w(f)d_p}$$

where

$$w(f) = 0.0185\frac{f^{2.5}G_{path}^{2.6}}{f^{1.5}G_{path}^{2.6} + 1.3\cdot10^3 f^{0.75}G_{path}^{1.3} + 1.16\cdot10^6}$$

This empirical expression describes the frequency dependence of ground absorption. For more details, see Defrance and Gabillet.[15]

The atmospheric refraction in downward-refraction conditions is taken into account by means of height corrections δz_s and δz_r applied to z_s and z_r. These height corrections are derived from the analogy between flat ground with curved rays versus curved ground with straight rays. Assuming a linear vertical sound speed profile with a stochastic part determined by the variance of the refraction index $\mu \ll 1$, this profile can be written as

$$c(z) = bz + c_0(1 + \mu)$$

The height correction terms are given by

$$\delta z_s = \frac{b}{2c_0}\left[\frac{d_p z_s}{z_s + z_r}\right]^2, \qquad \delta z_r = \frac{b}{2c_0}\left[\frac{d_p z_r}{z_s + z_r}\right]^2$$

where the refraction parameter, that is, the linear vertical sound speed gradient, is assumed to have a mean value $b = 0.07s^{-1}$.

This approach gives good results as long as there is only one reflection on the ground. To take into account multiple-reflection phenomena, the bound of $A_{ground,H}$ is replaced by

$$(1 - G_{path}')\left[-3 - 6(1 - 30(z_s + z_r)/d_p)\right]$$

for $30(z_s + z_r)/d_p \leq 1$.

Turbulence is also taken into account in the same way as refraction, that is, through a height correction δz_T added to z_s and z_r. It allows one to model the coherence loss between direct and reflected rays, in the case of propagation above flat ground. It will also enable one to shift the first interference toward low frequencies with a lower amplitude. This searched behaviour imposes δz_T to be equal for z_s and z_r. Comparisons with Daigle's results have led to the following expression of δz_T for a value $\langle \mu^2 \rangle = 2 \cdot 10^{-6}$:

$$\delta z_T = 6 \cdot 10^{-3} \frac{d_p}{z_s + z_r}$$

To summarize

$$A_{ground,F} = \max\left[A_{ground,F}^{unbounded}, A_{ground,F}^{floor} \right]$$

where

$$A_{ground,F} = -10 \log_{10}\left[4 \frac{k^2}{d_p^2}\left(\tilde{z}_s^2 - \sqrt{\frac{2C_f}{k}}\tilde{z}_s + \frac{C_f}{k} \right)\left(\tilde{z}_r^2 - \sqrt{\frac{2C_f}{k}}\tilde{z}_r + \frac{C_f}{k} \right) \right]$$

with

$$\tilde{z}_s = z_s + \delta z_s + \delta z_T$$

$$\tilde{z}_r = z_r + \delta z_r + \delta z_T$$

and

$$A_{ground,F}^{floor} = \begin{cases} (1-G_{path}')[-3-6(1-30(z_s+z_r)/d_p)] & 30(z_s+z_r)/d_p \leq 1 \\ -3(1-G_{path}') & otherwise \end{cases}$$

As mentioned in the Introduction, in some situations with an embankment close to the source, the mean ground plane approach leads to a significant underestimation of the ground effect. To address this problem, a corrective term has been introduced. It is limited to civil-engineered slopes at the vicinity of the road, and slopes of embankment between 15° and 45°. An additional image-source is introduced with respect to the plane of the embankment (see Figure 4.6). A Fresnel zone approach is used to compute the amplitude of the contribution at the receiver of the reflection on the embankment. This contribution increases as the ratio ε of the intersection of the Fresnel ellipsoid with the bank increases.

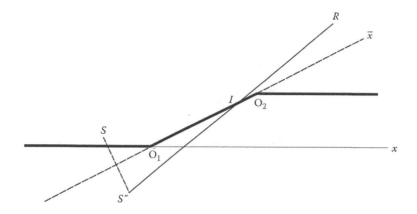

Figure 4.6 Additional image source in the case of an embankment close to the road.

The attenuation due to the embankment is defined by

$$A_{embankment} = -1.5\varepsilon(2 - G_{embankment})$$

where $G_{embankment}$ is the ground factor of the embankment. For more details on the definition of this correction, the reader shall refer to NMPB 2008[4] and Dutilleux et al.[11]

When it occurs, the contribution of this reflection is also taken into account in the ground effect part of the diffraction formulas introduced next.

Diffraction

One assumes a vertical barrier above a horizontal ground, with a source (S) on one side of the barrier, and a receiver (R) on the other one. Four paths must be considered: SOR, S'OR, SOR', and S'OR', where O is the edge of diffraction. S' is the image source and R' is the image receiver both with respect to the ground (see Figure 4.5). These paths are still valid in the more general case of two mean ground planes as previously introduced. The diffraction formulation used in NMPB 2008 accounts in first approximation for all these paths at once using a single formula:

$$A_{dif} = -20\log_{10}\left[10^{-\frac{\Delta_{dif}(S,R)}{20}}\left[1+\left(10^{-\frac{A_{ground}(S,O)}{20}}-1\right)10^{-\frac{\Delta_{dif}(S',R)-\Delta_{dif}(S,R)}{20}}\right]\right.$$

$$\left.\left[1+\left(10^{-\frac{A_{ground}(O,R)}{20}}-1\right)10^{-\frac{\Delta_{dif}(S,R')-\Delta_{dif}(S,R)}{20}}\right]\right]$$

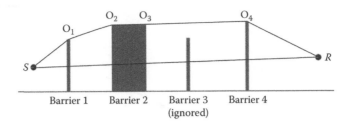

Figure 4.7 Sample definition of the convex hull in which multiple diffractions occur.

where

$$\Delta_{dif} = 10C_h \log_{10}(3 + 20N)$$

where N is Fresnel's number, and

$$C_h(f) = \min\left(1, h_0 \frac{f}{250}\right)$$

where h_0 is the larger of the two heights of the diffraction edge with respect to the two mean ground planes. f is the third octave median frequency of interest. $C_h(f)$ is a correction factor in order to better evaluate the insertion loss of low height barriers, like the 80 cm high continuous reinforced concrete gliders often used in France. When the height of the barrier increases, $C_h(f)$ tends be 1, and Δ_{dif} boils down to a classical formula for noise barriers.

When a potential diffraction is identified along a propagation path, the path length difference between the ray following the convex hull and the direct ray is computed. (See Figure 4.7.) If this diffraction is above a threshold value, then A_{dif} is computed, otherwise no significant diffraction occurs and A_{ground} is computed. It must be emphasized that A_{dif} takes the ground effect into account.

Fresnel's number, N, is proportional to the path length difference, δ. δ is computed with straight paths in homogeneous conditions and curved paths in downward-refraction conditions.

Multiple diffractions (see Figure 4.1) on several thin barriers, one or more thick barrier or earth berm, and one or more buildings are addressed in a simplified way in NMPB 2008. The additional hypothesis is that no ground effect occurs between the first and the last edge of diffraction.

Perhaps of less significance for linear infrastructures like roads, NMPB 2008 also provides a treatment of diffractions on vertical edges (see Hamet et al.[6] for more details). Combinations of diffractions on vertical and horizontal edges are not covered by the method.

Reflections on obstacles

Reflections on obstacles like noise barriers of building façades are dealt with by image sources. If one assumes a reflection on a surface whose absorption coefficient is α_r, the sound power level of the source is modified accordingly:

$$L_{w'} = L_w + 10\log_{10}(1-\alpha_r)$$

Complex configurations

For trenches, tunnel mouths, and partial covers, NMPB 2008 still refers to *Guide du Bruit*.[2] Albeit in theory, ray-tracing software could directly handle these configurations and identify the paths between a source and a receiver, the order of reflexion required to obtain correct sound levels is excessive and makes this approach impractical. Therefore, such configurations are addressed by the use of equivalent sound sources.

Total cover

A receiver on a façade located at the vicinity of a total cover will receive three different contributions (Figure 4.8):

- The energy radiated by the cover itself. The insertion loss introduced by the cover must be added to the other attenuations along the path from source to receiver:

$$L'_{w/m} = L_{w/m} - \left(R + 10\log_{10}\frac{A}{l}\right)$$

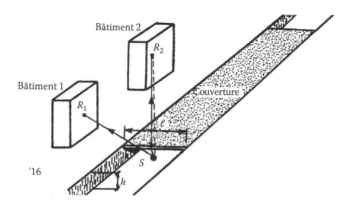

Figure 4.8 Total cover of finite length.

where A is the equivalent absorption area in a unit length of tunnel and l is the width of the cover.

- The energy from sources outside the cover. These sources must be handled exactly like the ones of a standard line source.
- The energy of an equivalent source representing the tunnel mouth effect:

$$L_w = L_{w/m} - 10\log_{10}\frac{A}{S}$$

where S is the area of the tunnel mouth.

Partial cover

In general, it is not worth computing the energy radiated by the cover itself, since it is negligible compared to the energy radiated by the openings. One can identify several contributions:

- Energy from direct paths and reflected at the first order
- Diffuse energy radiated by the openings after multiple reflections
- Energy from the ends of the partial cover like for a total cover

In NMPB 2008, a partial cover is characterised in a vertical cross-section by its covered perimeter, S, above the road and its opened perimeter, Ω. If the real cross-section is not made of a vertical and a horizontal wall, the real section must be replaced by such a bounding-boxlike approximate section. The contribution of each source under the partial cover must be computed like any diffracted path on the edge of the cover. The contribution of the image source with respect to the vertical wall is also added and the attenuation is also the one of a diffracted path on the edge of the cover.

If the partial cover is one-sided then the equivalent point source for the diffuse field is located at the midpoint on vertical passing by edge of the cover (if the cover is two-sided, the midpoint is between the extremities of the opening), as shown in Figure 4.9.

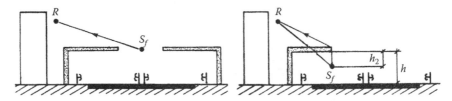

Figure 4.9 Cross-section of partial cover and position of equivalent sources for diffuse field.

For more details on special configurations, the reader shall refer to NMPB 2008.

Meteorological conditions

NMPB 2008 introduces the different ways to obtain occurrence probabilities necessary to combine a pair of sound pressure levels in homogeneous conditions and downward-refraction conditions in a long-term sound pressure level. As usual, the uncertainty increases as the amount of effort to collect and process meteorological data decreases. NMPB 2008 is delivered with precomputed occurrence probabilities distinguishing day, evening, and night periods for 41 meteorological stations distributed over the French territory (see Figure 4.10 for an example).[4] Since strategic noise mapping is a large-scale effort, with little time to delve into the details, it is likely that precomputed values will be used extensively in this context.

For these 41 French stations, 30 years of meteorological time series have been used. In another country, if long-term meteorological time series are available, the steps to calculate occurrence probabilities are as follows[16]:

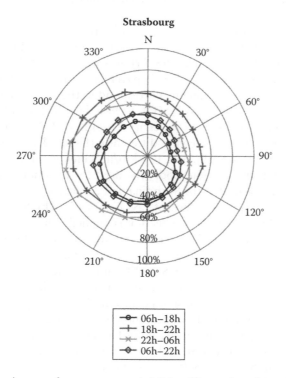

Figure 4.10 Sample rose of occurrence probabilities. **(See colour insert.)**

1. Use a micrometeorological model to compute gradients in a time series of temperature and wind speed gradients. For France, Choisnel's model has been used.[17]
2. Compute the hourly time series of sound speed gradient for each 20° sector.
3. Compute the histogram of these time series for each standard period day, evening, and night.
4. Divide the histogram in two classes with a threshold at 0.07 s^{-1} and compute the probability of downward-refraction conditions.

COMPARISON OF NMPB WITH HARMONOISE/IMAGINE

As mentioned in the Introduction to this chapter, NMPB 96 was the interim method for road traffic noise during the first round of application of the END. It is worth looking at the transition between the interim method and an IMAGINE-based CNOSSOS-EU, which is the preferred candidate of Europe to become the common mandatory method at the time of this writing.[18] As already explained, NMPB 96 is now superseded by NMPB 2008 so the comparison is provided with the latter method.

For the emission part we can compare the emission models. The correspondence between the two methods is only straightforward for categories 1 and 3 of CNOSSOS, since NMPB is based on two vehicle categories, light vehicles and heavy vehicles (>3.5 t). Results with NMPB are provided for pavement classes R2 ("average" pavements) and R3 ("noisy" pavements). Class 3 pavements roughly correspond to the interim emission data. CNOSSOS-EU category 1 and NMPB 2008 light vehicles on R2 pavement match very well. CNOSSOS-EU category 3 and NMPB 2008 heavy vehicles on R2 pavement behave somewhat differently (Figure 4.11). The difference of power level reach is 5 dB(A) at 100 km/h, which is quite a large model. With the IMAGINE emission mode, the need for regional corrections has already been emphasized in the Scandinavian countries.[19]

Here the total sound power level of CNOSSOS-EU of high and low sources is displayed and no directivity correction is applied. The assumptions comply as much as possible to the reference conditions of CNOSSOS-EU, that is, all correction factors are supposed to equal zero.

The propagation part of CNOSSOS-EU and the one of NMPB 2008 have been studied with respect to six measurement campaigns with combined acoustic and meteorological instrumentation. These campaigns have been carried out in the framework of the validation of NMPB 96 and of NMPB 2008. They feature embankments, noise barriers, hilly terrain, diffraction by an edge of a road platform, and road on the side of a valley. The maximum range from the road is 350 m, acoustic measurements were located at 2 and 5 m height above the ground. Forty-nine different acoustic measurement

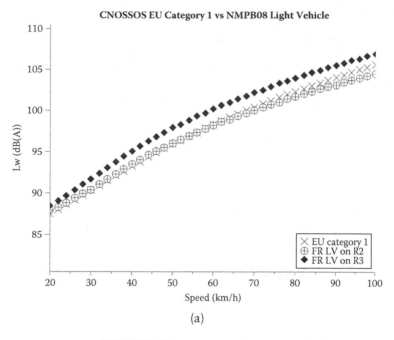

(a)

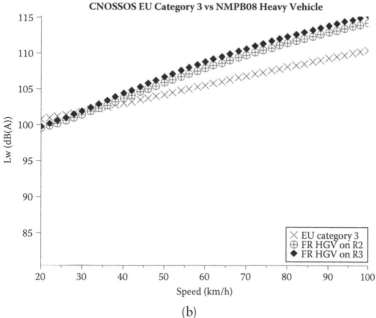

(b)

Figure 4.11 Comparison of sound power levels in dB(A) between CNOSSOS-EU and NMPB 2008 with class 2 and class 3 pavements.

points have been collected. From these measurements the postprocessing leads to a set of 310 validated day or night periods. For each period, a probability of occurrence of downward-refraction conditions is computed from meteorological data collected *in situ*. The line sources are located at 0.05 m height above the pavement. Computations are performed in homogeneous conditions and in downward-refracting conditions with a linear sound speed gradient of amplitude 0.07 s^{-1}. For each period, an L_D or L_N is computed taking the relevant probability of occurrence into account. The comparison is carried out on attenuations from a reference point located close to the infrastructure. For more details see NMPB 2008[4] and Dutilleux et al.[11]

Since CNOSSOS-EU is not available as software at the time of this writing, the closely related "Harmonoise DLL" point-to-point C library (release 2.019) distributed by CSTB has been used. Release 2.019 takes the output of IMAGINE project into account. The histogram of deviations is shown on Figure 4.12. Contrary to NMPB 2008 data, the Harmonoise ones

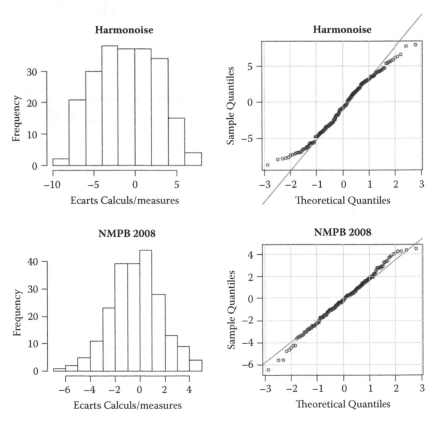

Figure 4.12 Histograms and quantiles versus Gaussian quantiles of measurement/prediction deviation for both methods.

are not normally distributed. The latter histogram is closer to a uniform distribution. The mean with a 95% confidence interval of deviations for Harmonoise is –1 dB(A) [standard deviation 3.7 dB(A)], for NMPB 2008 –0.2 dB(A) [standard deviation 2 dB(A)]. The slightly larger standard deviation for Harmonoise is related to a significant number of outliers. Better results may be obtained if more meteorological classes or a more detailed description of absorption of the ground is made. However, the amount of data used here is already quite detailed compared to what is available and used in END-related strategic noise maps and action plans.

CONCLUSION

NMPB 2008 is an in-depth revision of NMPB 96. Arguably, the most important change in the emission part is the separation of rolling noise and engine noise in the formulation. From the end user point of view, the most important change is in the description of road surfaces.

In the propagation part the most significant one is the formula for ground effect in downward-refraction conditions. The result is a more consistent formulation of the ground effect, with essentially the same formula for homogeneous and downward-refraction conditions. This formula is satisfying for a much wider range of source and receiver heights than in ISO 9613-2. By addressing the case of embankment slopes, the field of application of NMPB has been extended to the practical application of the estimation of the sound power level of an infrastructure from acoustic measurements instead of traffic data. Moreover, the diffraction formula can now cope with low barriers.

It must be emphasized that the emission part of NMPB 2008 is based on a very large experimental data set of ISO 11819-1 compliant measurements. The propagation part of NMPB 2008 benefits from a significant effort of experimental validation on real sites with nonflat topography and diffractions.

Compared to NMPB 96, NMPB 2008 exhibits a more moderate trend to the overestimation of noise levels in downward-refraction conditions. During the whole process of revision, attention has been paid to keep the method simple to use. As a result, while the method offers more possibilities to the noise consultant, the complexity of use and software implementation is not significantly increased. At least for transportation sources, if computation time matters, it is possible to limit the computations to octave bands without any significant change in the global level at the receiver.

NMPB 2008 has been compared with the so-called Harmonoise point-to-point DLL distributed by CSTB against experimental campaigns from the validation of NMPB and Harmonoise/IMAGINE. This exercise shows that the improvements expected from an advanced method like the outcomes of these European projects in terms of results are not systematic. Much more detailed input data is probably necessary for this method. Here misinterpretations in the implementation of the Harmonoise library used are unlikely, since its developer is one of the main designers of Harmonoise.

Therefore, when compared to ISO 9613-2 and the deliverables of Harmonoise and IMAGINE, NMPB 2008 appears to be a good trade-off between precision, CPU time, the skills required to handle the method properly, and the effort to collect the input data necessary for a simulation.

NMPB 2008 has been translated into AFNOR NF S 31-133:2011 standard.[4] This standard proposes a unified method for noise prediction of road, rail, and industrial sources. At the time of writing, the development of reference software libraries implementing this standard and the related emission values for road and rail sources is in progress. These libraries will be made available free of charge for noise prediction software vendors in order to reduce the discrepancies between two softwares implementing the method.

Adapting NMPB 2008 to the context of other countries seems feasible without any excessive effort. For road sources, other pavement formulations could be included on the basis of ISO 11819-1 compliant measurements. For rail sources, the source positions introduced should allow for the specificities of foreign rolling stock. The procedure for computing the probabilities of downward-refraction conditions is quantitative and straightforward as long as an hourly time series of meteorological data are available.

REFERENCES

1. Directive 2002/49/EC, 2002 of the European Parliament of the Council of 25th June 2002 relating to the assessment and management of environmental noise.
2. CETUR, 1980, *Guide du bruit des transports terrestres—Prévision des niveaux sonores*, in French.
3. Besnard F., Hamet J.F., Lelong J., Le Duc E., Fürst N., Doisy S., Dutilleux G., 2009, Road noise prediction 1—Calculating sound emissions from road traffic, SÉTRA, http://www.setra.equipement.gouv.fr/IMG/pdf/0924-1A_ Road_noise_prediction_v1.pdf.
4. Besnard F., Gauvreau B., Dutille G., Bérengier M., Defrance J., et al., 2009, *Road noise prediction 2—Noise propagation computation method including meteorological effects (NMPB 2008)*, Sétra, http://www.setra.equipement. gouv.fr/IMG/pdf/US_0957-2A_Road_noise_predictionDTRF.pdf.
5. NF S 31-133, 2011, Acoustics—Outdoor noise—Calculation of sound levels, AFNOR.

6. Hamet J.F., Besnard F., Doisy S., Lelong J., Le Duc E., 2010, New vehicle noise emission for French traffic noise prediction, *Applied Acoustics*, vol. 71, no. 9, pp. 861–869.

7. Hamet J.F., Pallas M.A, Gaulin D., Berengier M., 1998, *Acoustic modelling of road vehicles for traffic noise prediction—Determination of the source heights.* In 16th International Congress on Acoustics, Seattle, USA.

8. Golay F., Dutilleux G., Ecotière D., 2010, Source height determination for several sources at the same height, *Acta Acustica United with Acustica*, vol. 96, no. 5, pp. 863–872.

9. Toussaint L., Lefèvre H., Dutilleux G., 2010, *Emission acoustique des deux-roues motorisés: Scooters et cyclomoteurs*, 10th Congrès Français d'Acoustique, Lyon, France, in French.

10. Lefèvre H., Toussaint L., Dutilleux G., 2010, *Emission acoustique des deux-roues motorisés: motocyclettes*, 10th Congrès Français d'Acoustique, Lyon, France, in French.

11. Dutilleux G., Defrance J., Ecotière D., Gauvreau B., Bérengier M., Besnard F., Le Duc E., 2010, NMPB-Routes-2008: The revision of the French method for road traffic noise prediction, *Acta Acustica*, vol. 96, no. 3., pp. 452–462.

12. ISO 9613-1, 1993, Acoustics—Attenuation of sound during propagation outdoors—Part 1: Calculation of the absorption of sound by the atmosphere.

13. NMPB 96, 1997, Bruit des infrastructures routières: méthode de calcul incluant les effets météorologiques, version expérimentale, NMPB-Routes-96, CERTU, SETRA, LCPC, CSTB, in French.

14. Attenborough K., Li K.M., Horoshenkov K., 2007, *Predicting outdoor sound*, Spon Press.

15. Defrance J., Gabillet Y., 1999, A new analytical method for the calculation of outdoor noise propagation, *Applied Acoustics*, vol. 57, no. 2, pp. 109–127.

16. Ecotière D., 2008, Révision de la NMPB-Routes-96, calcul des occurrences météorologiques favorables à la propagation, Technical report LRS, in French.

17. Brunet Y., Lagouarde J.P., Zouboff V., 1996, *Estimating long-term microclimatic conditions for long range sound propagation studies,* 7th Long Range Sound Propagation Symposium, Lyon, France.

18. JRC, 2010, Draft JRC Reference Report on Common NOise ASSessment MethOdS In EU (CNOSSOS-EU), Version 2D.

19. Jonasson H.G., 2007, Acoustical modelling of road vehicles, *Acta Acustica*, vol. 93, no. 2, pp. 173–184.

Chapter 5

Railway noise

P. de Vos

CONTENTS

INTRODUCTION

Railway noise is the collective name for the noise created by the operation of rail-bound vehicles. Commonly, railway noise indicates the noise from trains running on main railway tracks; this mode of transport is known as *heavy rail*. A range of other sources of noise is linked to heavy rail networks, such as trains running or resting stationary in depots and shunting yards, in railway stations, together with fixed installations such as servicing equipment (washing, cleaning, fuelling stations), level crossing warning installations, and auxiliary power stations.

Opposed to heavy rail there is *light rail*, which is the collective name for all rail-bound rolling stock for urban and regional transport, with relatively lightweight vehicles, such as metros, streetcars, and light rail vehicles.

Currently, there is a development where the strict distinction between heavy and light rail gradually vanishes, as heavy vehicles are sometimes used in combined heavy and light rail networks. This occurs in passenger transport first but may be followed by certain types of (fast) cargo transport as well.

Rail traffic is distinguished as follows:

- Passenger transport with conventional speed (up to 200 km/h)
- High-speed (between 200 and 400 km/h) passenger transport
- Freight transport with speeds typically up to 80 or 100 km/h

Passenger transport runs mostly by a fixed time schedule; freight traffic often runs on demand, provided that there is track capacity available. The track capacity is then indicated as a "path."

Rolling stock

Passenger rolling stock consists of either multiple units (traction included) or of coaches pulled by an engine or locomotive. Both may be either diesel driven or electricity driven. Electric power for traction is provided by an overhead wire or sometimes by a third rail. Diesel traction is converted into electric power (diesel–electric traction) or hydraulic power (diesel–hydraulic traction).

Coaches and multiple units run mostly on *bogies*, which are units of four wheels that can rotate around the point carrying the vehicle. Most coaches have two bogies; in a multiple unit two adjacent coaches sometimes share the bogie (the so-called Jacobs bogie).

Freight wagons run on bogies or wheel sets (i.e., fixed axles with one wheel on either side). Freight wagons may have between two and six axles, depending on the load they have to carry. As we will see in the sections to follow, the wheel rail contact is the main source of noise for the majority of rail vehicles. Therefore, the number of axles per unit length of the vehicle is a highly important parameter to characterise the performance of the vehicle in terms of its noise generation.

The wheels

With few exceptions, such as part of the Paris metropolitan railways running on rubber tires, the vehicles have steel wheels, typically with a conical cross-section, consisting of a wheel web connected to a rim or tire and a

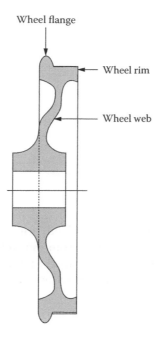

Wheel flange

Wheel rim

Wheel web

Figure 5.1 Wheel cross-section.

flange on the inner side of the track (Figure 5.1). Most modern wheels are monolithic, so-called monoblock wheels. For economic reasons, sometimes tired wheels are used, where the tire, once worn, can be exchanged. Today, the large majority of the wheels are monoblock wheels.

Railway wheels have a typical diameter of slightly less than 1 meter. The standard UIC* wheel has 920 mm. Smaller diameter wheels can be found in light rail and metro and in some specific heavy rail vehicles such as Ro-Ro lorry carriers.

The track

The track consists of steel rails, mostly joined together by wooden or concrete sleepers, either in a ballast bed or in nonballasted constructions such as slab track (Figure 5.2).

Heavy rail track consists of a standard rail, consisting of a foot, a rim, and a rail head. The current standard for heavy rail is UIC 60, the number indicating the mass in kilograms per meter length.

Streetcars usually run on grooved rails.

---

* UIC is the International Union of Railways, the standardising body for many railway issues. The UIC headquarters are located in Paris.

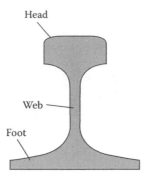

Figure 5.2 Rail cross-section.

Modern tracks are "continuously welded," that is, there are virtually no joints between sections of rail, apart from movable parts in switches and near bridges. In older parts of the network, jointed rail with many joints may give rise to the classic train noise (ka-boum, ka-boum).

SOURCES

The sources of noise for railway traffic can be distinguished according to the speed range (Figure 5.3):

- At low speed the *traction noise* or *engine noise* is dominant. Traction noise emerges from the main engine (exhaust, cooling, and ventilation for diesel engines; cooling, gearboxes, and transformer noise for electric engines). Its sound power output depends on the speed of the

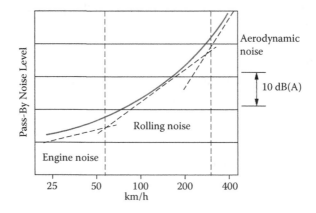

Figure 5.3 Pass-by noise level at different speed ranges.

train, generally between 0 and 20 lg v. Other parameters that have a significant influence in the sound power of the traction noise are

- Traction system type: Internal combustion engine/electric drive/other
- Number of traction units (motors)
- Rotational speed of the traction unit, load, and acceleration
- Type and number of cooling air fans and their rotational speeds
- Exhaust silencer performance (for internal combustion engines)

- At medium speed, that is, between 40 and 160 km/h, the *rolling noise* is the dominant noise source. Rolling noise is caused by surface irregularities in the wheel and rail contact area. These irregularities are indicated as "roughness." Rolling noise sound power increases with the speed of the vehicle typically as 30 lg v. Note that this relationship means an increase in pass-by noise level of approximately 10 dB for every doubling of speed.
- At high speed, from approximately 160 to 220 km/h, the *aerodynamic noise* is dominant. Aerodynamic noise is generated at any protruding objects at the train body, particularly at the pantograph, the bogies, intersections between coaches, and so forth. The sound power of aerodynamic noise is strongly dependent of the shape of protruding parts and increases with speed at typically 60 lg v.

In addition to these sources of noise, some other noise sources may be relevant at specific situations, such as

- Curve squeal, a high-pitch, almost pure tone noise, generated when a train passes through a (narrow) curve
- Brake screech, yet another high-pitch noise, originating from the interaction between the wheel and a brake block, obviously occurring only during braking
- Signalling noise, when a warning signal is produced by the train's horn

ROLLING NOISE AND ROUGHNESS

Out of these different generating mechanisms, rolling noise is by far the most important. For this reason, it has been the subject of thorough investigation during the last decades. Rolling noise is the result of vibrations in the wheel and rail, which are caused by mainly vertical dynamic forces due to minor surface irregularities in the rail and wheel contact area (Figure 5.4). These vertical surface irregularities are indicated as "roughness." The following paragraph, quoted from the Harmonoise final report,[1] represents the generation of rolling noise as an effect of surface roughness on wheels and rails.

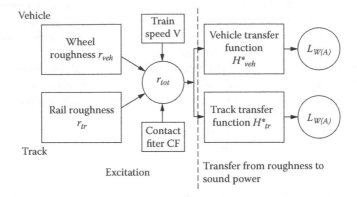

Figure 5.4 Scheme of generation of rolling noise. (Adapted from Categorisation of vehicles and tracks, Harmonoise report D11-WP2-HAR12TR-021107-SNCF10.)

Typical roughness amplitudes are in the order of a few microns (10^{-6} m) only, and have wavelengths of typically a few centimetres on the track and almost a decimetre on the wheel. Severe roughness on the track is indicated as "corrugation." Strong corrugation on the track may cause passenger discomfort due to vibrations in the coach. Strong corrugation is associated with noise levels alongside the track that are up to 20 dB higher than the average good condition track levels (Figure 5.5).

Figure 5.5 Periodic wear on rail head (initial rail corrugation).

Strong corrugation on the wheel is sometimes referred to as "polygonisation." Wheel roughness should not be confused with wheel flats, a type of wheel surface irregularities, which is produced when wheels slip. Wheel flats are another cause of the typical ka-boum–ka-boum sound of running trains. In modern braking systems there are antiblockage systems that prevent train wheels from slipping. In good maintenance, wheel flats are removed by "reprofiling" the wheels.

Track roughness is generated by periodic wear, caused by the dynamic properties of the track superstructure and its substructure. High track roughness can be mended by grinding the rail, but exceptionally it may grow back to unacceptably high roughness within 1 year; it may also take more than 10 years to grow back. It should be emphasized here that there is no evidence whatsoever that track roughness is the result of wheel roughness being imprinted on the rail or the result of braking or accelerating of trains. Rough wheels, however, do excite more vibrations in the track and could thus lead to higher periodic wear. The same applies to rough rails: when rails are smooth in the first place, there are few vibrations that will cause the rail roughness to grow. For this reason, some networks apply "preventive grinding" of new rails.

Wheel roughness has a distinct, proven cause: tread braking (Figure 5.6). This is the collective term for all types of braking that apply contact friction between a brake shoe and the wheel tread. Tread braking is the conventional type of braking. The friction between brake shoe and wheel tread transforms the kinetic energy of the train into heat that is partly radiated into the surroundings and partly absorbed into the brake shoe, the wheel tyre, and the wheel rim. The heat absorbed into the wheel causes the wheel tyre to expand and the surface of the circular tyre to "buckle." When the braking continues on the buckled wheel, periodic hot spots may occur, which then develop into worn spots.

Figure 5.6 Wheel with tread brakes.

Figure 5.7 Passenger bogie with disk brakes.

The conventional material for brake shoes is cast iron. About 99% of all freight vehicles are equipped with cast-iron brake blocks. Cast-iron brake blocks are known to generate high levels of roughness on the wheels. Alternatives for cast-iron brake blocks, such as composite and sinter-metal brake blocks, are known to generate lower roughness levels. These blocks either have a more favourable thermodynamic behaviour or a polishing effect. They thus lead to smoother wheel tread surfaces.

Modern braking technology is based on nontread braking, for instance, by means of disk brakes mounted on the wheel axle (Figure 5.7). Braking discs are found sometimes on the outer end of the axle, covering the wheel, but for higher braking performance a series of parallel braking discs is mounted on the inner side of the axle. Other, more exceptional braking types are drum brakes and magnetic rail brakes. Nontread braking leads to smoother wheel tread surfaces and therefore to lower rolling noise levels.

TRACK DECAY RATE

Apart from the roughness of the rail head, there is one other parameter that is very relevant to the track's noise performance: the track decay rate. This quantity indicates the loss of vibration energy by damping along the length of the rail. When this damping is high, the vibration wave does not travel very far along the rail, and this leads to a lower rail noise being emitted. The track decay rate is expressed in terms of decibels per metre (dB/m) and is a function of frequency. A distinction is made between lateral and vertical track decay rates. Typically, they are between 0.5 and 6 dB/m.

The track decay rate is dependent of the rail fixation and of the masses of rail and sleepers. When the rail fixation is stiff, the rail is held firmly, resulting in a high track decay rate. However, for reasons of comfort and to prevent damages, the rail needs a minimum amount of elasticity in its fixation points.

When the rail fixation is soft, the rail can vibrate more freely and the track decay rate is low. This leads to a higher noise emission from the rail. Rail fixation stiffness is to a certain extent determined by the "culture" of the rail design engineers. For instance, the typical track design in France includes a relatively soft rail fixation, whereas the track design in Germany and Austria is usually stiff fixation.

MODELLING ROLLING NOISE

In the previous paragraphs, it was stressed that rolling noise is the predominant cause of noise for rail traffic. In general terms, the sound production, E, produced by running trains at a given stretch of railway can be described by the following general equation:

$$E = a + b\lg v + 10\lg Q + c_t \quad \text{dB} \tag{5.1}$$

In this equation, a and b are constants that are dependent of the vehicle category, v is the vehicle speed, Q is the number of vehicles per hour passing on the particular stretch of railway, and c_t is a correction factor for the type of track the vehicles are running on.

Equation (5.1) is a very general equation. In interpreting noise prediction models, one should be aware of the different definitions used for the factor E. The usual interpretation is that E is the sound power level of a source, in dB re 10^{-12} W. However, this interpretation refers to point sources rather than line sources. Clearly, a railway line with running trains on it behaves as a line source. Thus, the time factor comes into the equation. One way to tackle this is to subdivide the total sound power level of a railway vehicle into sections of unit length, say 1 metre, and allocate these sources to a stretch of track of equal length, calculating the total amount of time that there is actually a train of that type present on that particular stretch of rail.

Obviously, that amount of time depends on the train's speed; when the train runs faster, the time it spends on that particular metre stretch of track is shorter. Therefore, when transferring sound production levels or sound power levels into equivalent sound reception levels, a factor c_{eq} comes into the equation, where

$$c_{eq} = 10.\lg\left(\frac{1}{v}\right) \tag{5.2}$$

where v is the train speed. It is important to note that the implementation of this correction leads to the fact that the equivalent sound level along the line relates to $\approx 20 \lg v$, whereas the pass-by level or the sound power level relates to ≈30 lg v. This explains the values for the factor b being around 20, as they can be found in some of the prediction models.

The general Equation (5.1) can be further scrutinized in different ways:

- Sometimes a distinction is made between trains running normally (traction switched on; no significant acceleration or deceleration) and trains with their brakes switched on. The latter usually produce a different rolling noise with a somewhat higher sound level. This should not be confused with brake screech, which is an entirely different source of noise with a typical high-pitch frequency spectrum.
- The total sound emitted by the sources of rolling noise, that is, the wheels and the track superstructure, is sometimes split over different source heights. For instance, the wheel source could be in the centre of the wheel, at 0.5 m above rail head, and the rail source could be at the same height as the rail head.
- Also, the lateral location of the sources may be of relevance. In some models, the source is located at the axis of the track under concern, whereas in other models the sources are located at the rail nearest to the receiver point.
- Finally, in more accurate models, the source output is modelled in terms of octave band or even third octave band levels, and in that case the values a and b in Equation (5.1) may assume different values for different frequency bands.

In the next section, Equation (5.1) is further elaborated, taking three current prediction methods as an example:

- The Dutch standard calculation method for railway noise, edition 1996, which was defined in Annex II of the Environmental Noise Directive as the recommended interim method for noise mapping.
- The Harmonoise/IMAGINE method, which was proposed as a candidate method for the CNOSSOS-EU set of methods, to be developed into the harmonized method for the 2017 round of noise mapping and action planning.
- The sonRail method, a Swiss development, building on the Harmonoise/IMAGINE method but developed further, representing the most recent state of the art model.

Rolling noise in the Dutch standard method

Edition 1996 of the Dutch "Reken- en Meetvoorschrift railverkeerslawaai" defines so-called emission numbers for eight different rail vehicle categories. Note that the emission number is not identical to, for example, sound power level per unit length or to the sound pressure level at any particular position along the track. It is merely a decibel number that should be taken as the input of the computation in order to arrive at the long-term average level.

The sound production is thought to be located at two different heights above rail head:

- The emission number $L_E{}^{bs}$, which is located at the height of the rail head
- The emission number $L_E{}^{as}$, which is thought to be located at the axle, that is, 50 cm above rail head

For most of the vehicle categories, the energy is distributed between the two source heights, such that 1 dB of the total energy is subtracted for the lower source and 7 dB is subtracted for the upper source height.

High-speed traffic is defined as the ninth vehicle category, with emission numbers for four source heights:

- 0.5 m above rail head for rolling noise and aerodynamic noise in the bogies
- 2.0 m above rail head for traction and ventilation noise
- 4.0 m above rail head for aerodynamic noise at the train's body
- 5.0 m above rail head for aerodynamic noise at the pantograph

To a certain extent, the eight vehicle categories are comparable to vehicle classes in other networks than the Dutch. For instance:

- Category 1 represents electric traction multiple units with cast-iron brake blocks.
- Category 2 represents electric traction multiple units and coaches with disk brakes and additional cast-iron block brakes (a Dutch specialty).
- Category 4 represents any freight vehicle (irrespective of the number of axles per wagon length).
- Category 8 represents electric traction multiple units and coaches with disk brakes only.

The total energy is calculated with an equation similar to Equation (5.1). The values of the parameters a and b and also the track correction factor, c, differ per vehicle category. The track correction c_t is equal to 0 dB for all vehicle categories on ballasted track with concrete monoblock sleepers. This is the reference track superstructure and the quietest of the track categories. Other track correction factors are given in Table 5.3.

Table 5.1 Values of Parameter *a* in Equation (5.1)

Category	Octave Band Centre Frequency in Hz							
	63	125	250	500	1k	2k	4k	8k
1	20	55	86	86	46	33	40	29
2	51	76	91	84	46	15	24	36
4	30	74	91	72	49	36	52	52
8	31	62	87	81	55	35	39	35

Values for *a* and *b* for eight octave bands and four vehicle categories given earlier are presented in Table 5.1 and Table 5.2. The values for *a* and *b* apply for a speed expressed in kilometres per hour (km/h).

There are more complicated relations for high-speed vehicles (due to four source heights) and for disk-braked electric multiple units (EMUs) with traction engines built into the axles (category 3). The latter requires a separation of engine noise and rolling noise. This separation is also made for diesel multiple units (DMUs, category 6).

The aforementioned parameters have been derived from multiple regression curves, based on a multitude of measured results, collected in the late 1980s. The measurements included several sites (so that a range of different track roughness levels would be included), different speeds, and different track superstructures.

Special corrections apply for rolling stock with brakes in operation. Usually, that is the case when trains are decelerating. The brakes cause additional friction noise (Note: Not brake screech!) that is accounted for by adding a correction factor to the rolling noise term, wherever the deceleration applies.

The Dutch regulation has been changed twice since 1996. New versions date from 2006 and 2009, respectively. The aforementioned values, based on historic data, have remained unchanged. The 2009 version includes two more vehicle categories:

- Category 10, light rail rolling stock
- Category 11, retrofitted freight stock with K- or LL-blocks

Table 5.2 Values of Parameter *b* in Equation (5.2)

Category	Octave Band Centre Frequency in Hz							
	63	125	250	500	1k	2k	4k	8k
1	19	8	0	3	26	32	25	24
2	5	0	0	7	26	41	33	20
4	15	0	0	12	25	31	20	13
8	15	5	0	6	19	28	23	19

Figure 5.8 Conventional ballasted track with wooden sleepers.

As there are only a few of these vehicles in service, these categories are less relevant for noise mapping.

The 2006 version of the regulation includes a complete description of a measurement method that should be applied in order to either allocate existing rolling stock of unknown category to either of the existing categories or to base a new category upon measured results. The approach is no longer to collect a large, statistical database of measurements under a range of conditions but to measure only under tightly defined conditions and correct the results to transpose these results to other conditions. The Environmental Noise Directive, in its Annex II, does not refer to this measurement regulation. It would be highly recommendable if it did.

In the Dutch "Reken- en Meetvoorschrift," track correction factors are given for nine different track types (see Figures 5.8 and 5.9). For most of the heavy rail tracks, the following track categories apply:

- Track Category 1, ballasted track with monoblock concrete sleepers
- Track Category 2, ballasted track with wooden sleepers
- Track Category 3, ballasted track with rail joints or switches and turn-outs

The correction is equal for all vehicle categories. Table 5.3 presents the corrections for the three track types given before. The 1996 method includes a further correction for the number of joints per unit length. Finally, indicative corrections are given for different types of concrete viaducts and bridges. For steel bridges, the additional sound produced by the steel construction shall be assessed by means of a measurement. This is due to the large variety of steel bridge constructions.

Figure 5.9 Rheda type slab track.

In later versions of the Dutch method, the measurement method to be applied is specified in a separate document, the Technical Regulation.

Propagation in the Dutch method

In this chapter about railway noise, the calculation of the sound propagation and the excess attenuation is not treated in detail. The railway line is normally split into sectors, each sector being determined by an opening angle looking from the receiver to the railway line. The railway itself is split into source points, each source point representing a certain length of railway line with constant noise emission (i.e., constant source data). The basic formula to compute the long-term average noise level per octave band, per sector, and per source point, reads as follows:

$$L_{Aeq} = L_E + \Delta L_{GS} - \Delta L_{EA} - \Delta L_{SC} - \Delta L_R - 58.6 \text{ dB(A)} \tag{5.3}$$

Table 5.3 Track Correction Factors c_t for Three Main Track Types

Track Category	Octave Band Centre Frequency in Hz							
	63	125	250	500	1k	2k	4k	8k
1	0	0	0	0	0	0	0	0
2	1	1	1	5	2	1	1	1
3	1	3	3	7	4	2	3	4

where

L_E = emission number (the dimensionless factor 58.6 illustrates the fact that L_E is not a real sound power level).

ΔL_{GS} = attenuation due to geometric spreading.

ΔL_{EA} = excess attenuation, including ground effect, air absorption, and meteocorrection (from downwind to long term average), but excluding geometric spreading.

ΔL_{SC} = attenuation due to screening. For noise reducing barriers, it is assumed that the inner side, directed toward the railway line, is absorptive.

ΔL_R = attenuation due to reflections against buildings and other objects other than the barrier alongside the railway.

Rolling noise in the Harmonoise/IMAGINE method

In the Harmonoise and IMAGINE research projects, a prediction method was set up using the same identical method for the sound propagating from the source to a receiver, no matter what the nature of the source would be. The consequence of that approach was that a well-defined indicator had to be used to describe the sound produced by the source. This was the sound power level in dB re 1 pW. For line sources such as a railway line or a road, the sound power level per unit length is used as the input quantity for the calculation of the long-term average level.

The sound power is distributed over five source heights:

- At rail head for the track contribution of the rolling noise
- 0.5 m above rail head for the wheel contribution to rolling noise, as well as traction noise and aerodynamic noise in the bogie area
- 2.0 m above rail head for traction noise
- 3.0 m above rail head for traction noise as well
- 4.0 m above rail head for traction noise and aerodynamic noise

For each vehicle, five different operation conditions can be distinguished:

- Rolling at constant speed
- Braking
- Accelerating
- Curving
- Stationary

For each vehicle, five source types are distinguished:

- Rolling noise and impact noise (joints)
- Traction noise

- Aerodynamic noise
- Braking noise
- Curve squeal

For each source type (p), source height (h), and frequency band (i), a given directivity in the horizontal plane is assumed. As a first approach, dipole directivity is assumed for most sources involved in rolling noise, given by

$$C_{dir,pih} = 10 \lg (0.01 + 0.99\cos^2(\pi/2 - \varphi)) \tag{5.4}$$

where φ is the angle in the horizontal plane between the propagation path and the source line.

Every source is characterised by its sound pressure level at 7.5 m from the track axis. A conversion formula is presented to calculate the sound power level per unit length of track from the 7.5 m sound pressure level. This conversion formula reads

$$\overline{A^h_{line,propagation(T_p)}} = 10 \lg \left\{ \frac{1}{4\pi r N} \sum_{n=1}^{N} \int_{\varphi_{n,min}}^{\varphi_{n,max}} 10^{\left[\Delta L(\varphi_n) - A^h_{excess}(\varphi_n)\right]/10} d\varphi_n \right\} \tag{5.5}$$

where

  r = measuring distance (7.5 m)
  n,N = nth subsource element of the N subsources of unit length, which constitute the line source
  φ_n = angle to the normal to the track of the nth subsource
  A_{excess} = total attenuation along the path φ_n, excluding geometric spreading

As an example, the following table presents the precalculated values of A for the two lower sources, at 0 and 0.5 m above rail head, for the main octave bands. (Note: The calculation is in third octave bands!)

	Octave Bands Centre Frequency in Hz							
	31	63	125	250	500	1k	2k	4k
A (0 m)	−8.7	−9.5	−12.4	−11.2	−13.3	−15.2	−15.7	−15.9
A (0.5 m)	−9.0	−10.3	−13.3	−15.3	−16.1	−15.6	−14.6	−15.0

The values apply to "neutral" weather conditions and a ground impedance between 200 and 200,000 kPa s/m².

For each source type, the 7.5 m level can be calculated using the forms given by the method. Here, only the algorithms for rolling noise are presented.

$$L_{peqi,roll} (h = 0 \text{ m}) = L_{rtot,net,I} (v) + L_{Hpr, nl, tr,I} + 10 \text{ lg } (N_{ax}/l_{veh}) \qquad (5.6)$$

where

$L_{rtot,net,I} (v)$ = total effective roughness as a function of train speed. The effective roughness is related to the direct roughness (as measured with an appropriate method) via the contact filter $A_3(\lambda)$.

$L_{Hpr, nl, tr,I}$ = the track and vehicle transfer function, where the track transfer function applies to the 0 m source and the vehicle (wheel) transfer function applies to the wheel. These transfer functions are independent of the speed. They are normalized to the axle density and are known from various measurements (STAIRRS project). As most wheel types used in heavy rail are rather similar, the same vehicle transfer function can be used for many different vehicle types. The track transfer functions depend on the usual track parameters such as rail type and dynamic stiffness of the rail fixation.

N_{ax}/l_{veh} = axle density (number of axles per vehicle length).

Once the rolling noise contribution has been established for 0 and 0.5 m above railhead, the other sources can be calculated:

- Impact noise, which is included into the rolling noise term by logarithmic adding of an additional roughness to the combined effective roughness.
- Traction noise. Most of these sources have to be obtained from measurements, for a range of predefined operation conditions.
- Deceleration noise, braking, and squeal. For the broad band braking noise, a speed-dependent correction factor is added to the rolling noise. For brake squeal, a time correction is added for the duration of the squeal noise for each braking operation.
- Curve squeal. For curve squeal, the method presents an indicative method to assess squeal noise as a function of speed and curve radius.
- Aerodynamic noise. The source is dependent on the speed and the source height. The basis data is acquired through measurement.

The overall assessment of the source terms according to the Harmonoise/IMAGINE method is a step-by-step approach, which is quoted from the IMAGINE report:

- Define railway source lines with end points (these are at each source height where sound is being created and must be acoustically homogeneous)

- Identify railways
- Identify rail vehicle type, track types, traction noise, rolling noise, aerodynamic noise
- Define operating conditions per unit of time (e.g., day, evening, night), number per hour, track roughness, speed, acceleration
- Define locations of source lines with end points
- Correction factors for directivity, curves, joints, bridges, and so forth
- Calculate sound power level (third octave bands, for each source height, per metre of source line, per hour, per D/E/N)
- Sum sound power levels per source height

In order to simplify and facilitate this step-by-step approach, a database structure was set up in the Harmonoise and IMAGINE projects. The database contains spectral information on source data, depending on a series of practical input parameters. For instance, the rolling noise term (including its spectral density) is assessed for a default value of the axle density, for three categories of braking systems (cast-iron blocks, K-blocks and disk brakes), for four different track types, for a given—measured or default—track decay rate, and track roughness.

The prediction method for the propagation of noise from the source to the receiver, as proposed in the Harmonoise/IMAGINE project, builds on the Nord2000 method that was developed in the Nordic countries in the 1990s. It includes state-of-the-art methods for ground absorption and screening and includes a range of different meteorological conditions. Detailed description of the propagation method for IMAGINE (which is identical for road and rail sources) is beyond the scope of this chapter.

ROLLING NOISE IN THE SONRAIL METHOD

In recent years, a large research project was carried out in Switzerland with the objective to bring the Swiss prediction method for railway noise, the so-called SemiBel method, up to date. The project and the resulting method are known as sonRail. The project treated both the source emission and the propagation. Although the propagation part is rather general and could therefore be applied to sources other than railways, the method is intended exclusively for railway noise.

The source model for sonRail builds on the methods proposed in IMAGINE. The sonRail method uses five source heights similar to the ones in the IMAGINE method. The sound power spectra are assessed for third octave bands from 100 to 8000 Hz.

The so-called primary sources are rolling-noise-related sources, which are calculated using effective combined roughness and transfer functions, similar to Equation (5.6). Traction and set noise are indicated as secondary noise sources. They are calculated on the basis of a database, with a speed-dependent function.

Wheel roughness values are given for disk-braked, K-block-braked, and cast-iron-block-braked wheels. Rail roughness levels are given for smooth rails, average rails, and bad rails.

There is a correction of rolling noise in (tight) curves, but other than in the IMAGINE method, there is no prediction for curve squeal.

The method was validated by means of extensive measurements in Switzerland between 2007 and 2009. The method includes a sophisticated method for sound propagation, including features like diffuse reflection against trees and hillsides.

THE CNOSSOS-EU METHOD

According to Article 6 of the Environmental Noise Directive, the European Commission shall lay down common noise assessment methods. Interim methods have been defined in Annex II of the directive. The development of a definitive common method has not been concluded at the moment that this chapter is drafted. The process of development of these common methods is called CNOSSOS-EU (common noise assessment methods).

The current situation (as of January 2011) is the following:

- The number of source positions might be reduced from five to four, combining the two higher sources into one.
- For wheel and rail roughness, it was decided to adopt the Harmonoise/IMAGINE approach and define separate values in third octave bands for wheel and rail roughness.
- The vehicle class description of the Schall 03 (German prediction method) should be adopted, which is simpler and more confined than the Harmonoise/IMAGINE method. It includes 13 different rail vehicle types. The source data in Schall 03, however, is in octave band, which may cause problems when the propagation has to be computed in third octave bands.
- For rolling noise, the IMAGINE approach should be adopted.
- For engine noise, aerodynamic noise, and squeal noise, the Schall 03 approach should be adopted.
- For braking noise, the IMAGINE approach should be adopted.

Where it says Schall 03, reference is made to the revised German method Schall 03 that was published in 2006. To date, it has no formal status, as it was not accepted yet to replace the previous Schall 03 version.

The current situation has not stabilized yet, and it is therefore not feasible to describe, with an acceptable level of certainty, the future common noise assessment method for rail noise mapping. The common methods will evolve during the year 2011 and a definite method is expected to be agreed and decided on in late 2012.

ALLOCATING VEHICLE CATEGORIES

For rolling noise, two main factors dominate the noise performance of the vehicle, namely:

- The number of axles per vehicle
- The brake type applied on the vehicle (cast-iron tread braking, composite tread braking, or disk braking)

In principle, it is feasible to categorise all vehicle types in a national fleet along these two characteristics. Additional deviating properties could be small diameter wheels or wheels with tuned absorbers.

Another approach would be to start from the type approval tests. For new vehicles, the Technical Specification for the Interoperability (TSI)–Noise would apply. This means that newly homologized vehicles would have to be measured. The measurements have to take place at a special track that needs to comply with requirements according to CEN/ISO 3095. This is a ballasted track with concrete sleepers and a roughness level that does not exceed a given upper limit. The results of these measurements could in principle be used as an input into a prediction method. However, the following remarks need to be made with respect to this approach:

- Usually there are many existing vehicles in a national fleet that have not been submitted to the TSI–Noise measurement procedure, as this procedure only applies to the homologation of new vehicles to be admitted to the European rail network.
- The measurements refer to CEN/ISO 3095 track quality. For normal practical situations, where the track quality is less, the track correction factor c_t has to be referenced to this high-quality track.

Most of the classical prediction methods for railway noise existing in various member states have adopted a very practical approach; pass-by noise levels of many different vehicles with different speeds have been collected at a range of different tracks. A statistical analysis of the results of such measurements produces different regression lines that are then considered to be characteristic for a particular vehicle/track combination.

The disadvantage of this approach is that a large number of measurements are required to achieve sufficient statistical confidence (typically around 150 measured pass-bys are necessary to fully describe one vehicle type). In addition, whenever a new vehicle type is included into the prediction method, a similar amount of measurement would be required for that particular new vehicle type.

In 2006, the Dutch government published the Technical Regulations for Methods of Measuring Emission of railway vehicles. These regulations have not been included in the annex of the EU Directive on the Assessment and Management of Environmental Noise, as they were published after the directive came into force. The regulations specify three so-called procedures for the assessment of the noise emission of rail vehicles, each intended to represent the emission factors required in a prediction method.

Procedure A is a simple and straightforward procedure. It includes measurements at a standard ballasted track with a specified upper roughness level and specified dynamic properties. The train should run at constant speed, several times for different speed classes. The results of these measurements are then compared to the noise emissions that have been established for the existing fleet and the vehicle under concern is ranked into the same vehicle category as the rolling stock type to which it fits best. The method has limited accuracy, as certain deviations between the new vehicle and the closest existing vehicle category are allowed.

Procedure B is the full method, allowing the assessment of all the necessary noise emission data at a single stretch of track. The track should comply with requirements with respect to its roughness and dynamic properties. The method allows for determination of traction noise (using maximum accelerated vehicles) and for source height determination using either microphone arrays or microphone positions close to the source.

For rolling noise, the method includes assessment of the effective rail roughness (according to EN ISO 3095) and measurement of the total effective roughness through a measurement of the vertical vibration of the rail head. From these two quantities, the effective wheel roughness is calculated and used as an input for the total effective roughness if the same vehicle was running on an average track with average national roughness.

Procedure C is the full method to assess the track correction factors for new, unknown types of track superstructures.

The methods specified in these regulations form the basis of a new, revised prediction method in the Netherlands, issued in 2006 in combination with the regulations. This is the version that is currently applied for legal procedures in the Netherlands.

ASSESSING TRACK ROUGHNESS

The track roughness is an essential element in the prediction of rolling noise. There are basically two ways to deal with this parameter:

- Most prediction models assume that a given track roughness is present anywhere on the track. This roughness level is then indicated as the "average network roughness." It is accepted, however, that local deviations of up to 20 dB may occur, due to local corrugation. The argument of rail infrastructure managers is that state of the art track maintenance will guarantee that this exceptionally high roughness is removed by means of rail grinding as soon as it is noticed, so that on the average the national roughness level applies. On the other hand, there may also be locations where the track roughness happens to be far below the network average, to the obvious advantage of people living near that spot. However, state-of-the-art maintenance would ignore this spot so its roughness might grow into the network average without maintenance interference.
- A far more precise but very laborious method would be to collect data on the actual track roughness levels on a periodic basis and feed these levels into a track quality database. This database could then be used to assess the actual rolling noise level, using the measured local track roughness in combination with the predicted wheel roughness, which is a function of the vehicle type. This method produces far more accurate noise levels along the track, but the drawback is that this prediction applies only to situations where the track data is available. Predictions for future noise levels would have to be made on the basis of a prognosis for the track roughness, with limited accuracy. The advantage of this method, however, is that in certain spots the infra manager gains an option to reduce noise levels by rail grinding.

Between these two methodical options is the option of the "especially monitored track."* This option is maintained only in Germany. It involves a track where the roughness is monitored regularly (by means of a monitoring vehicle actually measuring rolling noise) (Figure 5.10). As soon as the roughness exceeds a certain action level, the rail is ground to a lower roughness level. By this process, the rail infra manager can guarantee that the roughness, averaged over time, is lower than the typical network roughness. In the legal prediction scheme in Germany, there is a correction of −3 dB applied to a track that is designated as being "especially monitored."

---

* In German: *Besonders überwachtes Gleis or short BüG.*

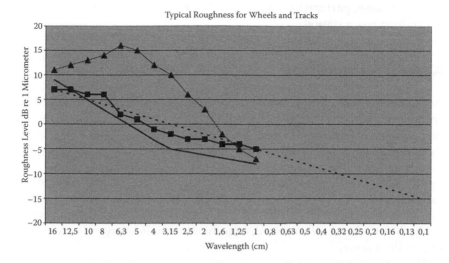

Figure 5.10 Typical roughness levels (rms dB re 1 micrometre) in dependence of the wavelength in centimetres (solid line: track roughness TSI Noise; dashed line: track roughness Dutch average; triangles: tread braked wheel; squares: disk braked wheel).

TRACTION NOISE

Traction noise is more difficult to model than rolling noise, as the differences between one vehicle and another are far more widespread than for rolling noise. There is an important difference between diesel-driven and electricity-driven vehicles. In diesel traction, both in locomotives and diesel-driven multiple units (so-called DMUs), the exhaust represents a significant noise source, which is usually located somewhere at the upper side of the vehicle. In electrical traction, the actual engine and possible gearboxes are often mounted near or directly on the driven axles, so this source is located close to the ground. This makes an important difference, for instance, for the efficiency of noise barriers and should therefore be considered in the modelling.

Both in diesel and electric traction, there is a need to cool the engine and elements of the traction line, so there are cooling fans included. Cooling fans, particularly axial fans with high rotational speed, are a significant source of noise, but their actual contribution depends on their design, which tends to differ from one vehicle to another.

In modern high-speed rolling stock, the electrodynamic braking energy is usually dispersed in an electric shunt, which turns the kinetic energy of the train into heat. The cooling fan for this braking shunt is a famous source of noise, particularly as it is usually in operation at full speed when the train enters a station.

AERODYNAMIC NOISE

Aerodynamic noise can be ignored for all rolling stock with commercial speed less than 160 km/h. In that speed range, the rolling noise is by far dominant over the aerodynamic noise. For speeds over 160 km/h, aerodynamic noise may be relevant and in some cases even dominant, for instance, when a noise barrier is involved. The noise barrier would efficiently screen the sources of rolling noise (as they are located close to the rail head) but would not screen some of the aerodynamic sources (as these are located near the top of the train, for example, near the pantograph). From 200 to 250 km/h upward, aerodynamic noise may even dominate over rolling noise.

The noise emission of aerodynamic noise sources needs to be assessed by means of measurements. The sound power produced depends on the location and shape of protruding parts on the train body, which may be located near or in the bogies (particularly the front bogie); near or at the pantographs; and near other parts such as door handles, window sweepers, and antennas.

The measurements are usually complex, because both the sound power and the location of the source needs to be assessed in dependence of the vehicle speed. These kinds of measurements are usually carried out with highly directional so-called array microphones. This type of microphone allows recognizing sources on a fast moving vehicle and assessing their sound power level independent of other sources.

Both for traction noise and aerodynamic noise, the sound production of a rail vehicle could be assessed in two phases:

- First, assess the rolling noise element by identifying the vehicle's brake system and using the standard equations for that type of brake system. This produces the rolling noise contribution to the total sound power.
- Second, assess the aerodynamic sources and traction noise sources by means of separate measurements and add these to the rolling noise source.

CURVE SQUEAL

Curve squeal is the noise that is produced when a rail vehicle runs through a curve. It originates from the lateral stick–slip phenomena in the contact area between wheel and rails. The process of stick and slip occurs because the two wheels on a fixed axle both have to run through the curve, where the outer wheel has to run through a longer path than the inner wheel. The stick–slip process is influenced by:

- The dimensions of the contact area between wheel and rail, which can be influenced, for example, by the profile of the rail head.
- The friction factor present in the wheel rail contact, which can be influenced by so-called friction modifiers. Usual friction modifiers are water (including rain), and also some lubricators that can be fed to the rail head (fixed lubricators) or to the wheel tread (lubricators fixed to the vehicle).

Curve squeal does not always occur. It depends on the weather conditions, the speed of the train, the diameter of the curve, and possibly other parameters whether or not curve squeal is generated. When it occurs, it produces a high pitch, almost tonal noise with very high levels. When it does not occur, it does not produce any noise at all, apart from the normal rolling noise. The effects of the curve squeal are limited to the direct surroundings of the curve under concern. Because of these two considerations (local effect and incidental occurrence) curve squeal is ignored in many prediction methods and certainly in the process of noise mapping. Local complaints may trigger the installation of local mitigation measures, such as lubricators or fixed water spray installations.

MODELLING SOUND PROPAGATION

In general, the noise prediction method consists of two elements: the source description and the propagation part. The source description has been treated in the previous paragraphs. The sound propagation is modelled in the same way as for other sources. This means that the following elements are included in the model:

- The distance attenuation caused by the sound energy being spread over a larger surface when the distance to the source increases. For line sources like a railway line, the distance attenuation, D, is related to the distance, r, between source and receiver in the following way:

$$D \div 10\log\left(\frac{1}{r}\right) \tag{5.7}$$

- The air absorption, which is influenced by the yearly average air humidity (which may be climate-zone dependent).

- The ground effect, depending on the areas where the reflected sound reaches the ground, and the impedance of that area, usually indicated as partially "soft" or "hard" ground.
- Reflections against buildings and other objects near the path between source and receiver point (except for the façade for which the incident sound level has to be assessed; the reflection into this façade is to be ignored).
- Wind and temperature effects leading to a convex or concave curvature of the sound path between source and receiver point.
- The screening effects of noise barriers along the track or of other objects that stand between the source and receiver point, the efficiency depending on the height of the diffraction edge(s) of that object.

For railway noise, all of these factors apply, but the last factor needs further attention. This is due to the fact that noise barriers in combination with rail traffic only perform as long as their inner side (rail side) is absorptive. The reason for that is the risk of multiple reflections between the barrier and the train, which may affect the diffraction such that more sound energy is radiated into the shadow zone behind the barrier. This phenomenon is sometimes indicated as "canyon effect."

The canyon effect can be prevented by having the inner side of the barrier highly sound absorptive. Alternatively, the barrier can be inclined instead of vertical, thus returning the reflected sound back into the ballast bed.

Another peculiarity of railway noise modelling lies in the acoustic properties of the ballast bed. Usually, this is a small area close to the source, but since the source is very low (at least it is for the track contribution of rolling noise), the area may have significant influence on the propagated sound. The typical track built-up is as follows: directly next to the track is usually a small heap of ballast stones, then there is a small inspection path (usually split stones), and then there is the soft and grassy shoulder of the earth wall carrying the track. Sometimes the various ground effects occurring in this area have been implicitly included in the source description of the wheel and rail. Otherwise, these sources are modelled as line sources without ground effect (this is the preferred way), but then the ground effects need to be modelled with care.

BRIDGES AND TUNNELS

Railway bridges and viaducts exist in a large variety of practical appearance. They serve to cross valleys, roads, other railways, and waterways. One distinction relevant for noise is between concrete, brick, or composition structures on the one side and metallic structures on the other. Metallic bridges are known to be loud, but their noise characteristics depend strongly on the construction and the track fixation.

In mountainous areas, steel bridges crossing valleys usually have a conventional ballasted track with wooden sleepers continuing on the bridge. The bridge has the shape of a trough fixing the ballast. To some extent, the ballast serves as a vibration insulator for these bridges, so that their noise radiation is modest.

Whenever a steel bridge crosses a waterway, bridges have to maintain a minimum height for vessels passing underneath and a maximum height to save cost in the track elevation. These bridges often have a direct fixation of the rail onto the bridge. This certainly applies to movable bridges that have to open to allow shipping to pass. Obviously ballast cannot be used on cantilever bridges!

The direct rail fixation serves as a shortcut for vibrations emerging in the rail. The vibrating bridge is usually a very efficient sound radiator, particularly for low frequencies. For receiver points close to the bridge, noise levels have been reported that can be up to 15 dB higher than noise levels at equal distance from a normal track with the same traffic density and rail roughness.

Their impact, however, is limited to the direct surroundings of the bridge. Therefore, for reasons of statistics, it is not absolutely necessary to include railway bridges in the noise mapping operation. However, for reasons of credibility with respect to the residents, it is recommended to pay attention to the increased noise radiation of metal bridges.

There have been attempts to predict, on the basis of constructional details, the additional noise radiation from steel bridges. Sophisticated prediction methods have been applied for new bridges, on the basis of finite element modelling (to assess the eigenfrequencies of the bridge structure) or statistical energy analysis. For mapping purposes, it would be recommendable to assess the additional noise radiated by the bridge by means of a comparative measurement, carried out both near the bridge and near the adjacent ballasted track. In doing so, two issues have to be emphasized:

- The roughness of the track on the bridge and the track on the ballasted track has to be at least in the same order or has to be irrelevant compared to the wheel roughness. If not, the comparison is not valid.
- When modelling the bridge, one has to be aware of differences in noise radiation properties between the bridge and ballasted track. The ballasted track behaves as a dipole line-source, whereas the bridge may behave as a monopole point source.

Tunnels are hardly ever included in noise mapping exercises. Obviously, when the train runs in the tunnel, the noise impact is zero. Sometimes tunnel openings represent a distinct source of noise, due to the reflections of noise inside the tunnel and the resulting diffuse noise field in the tunnel opening. It is left to the creativity of the modelling expert to assess noise levels in the vicinity of tunnel openings with an acceptable level of accuracy. The noise radiated from the tunnel opening can be reduced by the application of an absorptive lining at the tunnel walls and ceiling, provided that sufficient length before the actual opening is treated.

CONCLUSION AND OUTLOOK

The production of noise maps for railways has proven to be feasible with the methods currently available. Nevertheless, significant efforts were necessary for member states that did not have a national method in place. The assessment of noise creation factors for a range of different vehicles under different operation conditions and on different track types may involve a huge effort.

The interim method referred to in the Environmental Noise Directive is a Dutch method that has been since revised. In the revision, guidelines were included for the assessment of creation factors for vehicle types that were not in the calculation scheme. It is recommended to implement these guidelines in the Environmental Noise Directive, so that member states can use the guidelines when assessing the noise creation factors for their national fleet.

In the near future, the common European methods (CNOSSOS) will become available. Once that is the case, the assessment of noise creation factors will be more straightforward, as these will be based on a limited number of specific descriptors.

REFERENCES

1. Categorisation of vehicles and tracks, Harmonoise report D11- WP2- HAR12TR- 021107-SNCF10.
2. Rail noise database and manual for implementation, IMAGINE report IMA6TR-061015-AEATUK10-D12-13.
3. Technical regulation for methods of measurement emission 2006, edited by CROW, The Netherlands, December 2006.
4. Calculation and Measurement Scheme railway noise 1996, Dutch Ministry of Housing, Spatial Planning and Environment.
5. Annex 4 Railway to the Calculation and Measurement Scheme Noise 2006, Dutch Ministry of Housing, Spatial Planning and Environment.
6. Technical Specification for the Interoperability, Noise, Commission Decision of 23 December 2005, C(2005) 5666.
7. CEN/ISO 3095(2005), Railway applications—acoustics—measurement of noise emitted by railbound vehicles.
8. Thron, T., and Hecht, M. The sonRail emission model for railway noise in Switzerland, *Acta Acustica*, 92 (2010), 873–883.

Chapter 6

Industrial and harbour noise

J.R. Witte

CONTENTS

INTRODUCTION: INDUSTRIAL NOISE

In Directive 2002/49/EC of the European Parliament and of the Council of 25 June 2002 it states (Annex IV) that strategic noise maps for agglomerations shall put a special emphasis on the noise emitted by (see Figure 6.1)[1]:

- Road traffic
- Rail traffic
- Airports
- Industrial activity sites, including ports

Figure 6.1 Container terminal at Copenhagen with electric cranes, ships, and trucks.

Industrial noise sources are different from transport noise sources because of its constant presence at the same position. Transportation noise sources are based on a large number of cars, trains, or airplanes, divided into subcategories, and depending on speed, pavement, and so forth. But industrial noise may be just one fan operating for over 20 years. Industrial sources range from this fan (sizes less than one metre to several metres in diameter) to all kinds of equipment (for instance trucks, lifting trucks, and crushers), and complete refineries and power plants.

The perceived noise levels from industrial noise may also differ from general transportation noise. Tonal components like the hum of a power transformer may give rise to complaints but also impulsive noises from hammering or container handling. Especially the beeps of warning signals tend to annoy people.

When one models one fan, the location of the source is fixed. Together with an assumed nearby building, screening through lateral diffraction can be of great influence.

Knowing this background one can understand that industrial noise should be modelled carefully. In the next sections we go into details on how to obtain sound power levels of the industrial sources and next how to make an industrial noise model.

The next chapters lean heavily on the IMAGINE reports on industrial noise and on the NoMEPorts project.[2,7] Input on this subject was provided by CSTB (France), deBAKOM (Germany), DGMR (Netherlands), EDF (France), Kilde (Norway), and Muller-BBM (Germany).

HOW TO MAKE A STRATEGIC NOISE MAP ON INDUSTRIAL NOISE

The next steps should be taken to make a strategic industrial noise model:

1. Gather relevant sound power data
2. Gather operating times
3. Model industry and the surrounding area
4. Calculations

An extra step is optional but is highly recommended: validation of the noise model and map.

STEP I: INPUT OF SOUND POWER LEVELS

Different types of sources

Many industrial sound sources are complex in terms of geometry, sound-generating mechanisms, and radiation properties. It is therefore necessary to simplify the real, physical properties into a source model that can realistically be handled in a calculation process. The simplification must be done in such a way that the accuracy of the predicted sound level in the surroundings is acceptable.

General remarks

Noise data for industrial sources such as electric motors, pumps and compressors (Figure 6.2), fans, furnaces and boilers, coolers, piping and valves, stacks and flares, construction and building machinery, and many others can be obtained by means of direct noise measurements or by using default values (vendor data, prediction based on rules of the thumb) and available noise source databases (e.g., IMAGINE database, SourceDB). Direct noise measurements, using established techniques and specialised equipment and software, are considered to be the most accurate option. Measurements though are time consuming and often technically complicated (an accurately measured source should be isolated from background noise). The use of default values and databases offer an easier but less accurate approach. Validation of this type of data can be performed by means of measurements for only a small sample of the dominant sources from the complete noise data set.

Measurements

In principle, taking measurements of the source's noise emission will yield the highest accuracy of the sound power levels calculated from the

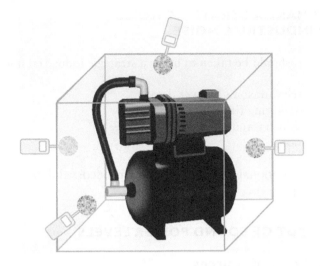

Figure 6.2 Simplified view on measuring noise on a surface near a compressor.

obtained results if the measurements are performed correctly. In contrast to this, the other methods mentioned in the previous section will inevitably require a compromise between high accuracy of the obtained noise data on one hand and the demand for as few source input/descriptive parameters as possible on the other. Because of the large variety of industrial noise sources and particularly in their operating conditions an individual state of noise control to many descriptive parameters would become impractical.

There are various possibilities to take noise measurements that make use of different methods and equipment, and produce results of different quality and degree of detail and accuracy.

However, for noise mapping it is essential that the input data sets are consistent with one another and with the noise propagation calculations to be used (for instance, 1/1 octave or 1/3 octave bands). This must be ensured for all input data, in spite of being acquired by a large number of different end users for a huge variety of noise sources in often completely different locations, situations, and operating conditions.

Generally, the measurement methods use the following principle: perform one or more noise measurements (1/1 octave or 1/3 octave bands) relatively near the object of interest and transform the mean measured pressure levels into power levels by summing the surface area that is enclosed by the measurement points including the measured object. This power level may be corrected for reflections in nearby walls, the near field, background noise, and so forth.

A compilation of existing international and national standards on noise measurements has been set up by the IMAGINE project that provides basic information and criteria that can be used to decide which method is best applicable to the specific measurement task. This compilation—the "Measurement Methods Report"[2]—also points out shortcomings, limitations, possible improvements, and special aspects with respect to the requirements on noise mapping input and gives general hints and warnings that can be of use in this context.

A fact sheet is drawn up for each standard or guideline. Each fact sheet is a one-page summary of the most important characteristics, with respect to noise mapping, of the measurement method described in the standard. Among other entries, the most important information given in each fact sheet includes:

Title and number of the standard/guideline
Short description
Method category
Accuracy of the results
Needed equipment
Preconditions/requirements
Expenditure (with respect to time needed, costs, etc.)
User group (required expertise: expert/layman)
Shortcomings
Remarks
Related standards

In compiling the measurement methods fact sheets it was the intention to provide a set of measurement methods that is complete enough but does not contain too many redundancies. A considerable number of international and national standards from several countries have been reviewed and many of them have been included in the database. Nevertheless, the database is not intended to be a complete and comprehensive compilation of all existing national and international standards on noise measurement methods.

Most of the standards included in the database were not originally intended to provide methods by which to acquire input data for noise mapping and action planning. Consequently, using standardised measurement methods for that purpose requires, to a different degree, "adapting" the procedures described in the standard to meet the requirements on input data for noise mapping purposes.

Unfortunately, this means that some of the advantages associated with standardisation are lost: To obtain the desired result, it is no longer sufficient to strictly adhere to the requirements set up in the standard and to exactly follow the procedures described in it. Also, in adjusting the measurement method in the standard in a way as to gain input data for noise mapping, special aspects and limitations may have to be taken into account. As these limitations may not be relevant or obvious for the purposes the standard was originally intended for, they may not explicitly be mentioned in the specific standard.

Accordingly, to provide the user with the necessary background knowledge, a variety of additional and complementary information in the context of taking measurements for noise mapping purposes can be obtained during measurements. Examples are

- Quantities and parameters that are crucial for successful noise mapping and that must be assessed during the measurements (e.g., dimensions, directivity, orientation and location of the source; surroundings; operating conditions and working hours of the source; meteorological data at the time of the measurements).
- Importance of considering the respective relevance of the individual sources for the overall noise emissions in plants consisting of many individual noise sources to avoid unnecessary costs and excessive amounts of data.

Although taking professional noise measurements is certainly the best and most accurate way to gather the relevant noise emission data of industrial noise sources, it may also be a time-consuming and costly way, especially for large industrial areas with many sources of different kinds. For example, performing noise measurements to collect only the relevant sound power input data for noise mapping in a petrochemical plant of 25.000 m^2 will take at least 4 man-days.

The fact that a single wrong result for an important individual noise source may have already led to a useless and ambiguous noise map and thus to wrong conclusions in the action planning following the mapping process, it justifies in many cases the efforts associated with noise measurements.

More information can be found in the IMAGINE report IMA07TR-050418-MBBM03 "Measurement Methods."[3] The SourceDB database, described in the next paragraph, contains the fact sheets mentioned earlier.

Database

If there is not enough time (or means) to obtain the relevant sound power data by measurements, the sound power levels might be obtained from manufacturers of the equipment, for instance, the CE-label. These sound power levels are of course for very strict operating conditions, probably described by the measurement method used.

If the manufacturers cannot supply these sound power levels, one can use data from a database.

The source database will give the user the sound power levels, including a mean third octave band spectrum, for individual sources and whole plants based on measurements and formulae with a small number of parameters. This information is based on the knowledge of noise experts throughout Europe.

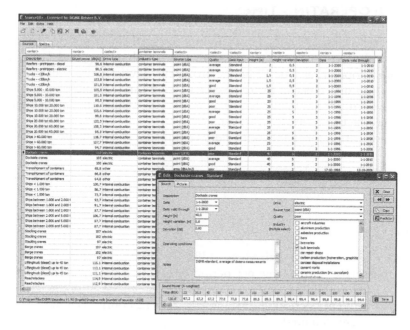

Figure 6.3 Screen dump from SourceDB.

For purposes of convenience, DGMR has developed a software tool, called SourceDB, for easy access and filtering of data. SourceDB (Figure 6.3) is distributed free of charge through the DGMR Web site (www.dgmr.nl).

Next to the information on sound power levels, the fact sheets about noise measurement methods (see also report IMA07TR-050418-MBBM03 "Measurement Methods") to obtain sound power levels are included as well. More information can be found in the report IMA07TR-050418-DGMR02 "Description of the Source Database."[4]

An indication of the accuracy for different methods for acquiring sound power data is given next, based on the Good Practice Guide[5] (see Figure 6.4).

Acquiring of Sound Power Levels Including Working Hours			
Method	Compl.	Acc.	Cost.
Measurements	⬠	2 dB	⬡
Measurement of dominant sources, extended up with the source DB or equivalent sound power database	◇	3 dB	◇
Use of sound power database only, no knowledge of working hours	△	> 5 dB	△

Figure 6.4 Toolkit estimating sound power levels.

STEP 2: OPERATING TIMES

The yearly averaged active hours are an essential input for the determination of long-term averaged noise indicators such as L_{den} and L_{night}. If a machine is operating 1 or 12 hours during the daytime, this will result in about 11 dB difference in the received L_{den} from this machine.

The working hours should be given for the day, evening, and night period. This must be based on a yearly average.

The correction for the working hours is calculated as follows:

$$C_W = 10\log\left(\frac{t}{T_0}\right) \text{ (dB)} \tag{6.1}$$

where

C_w = correction for active hours (dB)

t = active source time per period based on a yearly averaged situation (hours)

T_0 = reference period of time (hours) (day: 12 hours; evening: 4 hours; night: 8 hours)

For the more dominant sources, the yearly average active hours correction should be estimated at least within 0.5 dB tolerance in order to achieve an accuracy of 1 dB at 100 m. The yearly average active hours correction should be added to the sound power level.

Dominant sources are the sources that contribute most at the nearby houses. These dominant sources may range from sources with a low sound power level but located very close to this receiver to high sound power levels corrected for the working hours at larger distances. Also screening of noise will play an important role for the distinction of a dominant source. So, after the first calculation and defining the areas with a high contribution for industrial noise, extra care could be taken in estimating the yearly averaged active hours, first for the most dominant sources. As a rule of thumb, the most dominant sources are the top ten contributing sources at a certain receiver.

In Table 6.1 the yearly averaged active source time, t, is calculated as follows:

$$t = \frac{\sum days * percentage}{365} T_0 \text{ (hours)} \tag{6.2}$$

If no data is available on active source time, an estimate of the overall active hours must be obtained through contact with the company.

Table 6.1 Estimating the Yearly Averaged Working Hours and the Correction (C_w) for a Lift Truck

Period	Number of Days Working for a Certain Percentage of Hours in the Specific Period				t (h)	Correction C_w (dB)
	100%	50%	25%	0%		
Day	43	81	100	141	3.6	5.3
Evening	2	5	30	328	0.1	14.8
Night	0	0	60	305	0.3	13.9

This will relate to active hours during the weekends and holidays as well as the active hours during standard days and during overtime. A company that only works for eight hours during the daytime and does not always work during weekends and holidays (for instance 20 days) will result in approximately a total of yearly averaged active hours of about 5.2 hours or a correction $C_w = 3.6$ dB.

A petrochemical plant will be working almost 100% of the time. Only critical or planned stops will vary the noise levels. So $C_w = 0$ dB.

STEP 3: MODEL INDUSTRIAL AND SURROUNDING AREA

Geometry

Application of modelling guidelines to industrial noise:

Buildings and terrain between the industrial area and the receiver points must be modelled according to the overall modelling principles as used for traffic noises:

- If the industrial site is surrounded by walls or earth berms, it is highly recommended to take these into account as they may provide significant screening of the noise.[4]
- If the plant is characterised by the overall sound power method (L_w/m^2), no buildings and installations need to be modelled inside the plant area, the effect of obstacles being taken into account by means of a correction term for scattering. The source height shall be the same for the whole area and representative for the main noise sources on the site. It is recommended to take into account a large variation (e.g., up to 100%) on the estimation of the source height.

As an extension, the industrial area can be modelled by a set of distinct areas and sound powers and source heights assigned as a function of the specific activities for each area.

If the industrial area is modelled as a set of individual noise sources, the buildings and installations must be modelled according to a level of detail compatible with the used propagation models. Reflections and screening by buildings and screens must be integrated into the calculation scheme. Next to this, effects of lateral diffraction and scattering shall be taken into account.

Source modelling

The term "source modelling" here refers to the process of defining the characteristics that are needed for the calculation of "received," average, long-term sound levels in the vicinity of residential buildings, hospitals, schools, offices, recreational areas, and other places where the sound interferes with the activity of the place.

For road and railway sources that imply large numbers of independent sources, the errors introduced in source modelling are often small compared with the uncertainties attached to changing operating conditions and propagation effects due to meteorological conditions. For industrial sources, correct modelling of source positions and sound powers may be the dominant source of uncertainty (see "Modelling noise inventories").

The acoustic properties of any machine or piece of equipment producing sound can usually be defined in terms of four key parameters:

- "Type" of sound source (point, line, area)
- Source height(s)
- Total sound power produced by the source
- Spatial distribution of the sound radiation (the "directivity" of the source)

These parameters, and particularly the last two, are frequency dependent.

The choice of source type requires some comments: The "point source" does not exist in the real world. All machine have a finite size; they are often large and often consist of many components. But when the distance from the source to the receiver is much larger than the machine dimensions, the sound can still be assumed to radiate from a single point on the source. The next step is to choose a representative source height.

A "line source" is a source with one dimension much greater than others, when this dimension is large compared with the source to receiver distance. A line source can also be a point source that moves along a fixed route, for example, a lifting truck. The sound power will be given per unit source length (dB/m). A line source is often modelled as a series of uncorrelated point sources.

The radiated sound power from each point corresponds to the partial source length ("element length") that it represents. The distance between the point sources should always be smaller than the distance to the nearest receiver position divided by a factor greater than or equal to 1.5.

An "area source" is a source with large dimensions compared with distance to the receiver. It can be the surface of a building, an industrial site with many (often) complex sources distributed over an area, an area with mobile sources not moving along fixed paths, and so on. The source power will be given per unit of source area (dB/m^2). An area source can alternatively be modelled as a series of point or line sources. A building or an industrial site can consist of many area, line, or point sources with different sound power levels, heights, and so forth.

Sources treated as point sources may have large physical dimensions, and all frequency components may not be generated at the same position. Heavy construction machines consisting of an engine, exhaust, fan, hydraulics, and material handling can be mentioned. Stone-crushing units consisting of several different processes are another example. There are, of course, many ways of handling this. One is to divide the source in a number of point sources with a complete set of source spectra for each one. A simplified alternative is to define the overall source in terms of sound power, directivity, and source height for each 1/3 octave band.

The source height must be chosen at the centre of the source. The dimension of the source may also be an input parameter for the model (see IMAGINE project[2]).

In cases when it is difficult to determine the spatial distribution of the sound radiation, the sound power in the direction of the receiver(s) should be used.

Modelling noise inventories

Once the sound power levels of the sources have been determined either by measurements or from literature, a model of the industrial site including buildings and other noise barriers has to be set up. Also, the terrain has to be modelled to account for screening effects. In a further step, all the sources have to be located in relation to the buildings. This has to be done with some care since screening of the noise sources strongly depends on the position of the source in relation to the screens.

Depending on the sound propagation software, line or area sources are divided into subsources to model the screening effects properly.

Generally, the distance between the point sources should always be smaller than the distance to the nearest receiver position divided by a factor greater than or equal to 1.5. If screening takes place, the distance between point sources will have to be smaller. No general rules apply on how many sources per area should be entered, but calculation of several cases with reductions of the distance between sources will give the solution on what distance to use.

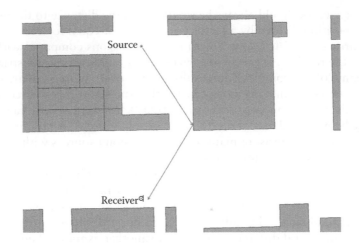

Figure 6.5 Example of dominant reflections.

The positions of the source have to be modelled carefully because of the sensitivity of screening effects based on distance to screen and screen edges. Especially if the model contains a few dominant noise sources, the "exact" location of these sources is important.

Another source of errors in the noise calculation is reflections from large facades. Depending on the location of the noise source and the receiver point, reflections can dominate the noise levels (Figure 6.5).

Modelling the buildings of industrial sites can also be a source of errors, since most sound propagation models do not account for horizontal screens, that is, screens that are not starting at ground level but at a certain height. Thus the screen will be "transparent" for a certain height, but will be treated in the model as being opaque.

Other noise sources beside the "stationary" sources are the traffic inside the plant like fork lifts or trucks, which can have a strong impact on the ambient noise levels especially during nighttime (e.g., backward driving warning signals).

Further points in modelling an industrial site are the variations in sound power levels due to different operating conditions and operation times of the noise sources. Usually noise measurements are carried out for periods from a few minutes up to a number of hours depending on the type of noise source and their typical operation cycles. In some cases, the measurements can cover only part of the operation conditions of the noise source. If, for example, a plant is processing different materials, measurements of these different conditions should be taken or at least the differences should be estimated.

At most modern plants a significant amount of the machinery is inside buildings. Noise radiated from buildings by façades (doors, windows, etc.)

are modelled, for example, by taking measurements inside the building and calculating the radiated sound power levels from the different parts of the façade by taking the size of the object and the sound insulation into account. For the sound insulation it is important to consider the installation of the façade. For example, if the façades are vibrating due to the rigid connection between machine and wall, the radiated sound power will be higher than expected from the theoretical values for the sound insulation. Also, poor fitting of walls results in acoustic leaks, which will further decrease the total insulation. For large façades this can result in errors in calculating the ambient noise levels.

STEP 4: CALCULATIONS

In order to calculate down to the 50 dB levels for the L_{night} around industrial areas, sometimes calculations over large distances have to be performed (End Annex VI: Data to be Sent to the Commission). This depends strongly on industrial activities and occupied area.

For instance, for a container terminal working 24 hours a day, the distance for the 50 dB contour (without screening or reflections) may be over 1 km away from the terminal. For large industrial areas (>3 km²) with mainly chemical plants, the contours may be 1.5 km away from the industrial area (see Figure 6.6).

The industrial interim method ISO 9613-2, which in many countries is used, does not have many parameters that are special for industrial noise

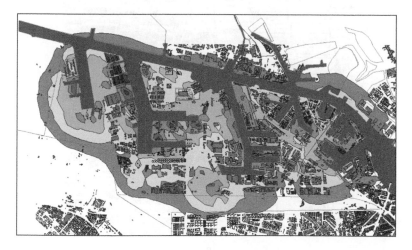

Figure 6.6 Contours of the industrial area Westport near Amsterdam, Netherlands. Contours of 50 dB are situated at 500 to 900 m from nearest industry. **(See colour insert.)**

calculations. Note that the averaged temperature and corresponding relative humidity is used for calculating the air absorption. If the number of reflections can be set, choose one reflection for normal cases where distance dominates the propagation. If large buildings dominate the propagation, more reflections might be necessary.

The geographical data has to be as simple as possible: only the outer contours of a physical object are relevant, the inner lines only slow the making and calculation of a model (e.g., modelling container formations; Figure 6.7). For instance, containers should not be entered one by one but as a contour around a line of containers with a height that is the average over a year.

(a)

Figure 6.7 The formation of containers (only use outer lines). (Courtesy of ECT.)

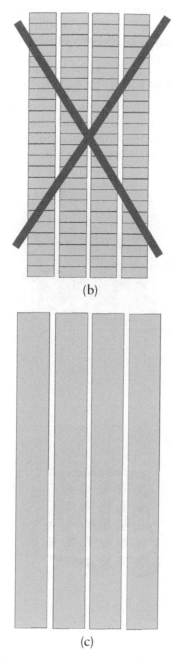

(b)

(c)

Figure 6.7 (Continued) The formation of containers (only use outer lines).

Reducing the number of sources may reduce input data as well as calculation time. Irrelevant sources are either low in sound power combined with the active hours or are far away from the area of interest. As a rule of thumb for the night period (T_0 = 8 hours):

If $L_{W,i}$ + 10 log(working hours/8) – 20 log($4\pi r^2$) < 30, neglect source i

If the model comprises many area sources modelled along a grid, the calculations points should not coincide with these grid points. This can be done by taking different spacing (do not use multiples of the spacing as well) in the source and the receiver grids, or use different origins (Figure 6.8).

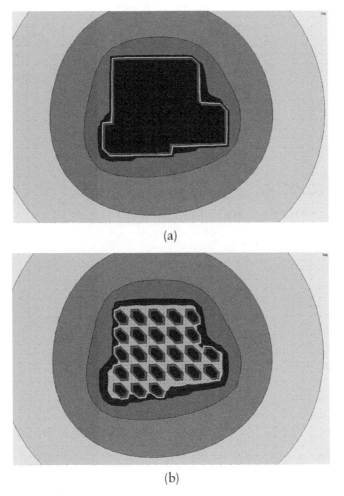

(a)

(b)

Figure 6.8 Different results for the same area source with different selection of grid spacing. From left 50 ´ 50 m, 40 ´ 40 m, 41 ´ 41 m while calculation grid is 100 ´ 100 m. **(See colour insert.)**

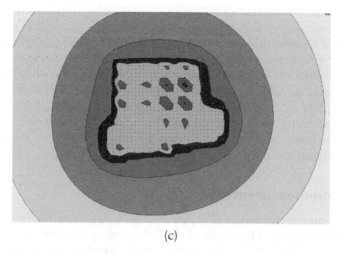

(c)

Figure 6.8 (Continued)

VALIDATION

There are several ways for validating a noise model. One option is the validation of the input data sets. Another option is measuring noise in selected locations and then attempting a comparison between the predicted and the measured noise levels. The validation by means of selected measurements could provide the means for accessing the accuracy of noise maps, but it cannot identify the causes of potential inaccuracies. The validation of the input data sets can be a more feasible tool to check where the problem generates.

Validation of input data

The validation of input data may result in a laborious task if it is to be performed on the complete data sets. Some kind of sampling would therefore be advised. Random sampling or sampling of the most significant input data could be selected (e.g., in terms of noise levels produced or effects to the exposed population).

When examining the reliability of noise data it is important to focus on both the source sound power level and its operational characteristics (e.g., timetables, volumes). The technical specifications of the machinery and equipment used could provide the sound power levels, and well-established noise emission databases (e.g., SourceDB) can also be of use. For more complex situations, specifically designed noise measurements can be performed. In that case the machinery or equipment under question can be isolated from other noise sources, and a set of measurements in selected distances from the machinery can take place. The measured values could then serve as an input to specialised software (e.g., Acoustic Determinator, Bruel & Kjaer) to estimate the sound power level of the source under question.

The examination of the operational characteristics of noise sources can be performed by selected checks and validation measurements. Another approach may be to cross-correlate information provided by different sources and verify their compatibility. For example, the number of ships that area berthed on a specific pier may be provided by both the terminal operator and by the port authority.

Validation measurements

Validation measurements can take place at selected spots of interest (e.g., near the housing areas or at the limits of the industrial area). It should be noticed that the goal of the strategic noise maps is to display the yearly averaged noise levels. Therefore, the validation measurements should be done long term or made during "selected" circumstances (usually favourable noise propagation condition) and then projected as the annual average of noise emission and propagation condition. Furthermore, it should be acknowledged that noise maps indicate trends more than actual noise figures and that their main function is to demonstrate problem areas. Nevertheless, it is considered useful to examine the noise mapping outcomes (predicted, estimated values) in line with some actual values. "WG-AEN recommends that wherever possible strategic noise mapping should generally be carried out by computation. However, it is recognised that noise measurement has many supplementary roles to play in the effective implementation of the END," according to the Good Practice Guide.[5] So, if many assumptions are made about sound power levels and active hours, noise measurements are the way to verify the model.

Different strategies can be applied for making noise measurements from an industrial site. If the noise levels are caused by stationary machines that run for years, the measuring time may be limited to half an hour near the residential areas. The best way to perform noise measurements is during meteorological noise favourable propagation circumstances.

Measurement protocol, based roughly on the Dutch "Guide for measuring and calculating industrial noise"[6]:

- Not too strong winds (between 2 and 6 m/s) from source area to receiver within 60°
- Measuring height: 4 m
- Number of measurements (at least 4 hours separated)
  - Distance between source area and receiver <150 m: 2
  - Distance between source area and receiver 150 to 1000 m: 3
  - Distance between source area and receiver >1000 m: 4

- Measurement time: depends strongly on variations of the noise level coming from the source area. At least 5 minutes of measurement time without the influences of extraneous noise. If constant extraneous noise cannot be prevented, the measured noise level has to be corrected.
- Measurement equipment should comply with IEC 651: 1979, type 1 with an omnidirectional microphone with windscreen.

To compare calculations with the industrial interim method ISO 9613-2, the meteorological term C_m has to be set to 0; this will calculate the noise levels for a downwind situation. If a HARMONOISE/IMAGINE calculation scheme is used, put in the meteorological situation as at the time of measurement.

If discrepancies between measurements and downwind calculations are smaller than 1 dB, the model may be called sufficient. If the differences are larger, corrections should be made.

Whole model validation

The IMAGINE project has made an effort in making a method for validating predicted noise models. This method is reported in report D5—Determination of L_{den} and L_{night} using measurements—IMA32TR-040510-SP08. In the following paragraph the introduction to this method is given.

This method describes how to determine L_{den} and L_{night}, as defined by the European Directive 2002/49/EC, by direct measurement or by extrapolation of measurement results by means of calculation. The measurement method is intended to be used outdoors as a basis for assessing environmental noise and for verifying the quality of predictions. The method can also be used for monitoring purposes.
The method is flexible and to a large extent the user determines the measurement effort and, accordingly, the measurement uncertainty, which has to be determined and reported in each case. Often the measurement results have to be combined with calculations to correct for operating or propagation conditions different from those during the actual measurement. In each case the long-term equivalent sound pressure level is calculated by taking into account the frequency of occurrence of the different operating and propagation conditions. For each of these conditions the sound pressure level is measured or calculated. In principle two different methods are described: long-term and short-term measurements. However, in practice, a combination of these will often be used. Short-term measurements involve measurements under specified source operating and meteorological conditions and the measurement results have to be used with a calculation method in order to

determine the L_{den}-values. Long-term measurements, on the other hand, involve measurements during a time long enough to include variations in source operating and meteorological conditions. Thus, the measurement results are more accurate and can be used with much less corrections than those of short-term measurements. This is a frame method, which can be applied on all kind of noise sources, such as road and rail traffic noise, aircraft noise, and industrial noise.

REFERENCES

1. Directive 2002/49/EC of the European Parliament and of the Council of 25 June 2002 relating to the assessment and management of environmental noise.
2. IMAGINE research project, www.IMAGINE-project.org.
3. IMAGINE report IMA07TR-050418-MBBM0, Measurement Methods.
4. IMAGINE report IMA07TR-050418-DGMR02, Description of the Source Database.
5. Good Practice Guide WG-AEN, version 2, 2006.
6. Guide for measuring and calculating industrial noise (in Dutch), Ministerie VROM, Zoetermeer, 1999.
7. NoMEPorts Good Practice Guide Technical Annex.

Chapter 7

Airport noise

R. Bütikofer

CONTENTS

INTRODUCTION

An aircraft noise calculation is considerably more complex than using a program with a specific name. Certainly, the concepts of acoustic models remain important. But there is a large amount of preprocessing and post-processing to be handled very similarly with any noise calculation program. The fact that a result can be only as good as the input data leads to the importance of the input data preprocessing. Decisions made by the user during preprocessing can easily have a greater impact on the uncertainty of the results than the acoustic model used. Therefore, this chapter on aircraft noise is not a technical manual on how to compute noise with a specific program, but it stresses common features and constraints of all aircraft noise calculation programs.

First, there is an overview on special aspects of aircraft noise calculations. The task to squeeze real airport operations into a few average input data sets, called scenario, is addressed next. Then the different generations of program-architectures and the consequences for the required sound databases are discussed. The main problem is and remains the availability of specific sound emission data. The four candidates for a European noise calculation program are briefly described. Then specific aspects of propagation in aircraft noise programs are discussed. The contour lines are generated in postprocessing. The quality of a noise calculation is characterised by the overall uncertainty, which depends on the scenario's complexity, the acoustic model used, and the completeness of the available sound data. User guidance is given in the application guide of DOC.29. To conclude, the revised third edition of DOC.29, published by ECAC in 2005 and used with the database ANP, represents the state of the art in Europe for aircraft noise calculations, unless a new sound emission database was established opening the road toward simulation programs.

The words *sound* and *noise* are used throughout the chapter. The meaning are as follows: *sound* describes a physical (measurable) property, whereas *noise* is used in a general meaning, sometimes including aspects of annoyance, but many times *noise* is used also instead of *sound*. Examples: sound emission, sound level, but noise contours.

OVERVIEW ON SPECIAL ASPECTS OF AIRCRAFT NOISE CALCULATIONS

Aircraft noise calculations differ from noise calculations for road or rail. This section gives an overview of the aircraft noise calculation elements.

Specific properties of aircraft noise calculations

The task of an aircraft noise calculation program is to characterise the overall noise situation around major airports over tens of square kilometres. The focus is on characterising areas, for example, those with high noise immission levels, not on qualifying the noise immission at one single apartment window. Such a single point could be measured directly with a monitor microphone.[32]

Large areas, no houses

For an international airport, the area under investigation may extend over 30 to 40 km in each direction. By consequence, the aircraft noise immission is calculated only in grid points with a grid spacing of 50 to 250 m. The grid points are then used for an interpolation to end up with contour lines of equal sound levels. There are three important implications:

- Aircraft calculations are bulky. An area of 30 by 40 km with a grid spacing of 100 m requires the calculation at 120,000 grid points. A finer grid spacing of, for instance, 10 m, would increase the number of calculations again by a factor of 100, without adding much extra value.
- Only terrain formations with dimensions greater than grid spacing may be accounted for in the calculation. Houses and sound barriers with dimensions of some metres are too small to be taken into account. (Usually the terrain altitude at the grid points for the noise calculation is interpolated from the grid points of the terrain information. Hence, topographic resolution is either that of the terrain grid or that of the calculation grid, whichever is larger.)
- Calculations are usually made for receiver points in a situation free of any obstacles like houses. Some programs assume a fixed receiver height (e.g., 1.2 m) above short cut grass.

Distinct aircraft types and their sound emission

Unlike road traffic with average vehicle types, aircraft traffic is characterised by distinct aircraft types with specific engines. Aircraft manufacturers tend to consider the sound emission data as very sensitive information that could be misused in commercial competition. Hence, source data is not readily available. The American Federal Aviation Authority (FAA) or EUROCONTROL may publish standardised data like the noise power distance (NPD) tables, which are derived from noise certification measurements according to ICAO, Annex 16.[26]

A key issue of any noise calculation program is the availability of sound source data for those aircraft types, which are acoustically dominant. As the sound emission may differ by several decibels for exactly the same aircraft type when it is equipped with different engines, the sound source data has to cover exactly that aircraft type–engine combination that operates at that airport for which the noise calculations are made.

What is considered to be aircraft noise?

Aircraft noise calculation programs usually consider only two operations: departures and landings, that is, (1) the aircraft movement from "start of roll" or "brake release" on the runway until the aircraft has left the area of the grid points; and (2) the aircraft movement from approaching the area of grid points until having landed and decelerated.

The following airport operations are excluded: taxiing of aircraft, the operation of the APU (auxiliary power unit) and all other noise sources from airport operation (e.g., truck, busses, heating/cooling devices, engine test stands). The acoustic impacts of these elements have to be assessed by other calculation tools not discussed here, if ever there is a need to do so. Note that national legislations may require including some of the aforementioned activities into the overall aircraft noise calculation.

Small aircraft (general aviation, propeller driven) are usually calculated with other simpler programs not discussed here. The general approach is the same, but specific considerations have to be made for individual source data and for defining the flight paths in absence of radar proof. The contributions of small aircraft to the noise emissions on an international airport are marginal.

Helicopters have rather distinct directional sound emissions. If they dominate, special programs apply, based on spectral, three-dimensional sound emission data. To have them included in the operation at an international airport, they may be roughly approximated by a similar sound emission description than a fixed-wing airplane.

Relevant distances for accurate noise calculations

The aim of aircraft noise calculation is to provide good noise data with low uncertainty within those noise contours relevant in a legal context. Depending on airport traffic, the noise contour with the lowest relevant level may extend several kilometres around the airport. At the locations of the lowest legally relevant contour line the aircraft is high up in the air and the distance from the aircraft to the receiver may be a few kilometres, where sound propagation through a turbulent atmosphere may increase the uncertainty of the result.

The aircraft is heard over much longer distances but this is out of scope for aircraft noise calculations.

"Air to ground" sound propagation

Close to the airport there can be a sound propagation "ground to ground" where all the specific effects of sound propagation close to the ground like barriers and curved sound propagation due to temperature gradients and wind may be important. For special situations, this has to be taken into consideration.

In general, however, the situation prevails where the aircraft is up in the air and the sound is propagating to residential areas below the aircraft. For this "air to ground" situation the sound propagation becomes very simple: it accounts for geometrical spreading and for air attenuation (plus "ground effects" discussed in the section "Specific Aspects of Propagation in Aircraft Noise Programs").

The effect of wind and of temperature gradients

In fact, wind does have an influence on source location rather than on propagation. It displaces the aircraft away from the expected flight path, resulting in a modification of the propagation distance from the new aircraft position to the receiver. If radar data is used for the description of the flight path, wind effects are accounted for.

Wind and temperature gradients (i.e., changes with altitude) may generate upward or downward bending of the sound propagation path. However, for a source high up in the air this has virtually no effect because in the absence of obstacles it does not matter if the propagation is straight or if it is curved.

Accounting for topography

Topographic issues are discussed in detail in the section "Specific Aspects of Propagation in Aircraft Noise Programs." The key words are receiver altitude, angle of sound incidence, shielding by hills, and the "nonvisibility" of barriers (e.g., houses) with smaller dimensions than the spacing of the calculation grid of typically 50 to 250 m.

Noise calculation of a single aircraft movement for a single receiver point

The kernel of any aircraft noise calculation is the estimation of the immission (usually the single event level L_{AE}) for a single aircraft movement (departure or landing). Figure 7.1 shows the typical components involved.

An aircraft noise calculation is much more complex than the acoustic calculation itself. Details may vary according to the acoustic model used, but there is always a considerable amount of geometric preprocessing: first, to estimate a set of coordinates describing the flight path, then breaking the flight path into segments, and finally deriving the parameters for the acoustic calculation for each segment (distance and emission and immission angles). If shading by the topography is enabled, the preprocessing

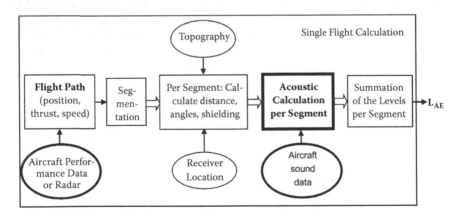

Figure 7.1 Basic structure for a single flight calculation for a given aircraft type and one receiver location.

also includes the generation of cross-sections of the terrain along the sound propagation path.

The result depends heavily on the user-defined input of the flight path and on the quality of the available aircraft sound database. These two factors will be discussed next in more detail.

Flight path information (position, thrust, speed)

The flight path describes the movement of the aircraft in the three-dimensional space. Usually the flight path is divided into

- *altitude profile,* which describes the altitude of the aircraft in function of the flown distance, and
- *track*, the projection of the flight path on the (flat) ground.

For departures, the altitude profile depends on (a) aircraft performance in function of airport altitude and temperature, (b) takeoff weight, (c) airline or airport specific departure procedures (e.g., "speed first," "altitude first," special noise abatement procedures), (d) use of derated power (also called flex power), and (e) the amount of headwind. For landings, the expression "altitude profile" is used to describe the descent. Here, the noise-relevant portion of the profile of the final approach is usually governed by the constant gliding angle of the ILS (instrument landing system).

The track depends on (a) the traffic control constraints, (b) the pilot's variations in following the nominal flight path, (c) operational constraints (i.e., initiate a turn only after having reached a specified height), and (d) side wind.

All the aforementioned effects are of no concern if direct positional data from the aircraft can be used, for example, the radar information from traffic control or (in general for research purpose only) the flight deck recordings from the aircraft.

There are two additional profiles needed, which are also given as functions of the distance flown:

- *Speed profile*—The ground speed of the aircraft affects the L_{AE} by the duration of the immission
- *Thrust profile*—For departures thrust is set according to flight procedures (reduced or flex thrust whenever possible) and, if maximum takeoff power was set, it is reduced after the initial climb. For landings, engine power may fluctuate for keeping the aircraft on the ILS guiding ray.

Aircraft sound database

It is relatively easy to devise a new version of a noise calculation program. The central problem is the availability of good quality sound emission data for exactly that aircraft–engine combination under consideration. A comprehensive, periodically updated sound database is the prerequisite for any harmonised European aircraft noise calculation. This will be discussed in the section "Database Characteristics."

The calculation of noise contours

The noise immission calculation for a whole area around an airport (Figure 7.2) is again more complex than the single-event calculation discussed earlier. First, the user has to reduce the real airport operations to a number of typical single-flight situations. The product of this process is the "scenario." A scenario indicates for specific time periods (e.g., day, evening, night) what aircraft type flew where and how often. The single-flight calculation then generates the L_{AE} values in the grid points for the whole area under consideration, using the appropriate aircraft sound source data from the internal database. This result is called a footprint. As the calculation time for a footprint may be rather long, footprint data is often stored for later reuse. Footprints have to be calculated for the most important aircraft types and for each departure route or landing approach route. The combination of, for example, the 20 most important aircraft types on 30 routes and subroutes adds up to several hundred footprints. To get the total noise immission, the individual footprints have to be added according to the number of movements within the various time periods of 24 hours (e.g., day, evening, night). Finally, the noise contours are calculated from the immission values in the grid points.

The quality of the final result depends on the way a scenario is defined. For a quick survey study, a coarse scenario might be appropriate, taking into account only a few aircraft types and a few flight paths, whereas for a precision calculation the scenario becomes rather complex.

Figures 7.1 and 7.2 illustrate that the acoustic model used in the calculations is only a small part of a complete aircraft noise calculation system, which includes tedious preprocessing to define the scenario and postprocessing to estimate the noise contours. For different programs, the form of the preprocessing may vary, while postprocessing is basically the same

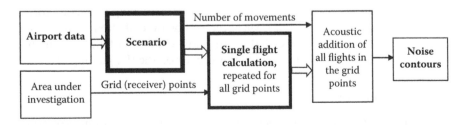

Figure 7.2 The components needed for the calculation of the noise contours around an airport.

for all kinds of programs. The "Applications Guide" of DOC.29 (2005)[13] addresses these topics.

Model features, quality requirements, and calculation times

Depending on the required quality, the number of calculations varies considerably. For a survey, it may suffice to group many aircrafts together and make only calculations for a few groups. Similarly, the spreading of flight paths may be modelled either with a few nominal flight paths or with many. And coarse grid spacing reduces the number of calculations sensibly. For the calculation of a major airport, the number of propagation calculations (one calculation per flight path segment, for each aircraft group and route, and for all receiver points) may increase from 10^5 for a survey up to 10^{12} for calculating every single flight within a year. Such a "full size" calculation was made, for example, by Empa for airport Zurich.[44] Using the Swiss aircraft noise calculation program FLULA2, it lasted 10 days on a 24 CPU Linux-cluster (10^{12} seconds make 32,000 years; hence one complete propagation calculation must be finished within a few microseconds).

New approaches to noise control like IMAGINE[27,28] promote a unified sound propagation handling for all kinds of sound sources. For aircraft noise, this may work for the aircraft on the runway. However, for air-to-ground propagation, this approach is impractical: the propagation models put forward for ground sources (like road or rail) cannot handle large angles of sound incidence up to the perpendicular sound incidence of a fly-over unless they are extended with some approximation algorithms, and, more important, such sophisticated algorithms are too time consuming for the millions of propagation calculations made for assessing the noise of one airport.

Noise metrics

There is a long tradition of various noise metrics. There was the NNI (Noise and Number Index) used in the United Kingdom and in Switzerland, the $L_{eq,4}$ used in Germany, the indice psophique used in France, and the NEF used in Canada. A special index is the NAT, the number above threshold, which counts simply the number of aircraft events exceeding a fixed sound level. In special cases, especially for small aircraft on airfields, it may be more practical to use the maximum sound level (weighted with the frequency filter A and the time constant SLOW): $L_{A,S,max}$. Today most countries have adopted an energy-based approach based on the sound exposure level (SEL; symbol L_{AE}). There are two calculation rules for the SEL of a single-aircraft flyby: In one case, the complete flight path contributes to the SEL, whereas in the other case only the portion of the sound event is considered, which is louder than $L_{A,S,max}$ –10 dB. The latter definition leads to sound levels that are 0.4 to 0.5 dB lower. All NPD data was calculated using the "10 dB down" rule.

Having estimated the overall L_{AE} from all flight events taking place within a specified period of time, the equivalent sound level L_{eq} is calculated from the L_{AE} by accounting for the time duration. The L_{eq} may be combined with specific correction factors, K, (e.g., penalties for night traffic) to produce a specific rating level (e.g., L_r, L_{DN}, L_{DEN}, L_{eq4}). The combined metric day–evening–night, L_{DEN}, is a weighted combination of the L_{eq} values for operation during daytime, evening (with 5 dB penalty), and night (with 10 dB penalty).[16]

Annoyance

The L_{eq}-based noise metrics are important for legal purposes and for planning issues. However, it must be kept in mind that the correlation of any noise metric to community response is not very strong. Other aspects like visibility of the aircraft, threatening by low-flying aircraft, living conditions, sound insulation of the apartment, and other noise immissions have a strong influence on the individual annoyance and thus how well the aircraft noise is accepted or not (see, for example, the DOC.29 (2005), 3rd edition, Volume 1[13]).

An example of using annoyance instead of sound levels is the Zürcher Fluglärm Index (ZFI). Using for daytime a level-annoyance relation and for nighttime the probability of additional awakenings, the local aircraft noise levels are combined with the number of people living in an area of, for example, 100 × 100 m to end up with the total number of highly annoyed people. The regional government of Zürich is obliged to induce countermeasures if the number of highly annoyed people exceeds a limit value.[50]

After this overview let us have a closer look at the tedious task of data preprocessing to generate a scenario.

THE SCENARIO

The scenario answers the question which aircraft type (or group of similar aircraft) flew (or will fly) how many times on which flight path within a specified time of the day.

The scenario is the result of mapping a real-world situation (e.g., the operation of an airport during one year) into simplified input data for the calculation. In Germany, this process was considered so important that it was regulated in the guideline "DES,"[11] now replaced by AzD.[5] Even if the acoustic calculation would be perfect, the old truth that the output can only be as good as the input keeps its full meaning with aircraft noise. It is not easy for the user to prepare a balanced input that is detailed enough to provide reliable results but that is not unnecessarily large. Further, not all required parameters are known to their full extent. It depends on the experience of the user to fill the gaps with reasonable assumptions. The new Applications Guide DOC.29 (2005), Volume 1[13] is the first document within the aircraft noise calculation literature providing user guidance, which may be also useful if other acoustic kernels than DOC.29, Volume 2 (2005)[14] are used.

The complexity of a scenario depends on the purpose of the calculation:

- Survey—Allowing a high level of uncertainty of the result, a simple scenario will do.
- Engineering—Allowing a moderate uncertainty, the scenario becomes rather complex.
- Precision—For minimal uncertainty, the scenario (and the number of calculations) becomes bulky.

The scenario also depends on the time period covered:

- Historic—For example, the noise contours of last year's traffic. Here all data on aircraft and their movements are known.
- Future—For a forecast noise exposure the uncertainty increases due to the following reasons: the calculation may consider aircraft not yet in use where the noise characteristics have to be estimated, the number of movements is a well-informed guess, depending on uncontrollable factors like growth of prosperity, and the landing and departure routes may be redefined in future, for example, due to political constraints.

Aircraft identification

The air traffic on the airport under consideration has to be analysed with respect to the aircraft types. As different carriers may operate the same aircraft type with different engines, the identification normally has to include the engine type. The airport authority may provide lists of movements. Based on the individual tail number each aircraft may be identified by looking up reference tables from the aviation authority, from commercial products or from the Internet. As an example, Empa maintains an updated list with more than 30,000 tail numbers of aircraft operating at Swiss airports.

Reducing the number of aircraft types in a noise calculation

Having identified all aircraft types, there arise two questions: (1) For which aircraft types are noise data available? (2) How can the number of aircraft types be reduced for a calculation? This leads to substitution (replacing an aircraft type without noise data by an acoustically equivalent type with noise data) and to grouping (clustering of aircraft with similar properties). Obviously, the way of grouping and substituting by the user of any noise calculation program has a direct impact on the final quality of a noise calculation. This topic is treated in detail in Chapter 6.4 of the Applications Guide (Vol. 1) of DOC.29.[13] For further details it is strongly recommended to consult this document. A free copy can be downloaded from the Internet.

To find out which aircraft types are important for a noise calculation, a ranking according to the acoustic energy may be made. The acoustic energy is based on the L_{AE} of a single flight (e.g., the NPD data for 305 m) and the number of movements. For an international airport there will be typically about 20 aircraft/engine types that generate over 80% of the acoustic energy. Those types shall be treated separately, each with its associated noise data from the noise database. The rest of the aircraft may be grouped and substituted in a few groups. Another approach is used in AzB,[4] where from the beginning the aircraft with similar acoustic properties are grouped together in predefined classes.

Whatever methods are used, the task of grouping and substitution has to be made for all aircraft noise calculation programs. The degree of grouping and substitutions, and hence the calculation efforts, depend on the required quality of the noise calculations (survey, engineering, precision).

Modelling of track dispersion

All departures on a given runway heading for a given route diverge after some kilometres. Common practise to model this is to define a centre track (the backbone) and on each side several sidetracks. The number of movements per track has to be evaluated from the original track dispersion.

At Empa another method has been used successfully for many years. Based on radar data, the average noise immission is calculated as follows: The individual noise immission (the footprints) for 60 to 100 randomly selected real flights are calculated and only the footprints are averaged and properly scaled. In this case, averaging is made on acoustic data rather than on track geometry. The advantage is a more realistic modelling of aircraft movements but the price is an increase in computer time by a factor of about 10. Details are explained in Bütikofer et al.[7]

Residents may argue that the simplified geometry of the centre track and sidetracks will not account for some aircraft flying along some unusual tracks. The ultimate solution is to make individual calculations with each of the available radar tracks. This full-size calculation was made for airport Zürich with 250,000 movements at Empa using a cluster of auxiliary programs around FLULA2. It is somewhat a tedious job to organise the data. Calculation time was about 10 days on a 24 CPU Linux cluster, but no doubt it is manageable, and it was also applied for airport Geneva. The main argument to perform a full-size calculation is a political one, namely, that the airport can assure a resident for having taken into account also that airplane that flew over his house. From the point of view of accuracy the benefits of a full-size calculation (taking into account exactly those aircraft movements that took place in the specific time of day interval) show up the shorter the time interval is (or more precisely, the fewer flights took place in the period considered). For example, in a 16-hour period of the day there is nearly no difference compared with the method of randomly selected single flights, whereas for a 1-hour period at night differences emerge.[44]

Altitude profiles

As mentioned earlier the altitude profile depends on various parameters of aircraft performance and operation. If radar data are available, the profiles may be calculated as an average geometric profile. Otherwise the altitude profile has to be estimated. For instance, SAE AIR 1845[42] provides formulas—repeated in updated form in DOC.29 (2005)[14] and implemented in INM[30]—to generate a profile starting with a flight procedure and using aircraft specific coefficients stored in the INM database and in ANP.[1] Some programs may also provide predefined altitude profiles. The relevant part of the landing profiles is usually governed by the ILS with a constant gliding angle and a straight approach to the runway.

Speed profiles

Together with the altitude profile, the speed along the flight path has to be known. Speed is closely linked to climb performance and thrust. Speed profiles can be estimated from radar data. Otherwise, default speeds of standardised flight procedures published by aircraft manufacturers and applied by pilots have to be used.

Thrust profiles or equivalent indications

During a takeoff and landing, the aircraft engines operate at different power levels and correspondingly emit various levels of sound power. Therefore, some indication on sound emission levels has to go along with the altitude profile. Despite the fact that the level of sound emission is the central factor that influences the immission level directly, little is known. Aircraft may accelerate on the runway with maximum power, but whenever feasible, less than maximum power is used to reduce engine wear. This is known as flex or derated power. For example, in the calculations made at Empa, maximum power is assumed if the ATOW (actual takeoff weight) is higher than 85% of the MTOW (maximum takeoff weight), and flex power for the lighter departures. One problem is that not all carriers report their ATOWs to the airport authority. In INM, the "stage length," that is, the runway length until liftoff, may be used for power estimation. And stage length is estimated from the distance to flight destination, that is, the fuel needed.

The "cutback" is a reduction of engine power after initial climb, for example, at 1500 feet (450 m) above ground. This cutback may be substantial for departures with maximum power, but small for departures with flex power.

The NPD (noise power distance) concept for source emission used in INM and in DOC.29 (2005) assumes that emitted noise is proportional to engine power. (For landing, this concept needs some adaptations.) A thrust profile is defined together with the altitude profile. This thrust profile is estimated based on formulas.

NUMBER OF MOVEMENTS

Noise immission depends on the number of movements per route and aircraft type. Depending on the purpose of the calculation, different numbers apply:

- Historic calculation—The exact number of aircraft movements that took place in the period under consideration (e.g., one year), split up into the time periods of the day (e.g., day, evening, night).
- Special rules—For instance, in Germany a situation may be assessed as if the whole traffic would use only one direction of the runway. This results in some kind of worst-case evaluation.
- Forecast—The calculation to predict a future situation is based on assumed numbers of movements.

Having set up the scenario, the next step is to look at the acoustic kernel.

THE ACOUSTIC KERNELS (NOISE CALCULATION PROGRAMS)

Aircraft noise calculations started in the late 1960, without computers. Calculations were straightforward and directly produced the A-weighted immission level. With emerging computer power the programs included more and more details but still used precompiled A-weighted data sets, which combined sound emission and a standardised propagation condition (e.g., the NPD database, defined for sound propagation at 25°C). If source emission is separated from propagation, the flexibility in handling various propagation situations is increased. But this requires extended spectral source data, increasing the computational effort to many spectral calculations instead of a single A-level calculation.

Characterisation of actual programs

There exist numerous noise calculation programs. Most of them were developed on the request of national authorities to handle aircraft noise in their national legislation. Today, some national programs are discontinued in favour of using the program INM 7.0; others have been adapted to use identical calculation algorithms. It has to be mentioned that there also exists very sophisticated programs used in research or by the aircraft manufacturers to answer specific questions on noise generation or to investigate quieter flight procedures, and so forth.

A "program" is originally a document describing the methodology plus a database. Examples are DOC.29[14] and AzB.[4] Here it is left to private organisations to implement it and eventually to sell it as a commercial product. In the case of INM,[30] the American Federal Aviation Authority provides a ready-to-use computer application and it is in charge of continuously updating the software and database.

For the purpose of estimating aircraft noise on major airports in Europe, the following programs are the most important.

DOC.29 (2005)

The ECAC (European Civil Aviation Conference) charged a group of international aircraft noise experts to update the old DOC.29. Work was performed in close collaboration with SAE A-21, the steering group responsible for INM and with FAA. DOC.29 uses NPD and performance data that are a subset of the INM database, maintained by EUROCONTROL as ANP[1] and accessible for interested users. An agreement between FAA and EUROCONTROL allows for generating new data for actual aircraft, especially Airbus. DOC.29 uses segmentation and the noise fraction algorithm, a generic lateral directivity, and the updated lateral attenuation similar to SAE AIR 5662.[43]

Prior to publication of DOC.29 in 2005, there was an international round-robin test with DOC.29 compliant program implementations from Norway (NORTIM), UK (ANCON2), and the United States (INM 7).

INM 7

The Integrated Noise Model has a long tradition. Version 7 incorporates all the features already mentioned with DOC.29. It is a ready-to-use software package, maintained and sold for a symbolic price by the American Federal Aviation Authority. Most important, it comes along with the NPD database and with a database of flight performance. The methodology of the program is developed in the international working group SAE-A21 and specified in corresponding SAE documents. Research by SAE A-21 for lateral directivity and lateral attenuation was included in version 7 of INM as well as in DOC.29.

The advantage of INM consists of being software ready to use. The only but important drawback is that the software package is a "black box" for the user with undisclosed source code. From one release to the next, internals may have been optimised or adapted, which could influence the results. This makes it not very suitable for legal requirements.

INM has a special feature to increase the number of grid points in areas with high changes in the sound levels, that is, close to the runways. Postprocessing (calculation of noise contours and graphic displays) is made with the stand-alone software package NMPlot.[38]

AzB

The AzB[4] was updated in 2008. It contains its own database of octave-band spectra for aircraft classes plus (longitudinal) directivity factors. It now uses segmentation. Starting with the sound power per octave band, emitted from a segment, propagation is calculated in each octave and the A-weighted level is only calculated at the immission point, which is at 4 m above ground. By default, grid spacing is 50×50 m. Special regulations are used for defining the scenario: calculations are made for classes of similar aircraft types, not for individual types; the number of movements relies on the six busiest months of a year and for runway usage special considerations apply. The AzD[5] prescribes standardised procedures for defining a scenario in Germany.

IMAGINE

IMAGINE was an EU-6RP research program continuing the EU-5RP program HARMONOISE. The goal was to develop a unified noise calculation for rail, road, and aircraft, using the same modules for propagation and at the receiver. IMAGINE is a simulation program, reproducing the spectral time-level history of an aircraft movement at the receiver point. For aircraft noise, this requires a new database for the sound power emission of the most important aircraft operating in Europe. An example was made on how to measure, process, and present source data, and existing IMAGINE propagation modules had been adjusted to also handle angles of sound incidence up to perpendicular. The sophisticated propagation calculations are rather time consuming. What would be needed to use IMAGINE on a large scale in Europe are some refinements in the acoustic software, a guideline on how to estimate altitude profiles for the scenarios (see, e.g., DOC.29 or INM), and, most important, a new spectral database for sound power emission for the dominant aircraft types. Measurements of such data are mainly a question of funding, which would amount to several million euro.

The EU interim model

It was intended that the updated DOC.29 3rd edition would be the platform for an EU model. In 2002 this new version of DOC.29 was not yet tested in practice. Therefore EU Directive 2002/49/EC[15] specified to use the DOC.29, 2nd edition (1997), which was the reaffirmed 1st version of 1986,[12] using a simple CPA structure (see following section). The directive tells to upgrade that model using segmentation, without specifying how to do it. The EU recommendation (6 August 2003)[16] gives more guidelines: The segmentation may be based on the noise fraction as described in the INM Technical Manual,[30] that is, the same approach as used in DOC.29 3rd edition. For the noise database to be used, the recommendation mentions the ANP[1] established in its final form in 2005 at EUROCONTROL , but specifies for the time being as a "default recommendation" the database from Austria[17] and from Germany.[18] Engineering enterprises implemented those guidelines, but there is no official reference implementation for compliance checks. The EU recommendation[16] acknowledges in paragraph 2.4.1 the revision of DOC.29 and suggests using the updated DOC.29 after release by ECAC, provided this is "appropriate and considered necessary." DOC.29 3rd edition was released in 2005.

Program generations

CPA (closest point of approach)

The earliest programs estimated the maximum sound level $L_{A,S,max}$ from tables for the shortest distance to the flight path. Correction terms were sometimes used to account for curved flight. The old program version of AzB used in Germany from 1975 until 2008 is based on this concept.[2,3] The L_{AE} is estimated by a theoretical approach from the $L_{A,S,max}$ level.

CPA with integrated database

The "integrated" database NPD (noise power distance) provides L_{AE} levels for a level flight, normalised to 160 knots (1 knot = 1.852 km/h). The straight flight segment with the closest distance to the receiver was used to determine the L_{AE}. Correction terms apply to account for speeds other than the 160 knots and usually for curved sections. DOC.29 1st and 2nd editions and the early INM used this technique.

Segmentation with integrated database

The problem of different power settings (initial climb from the runway, power cutback, continuous climb, etc.) and with different portions of the flight paths led to the formulation of the noise fraction algorithm. This algorithm assumes a mathematical sound power directivity of a $\sin^2$ with its main emission direction perpendicular to the flight direction. The advantage of this model was its analytical solution, which could be easily implemented in the calculation programs. The noise fraction allowed for dividing the flight path into straight segments. Each segment is first attributed the L_{AE} for a level flight at the corresponding distance and engine power, which is then reduced according to the noise fraction for the aspect angles under which the segment is seen. The method is described in detail in INM[30] and in DOC.29 (2005), volume 2.[14] The following programs use basically segmentation and NPD data: INM (from version 5 on),[30] DOC.29, 3rd edition,[14] ANCON2 (CAA, United Kingdom), NORTIM (Sintef, Norway), ÖAL 24 (Austria), Sverim (Sweden, made by Wyle, USA), SONDEO (Anotec Consulting, Spain), and NOISEMAP (U.S. Air Force).

Simulation

In the context of aircraft noise, the word *simulation* is used for programs that reconstruct (simulate) the level time history of a flyby at a receiver location as it could have been measured by a sound-level meter. They consider the aircraft as a point source, which is located consecutively at discrete locations along the flight path. From each location the immission level is calculated together with a time step during which this level will apply before the aircraft has moved to the next position, producing another level. As the flight paths are digitised in discrete points at, for example, 1 second separation of flight time, there are no restrictions for the geometry of the flight path. As the level-time history of the aircraft movement is reproduced, all the metrics that can be evaluated with a sound-level meter out in the field can also be calculated with the simulation. These are L_{AE} of the complete flight, L_{AE} only for the uppermost 10 dB of the event, $L_{A,S,max}$, NAT (number above threshold), and for comparison with monitor stations the L_{AE} above a predefined threshold.

Simulation models may account for the various sound levels emitted by the aircraft in different directions, the so-called directivity.

Older simulation programs like FLULA2 directly provide the immission level for a given distance and emission angle. Actual concepts are based on spectral sound power emission data with directivity and spectral propagation calculation according to the actual needs (more or less efficient propagation algorithms, specific temperature, soft/hard ground, height of receiver, topography, etc.). The following programs use simulation (series of point source positions) with various kinds of noise emission data: FLULA2 (Empa, Switzerland), DANSIM (Delta, Denmark), DIN 45'684 (for small airports, Germany), SIMUL (DLR Göttingen, Germany), SOPRANO (for research in EU-5RP "SILENCER"), GMTIM (airport Gardemoen, Norway), NMSIM (Wyle, USA), RNM (Rotorcraft Noise Model, NASA), and IMAGINE (EU).

A combination of segmentation and simulation is the program MITHRA-Avion (CSTB, Grenoble, France), which uses variable length segments, such that for each segment its endpoints are seen from the receiver location under the same aperture of angle.

A detailed description of the various concepts of aircraft calculation programs is given in Bütikofer.[6]

Database characteristics

The structure of the database depends on the architecture and the sophistication of the acoustic kernel. The database may list the A-weighted levels in function of distance and in function of engine power: the NPD tables defined in SAE AIR 1845[42] and used in INM[29] and as a subset in ANP[1] for use in DOC.29 (2005).[14] It may contain A-levels and directional information as used in FLULA2,[22,23,34] or spectral information as used in the old AzB.[2] It may contain spectral and directional information as used in the new AzB[4] and in a special version of FLULA2[33] and proposed by IMAGINE.[28]

The interaction between the content of a database and the calculation models is shown in Figure 7.3.

For research, various sophisticated models with limited databases exist. But this data is far from being complete to cover the most important aircraft types, and manufacturer's data is not available to the public.

The issue of sound data is the central challenge in aircraft noise calculations. Well-controlled measurements are expensive. A worldwide regulation would be required to allow manufacturers providing data in the same form their competitors would have to provide. The regulations on aircraft certification (ICAO, Annex 16[26]) provide the base today, which allows authorities to filter out NPD data from the undisclosed measurements. But more specific information would be needed in future.

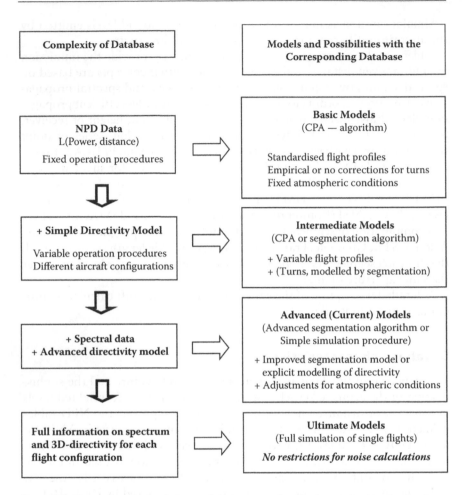

Figure 7.3 The more advanced the model, the more details are required in the database.

Source directivity

An aircraft has several noise sources. For departure, the engine is dominant; for landing approach there may also be important noise contributions from aerodynamic sources from the gear and the extended flaps. The noise from the jet engine itself has typically three components: fan noise emitting mainly toward the front, core noise emitting in all directions, and the jet noise from the hot gasses emitting laterally but toward the rear. The lateral sound emission is modified by reflections and turbulences on the wings (for wing-mounted engines) or by shading by the fuselage (for tail-mounted engines). Details on source directivity are found in the downloadable Deliverable 10 of IMAGINE, Work Package 4.[28] Old civil jet engines with a low bypass ratio and military aircraft have a pronounced level increase up to 15 dB for emission directions at 120° to 140° (front = 0°, rear = 180°). For modern civil aircraft with high bypass ratio engines, the longitudinal directivity shows only moderate level variations. Further, for a flyby, the main contribution to the SEL comes from the location where the aircraft is closest to the receiver, that is, for emission angles around 70° to 120°.

The real longitudinal directivity (front to rear) is only accounted for in simulation programs with appropriate source data. (Programs using segmentation with noise fraction assume a fixed mathematical model for directivity.) The lateral directivity (bottom to side) will be discussed in detail later.

SPECIFIC ASPECTS OF PROPAGATION IN AIRCRAFT NOISE PROGRAMS

In most aircraft noise calculation programs the propagation is rather straightforward, taking into account only a few distinct effects like spherical spreading, air absorption, and some kind of "ground effects." In this section, some specific terms used in several programs will be discussed.

Air absorption

Air absorption has to be applied to spectral data. In many programs used today, the spectral air absorption was calculated when generating tables for the A-weighted levels in function of the distance. The A-level tables are valid only for the temperature and humidity used in that processing. For NPD data, this is 25°C and 70% relative humidity (International Standard Atmosphere [ISA] +10°C, due to certification at that temperature).[26] For FLULA2 it is 15°C (ISA). Newer versions of INM and DOC.29, vol. 2 (appendix D) provide the possibility to approximate the NPD data for another temperature, using generalised spectra for the appropriate aircraft from the spectral class data tables included in INM[31] and in ANP.[1]

Lateral directivity

Specific measurements made in the working group AIRMOD for DOC.29 (2005) and in SAE A-21 revealed a lateral directivity of the sound emission of the aircraft, depending on propeller or jet engine, and for jet engines depending on the mounting of the engines at the wing or the tail.[20,40] Some findings on lateral directivity are also described by Krebs and Thomann.[35,36] Lateral directivity is a property of the individual source. However, the working group SAE A-21 decided that it would be too complicated to supplement the NPD data with lateral directivity. Instead, it was included in a generalised form in the revised lateral attenuation (see later).

Ground interference

The effect of ground interference is related to the height of the receiver point and the angle of sound incidence. Due to interference effects of the direct sound wave with that reflected on ground, the spectral shape of the measured sound at the receiver is modified. The effects are well understood based on the work of Daigle et al.[10] For example, the measurements for NPD data made at 1.2 m above ground show a typical dip in the one-third octave spectrum between 125 and 200 Hz. Hence, calculations using NPDs apply, strictly speaking, for a sitting person on a flat terrain. Fortunately, these interference effects have in most cases a small influence on the final result. This is reported in a case study from Bütikofer and Thomann.[8] These interference effects are the reason why the receiver height is now often set to 4 m. The physics is the same at 4 m, but the first dip of the interference effect is shifted toward lower frequencies and interferences at higher frequencies tend to level out within a one-third octave band.

Lateral attenuation/overall ground effect

Sound levels measured to the side of a flight path, especially at low angles of incidence, are usually lower than what was expected from propagation in the free air.

Empirical formulas for the A-weighted level to account for this effect are known as "lateral attenuation" or "ground effect." The lateral attenuation was defined in SAE 1751 (1981/1991)[41] and used in INM. Lateral attenuation was investigated in recent years.[21,25,46] Based on that, the working group SAE A-21 revised the recommendations for lateral attenuation and included the lateral directivity in the new standard SAE AIR 5662 (2005).[43] For the sake of not having to change the NPD source data, lateral directivity was included in a generalised form in the new lateral attenuation. This approach was adopted identically in DOC.29, volume 2 (2005). Lateral attenuation is considerably higher than what would be expected from the ground interference. It includes average experience of turbulences and partially acoustic shadowing as well.

Other programs use similar empirical formulas for the A-weighted levels to account for additional damping at low angles of sound incidence (usually below 15°). Recent research is reported by Krebs and Thomann.[36]

Topography

The topography has several effects on calculation:

- Taking into account the altitude of the receiver location will modify the distance to the aircraft and the angle of sound incidence. This is straightforward. It requires a digital map of the area and tools to extract the altitude at given coordinates. This adds only little load to computing time and it is included in many calculation programs.

- A difficult situation arises with high hills. For a receiver on the slope of the hill, the aircraft may be at a lower altitude down in the valley. Questions arise on definition of the angle of incidence (relative to horizontal results in negative angles; relative to the normal vector on an area element at the receiver location requires time-consuming calculations) and on the applicable "ground attenuation," which is usually defined only for flat terrain. (The acoustic knowledge to handle these situations is available, implemented, for example, since 1995 in the program NORTIM[39] and using most advanced models of sound propagation in sonRail[45] for the prediction of railway noise.)
- There may be shielding effects by hills. This is discussed in the next paragraph.

Shielding

As the grid spacing of receiver points is typically 100 m, only shielding of hills may be considered. Structures relevant for rail and roads like barriers or houses are nonexistent in aircraft noise calculations. To assess shielding, the flight path has to be divided into such short segments that a cross-section of the terrain from the middle of the segment to the receiver point is representative for the whole length of the segment. Computation time increases by factors to generate all these cross-sections and to find out where shielding applies. Nevertheless, there exists programs including shielding like NORTIM,[39] INM (version 6.2 and newer), and FLULA2.

Urban housing environment

As the noise contours of aircraft noise calculations describe the noise immission in the whole area around an airport, the "micro" details of the urban environment like shielding or reflection by houses are usually not taken into account (see preceding section "Shielding").

POSTPROCESSING

Postprocessing is very similar for all kinds of aircraft noise calculation programs.

Fleet mix and number of aircraft movements

Up until here the specific details of the noise calculation of one single aircraft on one single flight path have been described. If this calculation is repeated for all grid points, the result is called a "footprint." To end up at the average noise immission, such footprints have to be calculated according to the scenario for all aircraft specified and for all flight paths defined (see Figure 7.2).

In the scenario the number of movements per time period of the day associated with each aircraft group and air route have been specified. At each grid point, the footprints are now weighted according to the number of associated movements and summed. If all footprints are stored, the calculation of variants with alternative numbers of movements per aircraft type and route is a trivial summing up of the basic footprint data.

Noise contours and geographic information systems

The last step is the interpolation of the noise levels in the grid points to generate the contour lines. The shape of the resulting curve may depend on the method used (usually some kind of spline functions) and the degree of smoothing for curved lines. A widely used software package is the downloadable NMPlot[38] from Wasmer Consulting, which is also used within INM.

The noise contours may be exported to a GIS (geographic information system) for plotting the curves on a map or for performing additional calculations like estimation of the area enclosed within a contour line or estimating the number of people living within a contour of a specific noise level.

SPECIFYING THE UNCERTAINTY OF A NOISE CONTOUR

The concept of uncertainty put forward in 1995 with the GUM (Guide to Uncertainty in Measurements)[24] is a powerful tool to answer how good, trustworthy, or reliable is a calculation. For example, there is no need for huge calculation efforts if a very simplified calculation of the "survey" type with a high uncertainty will do to answer a specific question. On the other hand, to answer questions for land use where much money is involved, a "precision" calculation with a low uncertainty is adequate. Noise contours react clearly visible to small level deviations. As a rule of thumb, the deviation of a noise contour at levels of about 60 dB L_{eq} by one decibel will change the area enclosed in the noise contour by 20 to 25%. Therefore, *a noise calculation without indication of the uncertainty is meaningless.* The concept of uncertainty is discussed elsewhere in this book.

As an example, take the findings of the PhD thesis by Thomann.[47] He investigated the uncertainties for aircraft noise calculations using the program FLULA2 and also the uncertainties of monitor measurements to be used for comparisons with the calculated results. For the precision calculation using radar data and the complete airport statistics as input, a minimal standard uncertainty of 0.5 dB for daytime may be achieved for the areas close to the airport. The uncertainty will increase up to several decibels in regions far away from the airport with low aircraft noise levels.[48,49] Usually, the expanded uncertainty for 95% probability is used, which is in many cases twice the standard uncertainty, that is, the uncertainty (95%) of a precision calculation with an advanced calculation model at least 1.0 dB.

For calculations with other programs, the uncertainty is likely to be larger, especially if the altitude profile is based on a performance calculation and not on radar. For INM, some results were reported at Internoise 2009.[19]

For historic situations (situations in the past) the main factors adding to uncertainty in aircraft noise calculations are

- Incorrect distance—For distances from the aircraft to the receiver below 500 m, inaccurate radar data or unrealistic altitude profiles may result in errors of the propagation distance.
- Meteo—For distances from the aircraft to the receiver longer than 1000 m, the meteorological effects begin to contribute to uncertainty.
- Aircraft type—Errors in identifying the aircraft types and thus allocation to wrong groups increase the uncertainty.
- Inaccurate database—The same aircraft type may be operated with different engines, providing different noise levels. Often, the specific aircraft type does not exist in the database. Then it is up to the user to generate a new data set according to his best knowledge (which again has its uncertainty).
- Power settings—Noise emission depends on the power settings. Flex or derated takeoff procedures depending on the actual takeoff weight or specific procedures from aircraft carriers cannot always be modelled correctly in the scenario, due to lack of information.

For forecasts, additional uncertainties in the input parameters apply as mentioned at the beginning of the section "The Scenario."

APPLICATIONS GUIDE: DOC.29, 3RD EDITION, VOLUME 1

The topics discussed show that an aircraft noise calculation is much more than the mechanics of the acoustic model. The expertise of the user and how he specifies the scenario has probably a greater impact on the result than the model used. To provide assistance for the user, document DOC.29 (2005) volume 1 has been written. For free download, see www.ecac-ceac.org.[13]

Although aimed at the model described in DOC.29 (2005) volume 2,[14] this user guidance may be used with any model to clarify the expectations and the roles of model developers, users, politicians, and the public within the process of an aircraft noise calculation.

CONCLUSIONS ON AIRCRAFT NOISE CALCULATIONS

The two key points in an aircraft noise calculation are (1) the quality of the user input (scenario) and (2) the availability of precise acoustic source data for at least those 20 aircraft/engine types that dominate at the airport under consideration of the noise immission by the combination of number of movements and level of noise emission.

The uncertainty of the noise contours depends also on the calculation model used, but mainly on the quality of the input data (e.g., nominal tracks or real radar data) and on the degree of simplification made in the scenario according to the goal of the calculation: survey, engineering, or precision.

The number of propagation calculations for all aircraft types on all routes and for all receiver points may amount to many millions. Thus calculation efficiency is crucial in aircraft noise calculations.

A single flyby of an aircraft generates a level-time curve increasing from below ambient noise level to the maximum level and fading off again. The sound exposure level (SEL) may be based on the uppermost 10 dB—known as 10 dB rule—or on the complete event; the difference is typically 0.4 to 0.5 dB. NPD data uses "10 dB down," whereas simulation programs usually calculate the whole event resulting in systematically slightly louder noise contours.

The refinement of programs using segmentation tends to make the segments shorter and shorter to increase accuracy. Hence they are approaching the structure of the simulation programs.

For harmonised noise policies in Europe, it is up to the EU Commission to recommend an appropriate aircraft noise program for future noise calculations. The segmented model DOC.29 (2005) with the European database ANP is not at the front of research like many simulation programs, but it is the state of the art for rather fast computations and it has a comprehensive data base. It was designed by the experts on aircraft noise calculations in the international working group of ECAC and it was coordinated with the work in SAE. Besides segmentation, DOC.29 (2005) also includes the newest findings on lateral directivity and lateral attenuation. However, if political and financial power were available to build up a new, spectral sound power emission database, then, no doubt, a simulation program would be the first choice.

ACKNOWLEDGMENT

I thank Ullrich Isermann, DLR Göttingen, for reviewing and commenting so thoroughly on this chapter.

AIRCRAFT ORGANISATIONS

A-21	Working group of SAE; reviews aircraft noise activities and edits SAE Standards on aircraft noise
AIRMOD	Working group of ECAC: created DOC.29 3rd revision (2005)
ECAC	European Civil Aviation Conference, member of ICAO
FAA	American Federal Aviation Authority
ICAO	International Civil Aviation Organisation (Organisation of the United Nations)
SAE	Society of Automotive Engineers; see www.sae.org

REFERENCES

1. ANP (Aircraft Noise and Performance Database), www.aircraftnoisemodel.org.
2. AzB, Der Bundesminister des Inneren: Anleitung zur Berechnung von Lärmschutzbereichen an zivilen und militärischen Flughäfen nach dem Gesetz zum Schutz gegen Fluglärm vom 30.3.1971. GMBI 26, Ausg. A, Nr. 8, 162–227, Bonn, 10. März 1975.
3. AzB-Ergänzung Bundesministerium des Innern, Ergänzung der Anleitung zur Berechnung von Lärmschutzbereichen an zivilen und militärischen Flugplätzen – AzB – U II 4-560 120/43, Bonn 1984.
4. AzB, Anleitung zur Berechnung von Lärmschutzbereichen (AzB), Juli 2008, Anlage 2 zur Verordung der Bundesregierung "Erste Verordnung zur Durchführung des Gesetztes zum Schutz gegen Fluglärm" Drucksache 566/08, 2008.
5. AzD, Anleitung zur Datenerfassung über den Flugbetrieb (AzD), Juli 2008, Anlage 1 zur Verordnung der Bundesregierung "Erste Verordnung zur Durchführung des Gesetztes zum Schutz gegen Fluglärm" Drucksache 566/08, 2008.
6. Bütikofer, R. Concepts of aircraft noise calculations, *Acta Acustica united with Acustica*, vol. 93 (2007), 253–262.
7. Bütikofer, R., Thomann, G., Plüss, S. Track dispersion in aircraft noise modelling, Forum Acusticum Berlin, March 15–19, 1999.
8. Bütikofer, R., Thomann, G. Aircraft sound measurements: The influence of microphone height, *Acta Acustica united with Acustica*, vol. 91 (2005), 907–914.
9. Bütikofer, R., Thomann, G. Uncertainty and level adjustments of aircraft noise measurements, Internoise 2009, Ottawa.

10. Daigle, G.A., Embleton, T.F.W., Piercy, J.E. Some comments on the literature of propagation near boundaries of finite acoustical impedance, *Journal of the Acoustical Society of America*, vol. 66, no.3 (1979), 918–919.
11. DES–Der Bundesminister des Innern. Datenerfassungssystem für die Ermittlung von Lärmschutzbereichen an zivilen Flugplätzen nach dem Gesetz zum Schutz gegen Fluglärm vom 30. März 1971. DES, GMBI 26, Ausgabe A, Nr. 8, S 127–144, 10. März 1975.
12. ECAC.CEAC DOC.29 (1st edition 1986, 2nd edition 1997). Report on Standard Method of Computing Noise Contours around Civil Airports. European Civil Aviation Conference (ECAC).
13. ECAC.CEAC DOC.29, 3rd edition, volume 1: Methodology for Computing Noise Contours around Civil Airports, volume 1: Applications Guide. European Civil Aviation Conference (ECAC), 7 December 2005. Download at www.ecac-ceac.org.
14. ECAC.CEAC DOC.29 (2005), 3rd edition, volume 2: Methodology for Computing Noise Contours around Civil Airports, volume 2: Technical Guide. European Civil Aviation Conference (ECAC), 7 December 2005. Download at www.ecac-ceac.org.
15. Directive 2002/49/EC of the European Parliament and of the Council of 25 June 2002, relating to the assessment and management of environmental noise.
16. Commission Recommendation of 6 August 2003 concerning the guidelines on the revised interim computation methods for industrial noise, aircraft noise, road traffic noise and railway noise, and related emission data (notified under document number C(2003) 2807) (2003/613/EC).
17. ÖAL-Richtlinie 24-1 Lärmschutzzonen in der Umgebung von Flughäfen Planungs- und Berechnungsgrundlagen. Österreichischer Arbeitsring für Lärmbekämpfung Wien, 2001. (EU "interims model" data)
18. Neue zivile Flugzeugklassen für die Anleitung zur Berechnung von Lärmschutzbereichen (Entwurf), Umweltbundesamt, Berlin, 1999. (EU "interims model" data)
19. Noel, G., Allaire, D., Jacobson, S., Willcox, K., Cointin, R. Assessing the uncertainty in FAA's noise and emissions compliance model, Internoise, Ottawa, 2009.
20. Fleming, G.G., Senzig, D.A., McCurdy, D.A., Roof, C.J., and Rapoza, A.S. Engine installation effects for four civil transport airplanes: Wallops Flight Facility Study, Volpe Center, U.S. Department of Transportation, Cambridge, MA, 2003.
21. Fleming, G.G., Senzig, D.A., Clarke, J.-P.B. Lateral attenuation of aircraft sound levels over an acoustically hard water surface: Logan airport study, *Noise Control Engineering Journal*, vol. 50, no. 1 (2002), 19–29.
22. Pietrzko S.J., Hofmann, R.F. Prediction of A-weighted aircraft noise based on measured directivity patterns, *Applied Acoustics* 23 (1988), 29–44.
23. Thomann, G., Bütikofer, R., Krebs, W. FLULA2—Ein Verfahren zur Berechnung und Darstellung der Fluglärmbelastung. Technische Programmdokumentation, version 1.2, Abteilung Akustik, Eidgenössische Materialprüfungs- und Forschungsanstalt (EMPA), Dübendorf, Schweiz, 2005.
24. GUM: Guide to the expression of uncertainty in measurement. ISO/ENV 13005, Geneva, 1995.

25. Granoien, I.L.N., Randberg, R.T. Corrective measures for aircraft noise models, new algorithms for lateral attenuation, Joint Baltic-Nordic Acoustics Meeting, Mariehamm, Aland, 2004, http://www.acoustics.hut.fi/asf/bnam04/webprosari/papers/o41.pdf.
26. ICAO Annex 16—Environmental Protection; Annex 16 to the convention of international civil aviation; Volume I: Aircraft Noise, International Civil Aviation Organization (ICAO), Montreal, Canada.
27. Beuving, M., de Vos, P. Improved Methods for the Assessment of the Generic Impact of Noise in the Environment IMAGINE—State of the Art. Deliverable 2 April 2003, http://www.imagine-project.org.
28. Bütikofer, R. Default aircraft source description and methods to assess source data, December 10, 2004, http://www.imagine-project.org/bestanden/IMA4DR-061204-Empa-10_Aircraft_Source.pdf.
29. He, H., Boeker, E., Dinges, E. Integrated Noise Model (INM) Version 7.0 User's Guide, Federal Aviation Administration, Office of Environment and Energy, FAA-AEE-07-04, Washington, April 2007.
30. Boeker, E., Dinges, E., He, B., et al. Integrated Noise Model (INM) Version 7.0 Technical Manual, FAA-AEE-0801, January 2008.
31. Spectral Classes for FAA's Integrated Noise Model, Report No. DTS-34-FA065-LR1, Cambridge, MA, John A. Volpe National Transportation Systems Center, December 1999.
32. ISO 20906:2009, Acoustics—Unattended monitoring of aircraft sound in the vicinity of airports.
33. Krebs, W., Bütikofer, R., Plüss, S., Thomann, G. Spectral three-dimensional sound directivity models for fixed wing aircraft, Acta Acustica United with Acustica, vol. 92 (2006), 269–277.
34. Krebs, W., Bütikofer, R., Plüss, S., Thomann, G. Sound source data for aircraft noise simulation, Acta Acustica United with Acustica, vol. 90 (2004), 91–100.
35. Krebs, W., Thomann, G. Lateral directivity of aircraft noise, Acoustics 08, Paris, 2008.
36. Krebs, W., Thomann, G. Aircraft noise: New aspects on lateral sound attenuation, Acta Acustica United with Acustica, vol. 95, no. 6 (2009), 1013–1023.
37. Fleming, G.G., Senzig, D.A., McCurdy, D.A., Roof, C.J., Rapoza, A.S. Engine installation effects of four civil transportation airplanes: The Wallops Flight Facility study, October 2003, http://techreports.larc.nasa.gov/ltrs/PDF/2003/tm/NASA-2003-tm212433.pdf.
38. NMPlot Software and User Guide, free download at http://wasmerconsulting.com/nmplot.htm.
39. Olsen, H., Liasjø, K.H., Granøien, I.L.N. Topography influence on aircraft noise propagation, as implemented in the Norwegian prediction model. NORTIM. SINTEF DELAB Report STF40 A95038, Trondheim, April 1995.
40. Granøien, I.L.N., Randeberg, R.T., Olsen, H. Corrective measures for the aircraft noise models NORTIM and GMTIM: 1. Development of new algorithms for ground attenuation and engine installation effects. 2. New noise data for two aircraft families. SINTEF Report STF40 A02065, December 2002.

41. SAE AIR 1751 (1981/1991)—Prediction method of lateral attenuation of air-plane noise during takeoff and landing. Society of Automotive Engineers, reaffirmed 1991.
42. SAE AIR 1845 (1986)—Procedure for the calculation of aircraft noise in the vicinity of airports. Society of Automotive Engineers.
43. SAE AIR 5662—Method for predicting lateral attenuation of airplane noise. Society of Automotive Engineers, Draft 2005.
44. Schäffer, B., Plüss, S., Thomann, G., Bütikofer, R. Aircraft noise calculations for periods of day using a complete set of radar data, Rotterdam, NAG/DAGA 2009.
45. Wunderli, J.M. The sound propagation model of sonRAIL. EURONOISE 2009, Edinburgh, Great Britain, October 2009.
46. Smith, O.K., Ollerhead, J.O.B., Rhodes, D.I.P., White, S., Woodlyn, A.R.C. Development of an Improved Lateral Attenuation Adjustment for the UK Aircraft Noise Contour Model, ANCON, London, England, Civil Aviation Authority, Draft, February 2002.
47. Thomann, G. Mess- und Berechnungsunsicherheit von Fluglärmbelastungen und ihre Konsequenzen, Dissertation, ETH Zurich, 2007.
48. Thomann, G., Bütikofer, R. Quantification of uncertainties in aircraft noise calculations, Internoise, 2007.
49. Thomann, G. Uncertainties of measured and calculated aircraft noise and consequences in relation to noise limits, Acoustics 08, Paris, 2008.
50. Schäffer, B., Thomann, G., Huber, P., Brink, M., Plüss, S., Hofmann, R. ZFI, an index for the effects of aircraft noise on the population: Experiences, Proceedings of Inter-Noise 2010, Lisbon, Portugal, June 2010.

Chapter 8

The Good Practice Guide, Version 2

G. Licitra and E. Ascari

CONTENTS

INTRODUCTION

During the last 10 years, the European Commission has built up a European Union (EU) noise expert network whose mission is to provide assistance in the development and implementation of the European noise policy. Part of this network is the Working Group Assessment of Exposure to Noise (WG-AEN), which was formed in 2001 from two former working groups, the Noise Mapping and the Computation and Measurement group.

The role of WG-AEN was to assist the commission and the member states in the implementation of specific requirements of the Environmental Noise Directive 2002/49/EC of the European Parliament and of the Council of

25 June 2002 (hereafter, END). More specifically, WG-AEN, among other working groups, should help the commission in drawing up the guidelines on the assessment methods for the new noise level indicator calculations, and provide technical specifications for a study concerning the identification and development of a good practice in the field of noise mapping.

In December 2003 the WG-AEN produced the first version of a position paper titled "Good Practice Guide for Strategic Noise Mapping and the Production of Associated Data on Noise Exposure" (hereafter, GPG).[1] The aim of this position paper was to assist member states (hereafter, MS) and their competent authorities to produce noise mapping and the associated data required by the END. This guide focused on those requirements associated with the first round of strategic noise mapping (to be completed by 30 June 2007).

This guide was not assessed to assist software designers to be consistent with the requirements of the END, or to establish the role of geographical information systems (GIS) in noise mapping and the production of associated data, even if those issues are recognized as important factors.

The position paper avails of the results of a study funded by the UK Government's Department for Environment, Food and Rural Affairs (DEFRA), which, in May 2002, published an invitation to tender (Ref. EPG 1/2/41) titled "The Identification and Development of Good Practice in the Field of Noise Mapping and the Determination of Associated Information on the Exposure of People to Environmental Noise." Wölfel Meßsysteme Software GmbH & Co (main contractor) and Lärmkontor GmbH were contracted to carry out this study to identify the potential difficulties encountered during the practical implementation of the directive at a MS level; together with some MS representatives the list was set up.[2] Therefore, part one of the study, completed in October 2002, was a questionnaire-based exercise to identify general issues and technical challenges that member states may need to solve when they implement the END.

Part two of the study, completed in April 2003, was a first good practise guide and associated toolkits to address the key issues and challenges identified in part one of the study.

For WG-AEN, a challenging issue was to consider how much guidance should be provided. In fact, the guide attempted to find an appropriate compromise between the need for a coherent approach across European countries and the flexibility required by each MS to develop noise mapping that also fits to individual national needs.

The guide provides recommendations for dealing with the general issues and specific technical challenges identified by MS and it gives options for dealing with a number of the technical challenges by a series of toolkits, which constitute the last part of the guide.

Table 8.1 Legend of GPG Version 1 Toolkits

Level	Usability Colour Codes (Light to Dark Blue)	Accuracy Colour Code (Light to Dark Green)	Cost Colour Code (Light to Dark Red)
Low	△ ◇ ⬠	△ ◇ ⬠	△ ◇ ⬠
High	⬡	⬡	⬡

Source: European Commission Working Group Assessment of Exposure to Noise (WG-AEN), Position Paper: Good Practice Guide for Strategic Noise Mapping and the Production of Associated Data on Noise Exposure, Version 1, December 2003.

The guide can be divided in two big sections:

1. The first part of the guide tackles different issues through a brief explanation about the problem, a discussion, and recommendations.
2. The second part of the guide explains which degree of accuracy, cost, and complexity might be associated with the specification of a modelization parameter choice over the whole process through different toolkits having codes and ratings like the ones in Table 8.1.

Therefore, different toolkits were developed for each issue treated in the first part and here listed:

- Toolkits for input data—source-related issues
  - Toolkit 1. Road traffic flow
  - Toolkit 2. Average road traffic speed
  - Toolkit 3. Composition of road traffic
  - Toolkit 4. Train speed
  - Toolkit 5. Sound power levels of industrial sources
- Toolkits for input data—geographical issues
  - Toolkit 6. Building heights
  - Toolkit 7. Obstacles
  - Toolkit 8. Cuttings and embankments in the site model
  - Toolkit 9. Sound absorption coefficients α for buildings and barriers
- Toolkits for input data—meteorological issues
  - Toolkit 10. Occurrence of favourable sound propagation conditions
  - Toolkit 11. Humidity and temperature

- Toolkits for input data—demographic issues
  - Toolkit 12. Assignment of population data to residential buildings
  - Toolkit 13. Determination of the number of dwelling units per residential building and the population per dwelling unit
- Toolkits—miscellaneous issues
  - Toolkit 14. Determination of agglomerations
  - Toolkit 15. Area to be mapped
- Toolkit 16. Area outside the area being mapped

This was the structure of the first version of WG-AEN's Good Practice Guide: it did not tackle all the identified key issues, however, it was the first step toward providing a comprehensive document that is now available to scientists.[3]

In fact, at the beginning of 2004, the WG-AEN received a new one-year mandate, which included a requirement to collect and assess feedback on the content of the GPG and produce a second version before the end of 2004.

The toolkits of GPG version 1 were designed to give guidance on potential steps to be taken or assumptions to be made if the available data set fell short of the coverage or detail required for the large-scale noise mapping. Therefore, even if providing advice on decision making, there was at this stage no corresponding indication of the acoustic accuracy implications of making the decisions. This resulted in two serious consequences:

- The MS made decisions that introduced unknown uncertainty into the process so that both the MS and the EU Commission were uncertain about the potential accuracy and robustness of the results, even when the methodology followed the advice within the GPG.
- This lack of acoustic guidance within the GPG did not help MS making informed decisions on the relative importance of the various data sets, which should help focus finite funds in the procurement of missing data.

Therefore, WG-AEN committed to DEFRA a research project to determine the effects on the acoustic accuracy of calculated noise levels caused by following the advices contained within the GPG version 1. DEFRA, focusing on road traffic noise, studied the consequential acoustic accuracy in strategic noise map results of adopting the advice of the GPG.

The project[4] objectives are summarised as follows:

- Provide potential additional GPG toolkits for issues not included within existing guidance using a format compatible with the existing ones (for road surface type, road junctions, road gradient, ground surface elevation, ground surface type, and barrier height).

- Quantify the accuracy symbols within version 1 in toolkits 1, 2, 3, 6, 7, and 8 plus the new road surface Toolkit, considering their use in conjunction with CRTN and the recommended Interim Method for roads XPS 31-133.
- Provide practical guidance on the acoustic accuracy implications of following the recommended toolkits.
- Provide practical assistance to MS and professionals dealing with data handling and procurement across the European Union.
- Cooperate closely with WG-AEN to ensure that the views and requirements of the EC and MS are taken into consideration during the project.

In order to achieve these objectives, it should develop a testing methodology that can be used to assess the implications for acoustic accuracy of adopting the advice in the GPG; moreover, a quantitative testing methodology should be carried out using the proposed toolkits and the input parameters used within CRTN and XPS 31-133 in order to assess the ranking of data sets, and their effects upon the calculation method end result.

All these tasks were covered by this DEFRA study leading to the second version of the GPG. In the following paragraph its results are described before detailing the final version of the guide.

DEFRA ACCURACY STUDY

DEFRA's study started analysing accuracy requirements of the END: in fact, absolute accuracy is necessary to report absolute limits and number of people in discrete 5 dB wide bands. Moreover, absolute accuracy is extremely important to assess noise exposure across different groups and states and to draw up action plans. Thus, identifying the sources and magnitude of the potential errors within the noise mapping process is a key factor, especially to increase public perception of accuracy. Articles within television and print media may want to compare towns, states, and countries, so good results and robust recommendations for mitigation are needed to avoid credibility loss.

Therefore, this study considered all potential uncertainty sources within a system designed to reproduce a real-world environment, which can be summarised into four areas to be investigated:[4]

1. Estimation of uncertainty in model inputs and parameters (*characterisation of input uncertainties*)
2. Estimation of the uncertainty in model outputs resulting from the uncertainty in model inputs and model parameters (*uncertainty propagation or sensitivity*)

3. Characterisation of uncertainty associated with different model structures and model formulations (*characterisation of model uncertainty*)
4. Characterisation of the uncertainty in model predictions resulting from uncertainty in the evaluation data (*uncertainty of evaluation data*)

These issues are separately described in the following sections.

Different types of uncertainties

Input

To characterise input uncertainty, a study of each type of data has been carried out: in fact, uncertainty may arise from various sources including measurement, management, factoring, and assimilating of the actual information. In order to quantify the scale and distribution of these uncertainties, detailed analyses were carried out for each type of input data set.

A flowchart in Figure 8.1 shows how it is introduced into the noise mapping process: two types of input uncertainty were identified, one related to raw data and the other to data managing. It was also assumed that each input data set has a normal distribution of uncertainty, but the validity of this assumption was not evaluated in that study, nor it is yet assessed.

Sensitivity

Sensitivity analysis was carried out to understand how the variation in model output could be assigned to different sources of variations, and how the given model depends upon the information provided.

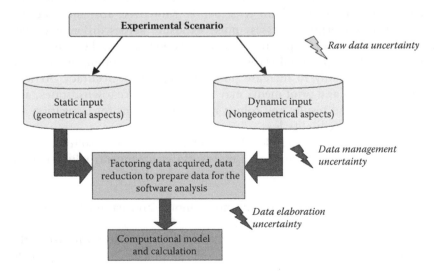

Figure 8.1 Uncertainty types in input data.

This task was centred on assessing the means by which uncertainties, errors, or assumptions within the input data sets propagate through the calculation tools to produce uncertainties or errors in the decibel results obtained.

This analysis was carried out for the CRTN and Interim Method but the recommendations set out within the toolkits proposed for the GPG version 2 referred only to the XPS 31-133 Interim Method and will be further detailed in the section "Sensitivity Analysis."

Model

The characterisation of model uncertainty should be carried out by noise models owners and developers. Comparative studies of the national methods would be useful together with error propagation analysis for each of them: in fact, it could help to determine a way to demonstrate "equivalence" for the END.

Another aspect of the model uncertainty is to investigate how the documented standard is transposed from a paper document into a 3D noise calculation tool, and how the additional simplifications, efficiency techniques, and assumptions introduced by software add further uncertainties in order to create usable calculation times. Therefore, the model uncertainty is introduced into the process in the following steps:

- Adapting physical scenarios to standard calculation methods
- Adapting standard calculation methods to software tools
- Using calculation tools together with user-controlled parameters to start the final calculation engine

Evaluation data

The issues surrounding uncertainties in environmental noise measurements was researched in detail by Craven and Kerry,[5] whose work concluded that for short term measurements a good result would be to obtain a spread of 5 dB(A) within measurements at the same site, for the same source, on different days.

Moreover, work within the Harmonoise project indicated that the uncertainties in the measured levels could be reduced by spanning measurements over a year and using meteorological and ground absorption factors.

The uncertainty is introduced into the measurement according to the flowchart of evaluation data presented in Figure 8.2. In particular, there are two kinds of processes introducing uncertainty: the one of measurements that could be evaluated according to existing guidelines,[7] and the other related to data evaluation due to variability of source and propagation conditions.

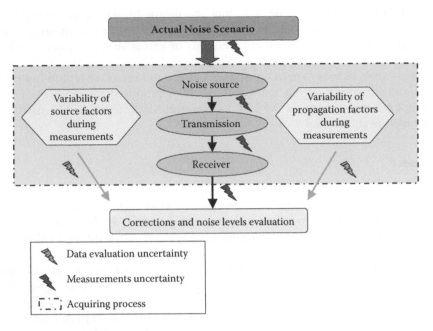

Figure 8.2 Uncertainty types in evaluation–data process.

Overall uncertainty determination

The aforementioned four kinds of uncertainty are interrelated to each other, as shown in Figure 8.3. They are related in a specific way to results evaluation. Therefore, the different types of uncertainties have been taken into account evaluating the decibel error in the noise mapping result, so that an absolute accuracy evaluation has been associated with each modelization choice in the toolkits.

Sensitivity analysis

To evaluate and quantify in terms of decibels these uncertainties, different methodologies were available to DEFRA: the simplest method is to run the model varying only one input parameter at a time and to compare the outputs with the nominal estimate. Although this approach has the advantage of being fast in design and execution, it does not allow a simultaneous exploration of input factors, so it cannot capture interaction effects.

Instead, multiple inputs could be simultaneously analysed using an error propagation model, which requires the consideration of input parameters "measured values" with their respective uncertainties. An error model provides an output uncertainty based upon the uncertainties of its respective inputs.

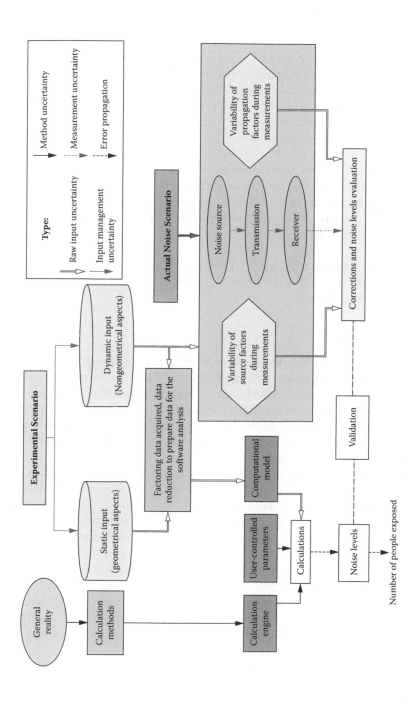

Figure 8.3 Uncertainty interrelation with result.

In environmental modelling there are two methods most commonly used for evaluating multiple inputs: Taylor series expansion (or finite order analysis) and Monte Carlo simulation. These techniques applied to the whole scenario, including the propagation path, would have required DEFRA to develop a highly complex analysis system due to a number of variables; however, a fast and practical guide was required and so a high detailed system was not within the scope of the accuracy study committed to DEFRA. Thus, two separate approaches were carried out to assess two different parts of the noise mapping process:

1. Analytical analysis techniques were used to assess the uncertainty propagation in the nonspatial factors of the calculation methods.
2. The spatial aspects of the calculation methods were investigated using a modelling-based approach with a series of test scenarios to investigate a specific effect.

There were some variations in how results analysis was performed according to the testing method used. The following paragraphs will discuss both methodologies.

Nonspatial effects

As already explained, the two most popular means of doing uncertainty analysis are the Taylor series expansion and the Monte Carlo analysis techniques. The first technique provides an approximate but direct assessment of potential error due to uncertainties contained within input parameters. The method consists of taking the first-order partial derivative of a function, which is the change rate of the function due to the input parameter at any value, that is, its sensitivity.

On the other hand, Monte Carlo simulation does not provide an analytical link between input and output uncertainties, but it consists of a statistical and probabilistic analysis of uncertainties. The idea of Monte Carlo simulation for uncertainty analysis is to calculate the outcome of a model using different input values, which are randomly sampled from a series of possible values (taken from a prior associated distribution).

Due to the nonlinearity of noise calculation methods, the use of the Taylor series expansion method was recognized to be unsuitable because it would have given approximate answers. In fact, the method creates a straight-line approximation of the function at a point, so it works very well for small errors and uncertainties, but in the case where a correction is highly nonlinear, the method breaks down significantly.

Therefore, analytical tests were performed using Monte Carlo simulation: the main assumption made in the whole analysis was that the uncertainties distributions for each parameter were considered to be normal

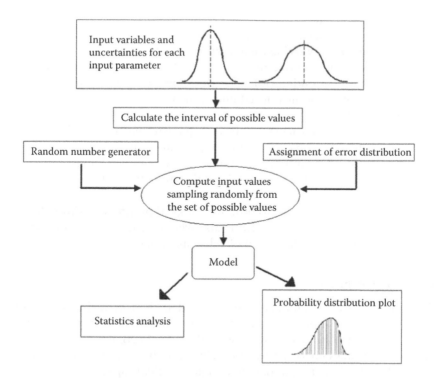

Figure 8.4 Flow diagram of Monte Carlo tools. (Adapted from S. Shilton, H. Van Leewen, R. Nota, Error Propagation Analysis of XPS 31-133 and CRTN to Help Develop a Noise Mapping Data Standard, Proceedings of Forum Acusticum, Budapest, Hungary, August 2005.)

curves. Overall, Monte Carlo simulation is essentially easy to implement and gives a high level of accuracy as long as the correct input distributions are known. The output can then be displayed in a histogram, which defines the probability distribution of the output allowing for the calculation of statistical parameters such as standard deviation and variance. Figure 8.4 shows the process flow of the Monte Carlo tools.

As the Monte Carlo approach was used on the nongeometric aspects of the methodology, this resulted in an investigation of the formulas used to calculate the source emission noise level by both the CRTN and XPS 31-133 methods.

The general approach was carried out in three steps:

1. General behaviour
   a. Investigate the general behaviour of the source emission function across a range of likely traffic flow values
   b. Identify scenarios to use within the subsequent tests

2. Single parameter
   a. Using the scenarios as the reference condition, run Monte Carlo simulations varying each input parameter
   b. Assess the resulting uncertainty in the calculated noise level
   c. Obtain a ranking order for the input data sets based upon the magnitude of introduced uncertainty
3. Multiparameter
   a. Select the three most significant input parameters, run Monte Carlo simulations varying all three input parameters simultaneously
   b. Assess the resulting uncertainty in the calculated noise levels
   c. Compare and contrast with single parameter tests

This approach was designed to produce a logical progression through each step, such that any important effect, discovered during the previous stage, could always influence the design of subsequent ones.

Geometric aspects

Analytical analysis techniques can be used if there is a direct relationship between the input data and the result but if the accuracy depends upon many variables, which relay with the actual geometry, an analytical approach becomes no longer feasible. Thus, another approach was used for the analysis testing of input data with a geometrical aspect.

The consequences on output accuracy of such data sets were examined by the use of a test map: different accuracy degrees on input were tested referring to a situation with a very detailed input data, known as the *crisp model*. Thus, the level of certainty was decreased stepwise, according to the tools provided by GPG and a series of *metamodels* have been produced. Each one was a copy of the crisp model for which the data within the crisp model, for a particular data set or attribute, was reduced in quality, or simplified, according to the suggestion of GPG tools. The crisp model and metamodels were then calculated using noise mapping software to obtain a number of grid results, which were compared to those from the crisp model.

For each studied input parameter, a number of metamodels were produced in order to create a spread of uncertainty. Each was then computed to obtain a series of uncertainty propagations, and finally the series of result sets were analysed together against the crisp model results to estimate the effect upon the accuracy.This method was quite simple but time consuming in order to achieve a good spread of results for each input uncertainty. For this reason only five scenarios were carried out for each input parameter under investigation. This did not lead to definitive results but it provided knowledge of the possible uncertainty to help authorities when using the GPG toolkits.

Accuracy grouping

The approach to accuracy constraints, defined within the GPG, was based upon the sensitivity testing carried out on an interim method. A reference "group" is assigned to the supplied data set, such that the potential output error is identified. Five different groups were defined to establish accuracy needed for the input data to obtain a decibel error on the final result:

- Group A is aimed to have very detailed input data. This group is suitable for detailed calculations or for validation.
- Group B is aimed to manage the input specifications such that potential errors in each parameter produce less than a 1 dB error.
- Group C is aimed to manage the input specifications such that potential errors in each parameter produce less than 2 dB of error.
- Group D is aimed to manage the input specifications such that potential errors in each parameter produce less than 5 dB of error.
- Group E is assigned when requested limits desired for Groups A, B, or C cannot be achieved with confidence, in this case it is recommended to improve data quality.

It should also be noted that the multiparameter sensitivity testing indicated that the combined effect of many parameters each in error, results in an error of higher magnitude. For example, managing to contain each data set within Group C could lead to an overall calculated level with an uncertainty in the order of Group D. Figure 8.5 shows an example of the recommendations for the uncertainty values (expressed as percentage errors on

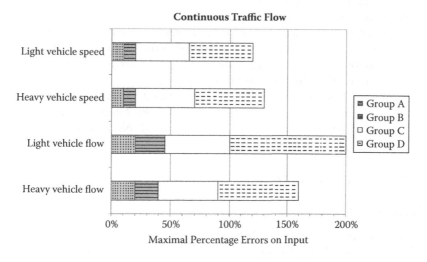

Figure 8.5 Recommended accuracy for inputs to obtain different decibel errors in noise maps.

Table 8.2 Order of Importance for Input Parameters in Noise Emission Calculation

Ranking	If Percent of Heavy Vehicles >30	If Percent of Heavy Vehicles <30
1st	Heavy vehicle speed	Light vehicle speed
2nd	Heavy vehicle flow	Light vehicle flow
3rd	Light vehicle speed	Heavy vehicle speed
4th	Light vehicle flow	Heavy vehicle flow
5th	Road gradient	Road gradient
6th	Road surface	Road surface

Table 8.3 Legend for Numerical Accuracy Toolkits

Level	Complexity Blue Code Gradations	Accuracy Code	Cost Red Code Gradations
Low	△	> 5 dB	△
·		4 dB	
·	◇	3 dB	◇
·	⬠	2 dB	⬠
·		1 dB	
High	⬡	< 0.5 dB	⬡

Source: European Commission Working Group Assessment of Exposure to Noise, Good Practice Guide for Strategic Noise Mapping and the Production of Associated Data on Noise Exposure, Version 2, August 2007.

the true value) to be used in order to assess the quality of an input data set for noise mapping purposes.

Moreover, results of Monte Carlo simulations on the Interim Method not only evaluated the sensitivity of the decibel error in the calculated result but also ranked them according to relevance of different input parameters. Table 8.2 highlights the ranking of importance.

Thanks to these testing procedures, quantified toolkits have been proposed and implemented in the new GPG version 2 together with previous toolkits where quantification was not required. New toolkits are based upon the legend in Table 8.3.

GPG VERSION 2 (2007)

Main structure

After DEFRA's accuracy study, the Good Practice Guide version 2 was published (last release 13 August 2007).[3] The main structure of the position paper is still the same as the previous version: discussion and recommendations about each issue arisen in the implementation of the END are widely presented in the second chapter; implications for accuracy using some of the toolkits provided are described in chapter 3; 21 toolkits are presented in chapter 4 (see following sections for further details).

Some of the subjects treated in chapter 2 are directly related to the tool-kits, but other issues are simply discussed without further assistance: issues are divided into general, source-related, propagation-related, and receivers-related ones. The following is an index of the issues discussed.

General Issues
- Strategic noise maps (and mapping)
- Assessment methods
- The role of noise measurement
- Area to be mapped
- Sources outside the agglomeration area being mapped
- Relevant year as regards the emission of sound
- Average year as regards the meteorological circumstances
- Reviewing strategic noise maps
- Special insulation against noise

Source-Related Issues
- Road
    - Road traffic models; traffic flows and traffic speeds
    - Major roads with less than 6 million vehicle passages per year on some sections
    - Low flow roads in agglomerations
    - Speeds on low flow roads in agglomerations
    - Geographical errors in road alignment
    - Road surface type
    - Speed fluctuations at road junctions
    - Road gradient
    - Determination of the number of road lanes
    - Assignment of flows and speeds to different lanes of multi-lane roads
- Railway
    - Calculation of railway noise
    - Rail roughness
    - Trams and the sound power levels of trams and light rail vehicles
    - Train (or tram) speed
    - Major railways with less than 60,000 train passages per year on some sections
    - Noise from stopping trains at stations
    - Geographical errors in rail track alignment
    - Assignment of train movements to different tracks in multi-track rail corridors

- Others
  - Helicopter noise
  - Noise from aircraft activities other than aircraft movements and noise from other sources at airports
  - Sound power levels of industrial sources

Propagation-Related Issues
- Ground surface elevation
- Ground surface type
- Barriers
- Building heights
- Simplification of building outlines
- Merging of heights on individual buildings and buildings of a similar height
- Tunnel openings in the model
- Sound absorption of building façades and barriers
- Consideration of meteorological impacts and favourable sound propagation conditions

Receiver-Related Issues
- Calculation height
- Most exposed façade
- Quiet façade
- Assessment point (grid spacing, contour mapping, and reflections)
- Assignment of noise levels to dwellings
- Assignment of population to dwellings in residential buildings
- Dwelling
- Determination of the number of dwelling units per residential building and population per dwelling unit
- Quiet areas in an agglomeration
- Quiet areas in open country

Other relevant arguments are treated in appendices: introduction to the use of geographical information systems in noise mapping, WG-AEN's proposals for a research project concerning quiet areas, and an in-depth section regarding the DEFRA study upon understanding sources of uncertainty in noise modelling.

Toolkits with numerical or qualitative uncertainty

The main issues of the toolkits are evidenced in Figure 8.6: the toolkits that present numerical accuracy are underlined and the ones added in the new version are marked as "(new)." It must be noticed that each toolkit is divided into different tools according to the level of detail of the available input. Then, for each kind of input, a different accuracy group is assigned

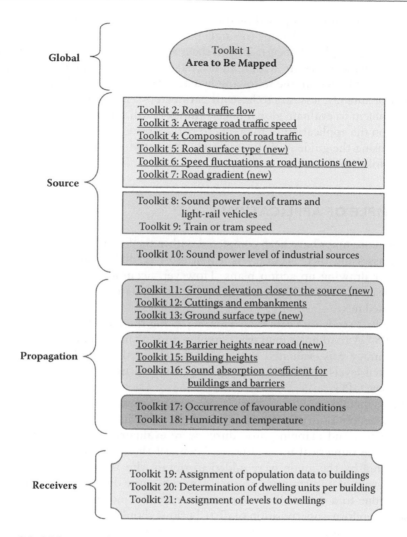

Figure 8.6 GPG version 2, toolkits overview.

for each modelling choice. Thus, using this guide, the following steps are required for each kind of input parameter:

1. Identify the appropriate toolkit
2. Evaluate the detail of the available input
3. Choose the tool that is suitable to the available input
4. Analyse all the different suggested procedures to process the input
5. Choose a procedure and assign uncertainty level as suggested by the guide

All uncertainties proposed by the guide are to be considered as possible variations on the results due to single parameters; however, planners have to evaluate the global uncertainty on the map. It was already explained that if all the parameters are within an uncertainty group (say, Group B), then results may be within the following less accurate group (i.e., Group C).

Unfortunately, it is rare to have all parameters within the same group, so the problem to evaluate global uncertainties is still unsolved. Some experiences on the application of the GPG to real maps demonstrate the possibility of using the guide in different ways; some studies are briefly described in the following section.

EXAMPLE OF APPLICATION

Many noise maps have been completed within the first round of noise mapping: assessing their accuracy is a matter of credibility and it is mandatory for drawing up action plans. However, accuracy studies performed on noise maps are still few: some Italian and Spanish studies are briefly discussed next.

The first study on evaluating accuracy using the GPG was done in Italy in 2008.[8] A road noise map was completed for a small city of Tuscany and its accuracy was evaluated both using the classical method of comparing predicted levels and measurements in a sampled number of points, both using the GPG.

In particular, predicted levels were compared to more than 150 measurements obtaining the distribution of the differences. On the other side, parameters and mapping procedures were evaluated with the toolkits to assess the numerical uncertainties. Then, all the numerical uncertainties suggested by the guide were added to obtain the "total" uncertainty. A square sum of uncertainties was performed because distribution of possible errors due to a single choice was supposed to be independent from other ones: therefore, superposition of distributions should be a normal distribution with the total uncertainty obtained squaring single uncertainties.

$$\sigma_t = \sqrt{\sum_i \sigma_i^2}$$

It resulted that more than 80% of compared points reported a difference smaller than this total uncertainty. This study evidenced that the GPG could be used as a way to evaluate combined uncertainties to assess accuracy of final results. Moreover, looking at single parameter uncertainty, it arises that the source-related parameters should be more reliable to increase accuracy. In fact, the same study was updated in 2009, using a traffic model to predict flows, as suggested in the toolkits, and a new uncertainty evaluation has been carried out: this new stage confirmed the reliability of values predicted by the GPG, having the 95% of compared points (more than 200) with a difference smaller than GPG value (which was of course improved too). This means that the guide was used not only to assess accuracy but also to identify those parameters, as to which good definition could improve accuracy, itself.

Another approach was used in the Spanish study[9,10] of noise maps of Buenos Aires Macrocenter carried out in 2009. In this work, the evaluation of uncertainty started considering the different groups of uncertainty for each parameter and evaluating the order of merit according to GPG suggestions. With this ranking, the uncertainty of the most relevant parameter was compared to deviations between simulated and measured values. Measured values were considered with their uncertainty according to the GUM.[7] Also, in this case 90% of compared points reported deviations smaller than the one previewed by the GPG (most relevant parameter uncertainty). Moreover, this study suggested evaluating simulation uncertainty subtracting measurements uncertainty contribution from the empirical distance between simulation and measurements. In fact, the total uncertainty was considered as a result of independent contributions.

Given that the combined uncertainty of the measurement process (and therefore the expanded uncertainty) can be analytically calculated and the total expanded uncertainty of the entire process is empirically determined depending on the coverage factor established (i.e., $k = 2$), the expanded uncertainty due exclusively to the simulation process can be calculated using the following equations:

$$U_T = u_{cT} \cdot k \quad where \quad u_{cT} = \sqrt{u^2(M) + u^2(S)}$$

$$u(S) = \sqrt{u_{cT}^2 - u^2(M)} \rightarrow U(S) = k * u(S)$$

where
   U_T is the total expanded uncertainty of the noise map, empirically determined.
   u_{cT} is the total combined uncertainty of the noise map, calculated from U_T.
   $u(M)$ is the combined uncertainty due to measurement process, analytically determined.
   $u(S)$ is the combined uncertainty due to simulation process, calculated from u_{cT} and $u(M)$.
   $U(S)$ is the expanded uncertainty due to simulation process, calculated from $u(S)$.

With this methodology, an indication of the most relevant parameters that contribute to total uncertainty is determined. In fact, even if there are some parameters of Group E (very high uncertainty), their order of merit was not relevant on final uncertainty.

It should be noted that these studies were carried out simultaneously so no efficient integration of both approaches have been developed until now.

CONCLUSIONS

The Good Practice Guide established ranking of importance of different mapping parameters and provided guidance for first round mapping of the END, especially for those conditions where knowledge is lacking.

New developments are attended due to the adoption of a common assessment method[11] for future maps: new calculations will be carried out to evaluate sensitivity of new methods and to update the guide. However, the great work done by scientists will remain as a milestone in the development of noise maps production to ensure their credibility.

REFERENCES

1. European Commission Working Group Assessment of Exposure to Noise (WG-AEN). Position Paper: Good Practice Guide for Strategic Noise Mapping and the Production of Associated Data on Noise Exposure, Version 1. December 2003.
2. E. Wetzel, C. Popp. The Identification and Development of Good Practice in the Field of Noise Mapping and the Determination of Associated Information on the Exposure of People to Environmental Noise Contract: EPG 1/2/41 Wölfel project number: P506/01 Final Report, October 2002.
3. European Commission Working Group Assessment of Exposure to Noise. Good Practice Guide for Strategic Noise Mapping and the Production of Associated Data on Noise Exposure, Version 2. August 2007.
4. Department for Environment, Food and Rural Affairs (DEFRA), Research Project NAWR 93: WG-AEN's Good Practice Guide and the Implications for Acoustic Accuracy. May 2005.
5. N.J. Craven, G. Kerry. A Good Practice Guide on the Sources and Magnitude of Uncertainty Arising in the Practical Measurement of Environmental Noise. DTI Project: 2.2.1—National Measurement System Programme for Acoustical Metrology, University of Salford, October 2001.
6. S. Shilton, H. Van Leewen, R. Nota. Error Propagation Analysis of XPS 31-133 and CRTN to Help Develop a Noise Mapping Data Standard. Proceedings of Forum Acusticum, Budapest, Hungary, August 2005.
7. JCGM 100:2008, GUM 1995 with Minor Corrections: Evaluation of Measurement Data—Guide to the Expression of Uncertainty in Measurement, September 2008.

8. G. Licitra, G. Memoli. Limits and Advantages of Good Practice Guide to Noise Mapping. Proceedings of Forum Acusticum, Paris, France, July 2008.
9. M. Ausejo, M. Recuero, C. Asensio, I. Pavón, J.M. López. Study of Precision, Deviations and Uncertainty in the Design of the Strategic Noise Map of the Macrocenter of the City of Buenos Aires, Argentina. *Environmental Modeling and Assessment* 15 (2010), 125–135.
10. M. Ausejo, M. Recuero, C. Asensio, I. Pavón, R. Pagán. Study of Uncertainty in Noise Mapping. Proceedings of Internoise, Lisbon, Portugal, July 2010.
11. Draft JRC Reference Report (Contract no. 070307/2008/511090). Common NOise ASSessment MethOdS in EU(CNOSSOS-EU) Version 2, May 2010.

Chapter 9

Uncertainty and quality assurance in simulation software

W. Probst

CONTENTS

GENERAL ASPECTS

The basic task in a noise mapping project is the calculation of a noise indicator like L_{den} or L_{night} at defined locations in a given environmental scenario. Uncertainty describes the possible deviation of the level determined from a "true value"; it can only be quantified if this latter ideal result is clearly defined.

People exposed to noise are interested in measures to minimise its harmful effects. It is certainly assumed that the noise indicators correlate with these effects and it is therefore obvious that effects should be minimised if the value of the indicator is reduced to the possible minimum. From this point of view the "true value" is the value of the noise indicator determined with an "ideal measurement" where uncertainties in the measurement process are small enough to be neglected.

According to the state of the art it must be accepted that a perfect noise prediction with negligible uncertainty can never be ensured. Many approximations are included in the models to describe sound emission and sound propagation, and there are even physical phenomena encountered in measurements where existing scientific knowledge is not sufficient to understand them theoretically or to describe them mathematically correct. But there are even well known physical dependencies that cannot be implemented in software because calculation times would explode.

However, scientific research does not stand still, the knowledge about sound emission and propagation is growing continuously and the increasing number of relevant publications partially driven by public financed projects is impressive.

In calculation of sound propagation it can be distinguished between more scientifically based models (SM) on the one side and engineering models (EM) on the other side, even if this classification is somewhat misleading, because the engineering models are also based on scientific research (Figure 9.1). An additional distinguishing feature may be that the methods in the SM group are generally not standardised but realized in individual

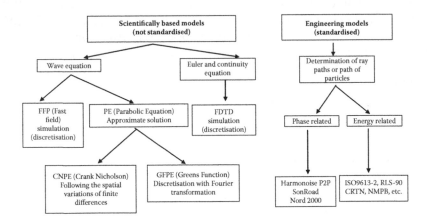

Figure 9.1 Scheme to classify the methodologies to calculate sound propagation.

software packages, whereas the methods of the EM group are—or should be—unambiguously defined and standardised. The correct implementation in software packages can be checked by methods of quality assurance. An example is the procedure described in the German Standard DIN 45 687.[1]

The models more scientifically based are tools for experts and some knowledge about the underlying principles is necessary for applying them correctly.

These methods cannot be used for noise mapping projects on a large scale or with complex structured environments like agglomerations. They may be more accurate in the aforementioned sense in some cases with relatively simple and small model scenarios, but requirements about necessary memory and calculation time would quickly exceed all practically acceptable limits if applied with realistic environments.

Engineering models are generally clearly described in guidelines or standards. They are designed to assure high precision (identical results if applied by different experts in the same case), and if implemented in software they can even be used by persons not experts in acoustics. In some countries with long tradition of legally based and strongly controlled requirements with respect to noise limits for each of the source types—road, railway, industry, and aircraft—such an engineering method is legally fixed and has to be used. Due to their relatively simple and transparent structure the application range of engineering models can in many cases be extended to large scenarios and therefore they have been and are applied in noise mapping projects.

The limits between these two mentioned classes of models are not sharp and from time to time, due to progress in hardware performance and software techniques, some elements of the more scientific-based models are integrated, for example, in standards and in engineering models.

Standardised models cannot continuosly be adapted to the most recent findings in acoustics; this is not even worthwhile because planning reliability is necessary in all projects going far into future. Therefore, models to be used in noise mapping and to control legal requirements may not always reflect the most detailed description of the environment and the best possible state of the art in the calculation of sound emission and sound propagation. But they should be precise to ensure comparable results if applied by different persons. Furthermore, they should not react too sensibly on parameter variations caused by the unavoidable uncertainties of input parameters. Last but not least, they should be so transparent that the reason for unexpected results can be detected. This traceability is a very important feature of models to be implemented in software and applied for large-scale noise mapping projects.

The decision about the methodology to be used as the legally fixed engineering method for noise mapping purposes is important. It needs a thorough balancing of these aspects: accuracy, precision, transparency or traceability, and robustness. If such a decision has been made it is recommended to thoroughly check the influence of the main parameters and to add a chapter about uncertainty to the technical specification (standard or guideline) describing the calculation method.

In individual noise calculations or noise mapping projects this existing knowledge about the uncertainty of the prescribed calculation method itself in relation to the true value "ideal measurement" is preconditioned and must not be checked further. The decided and often legally fixed calculation method is the "truth" in all such noise-related investigations.

This true value results from an exact application of the defined calculation method without any further approximation. The accuracy of the method itself is not further questioned, because there is no freedom to modify it. If the influence of any action (e.g., the geometrical simplification of a very complex ground model) on the uncertainty shall be examined, the results based on the original ground model are compared with those obtained with the simplified model, but in both cases applying the same calculation method. A quantified uncertainty is therefore only meaningful in relation to the calculation model that has been used.

THE CHARACTERISATION OF UNCERTAINTY

When quantifying the uncertainty of a predicted noise level, one has to determine the possible deviations from the true value. Such a quantification can be obtained by checking this deviation in a number of cases and describing the basic population alternatively by a lower and an upper limit of these values or by a mean value and a standard deviation.

Characterisation by quantiles $q_{0,1}$ and $q_{0,9}$

According to DIN 45 687,[1] the spread of possible results due to not negligible uncertainties shall be quantified by an interval including 80% of all possible results. Generally it is recommended to use a sample not smaller than 20 values to characterise the measure of dispersion of all these possible values; the more the better. The lower and upper boundary of this interval is defined by the 0,1 quantile $(q_{0,1})$ and the 0,9 quantile $(q_{0,9})$.

These quantiles can be determined by sorting the values of the sample as an ascending sequence with consecutive numbering R. Depending on the size of the sample, the numbers R of the quantiles $q_{0,1}$ and $q_{0,9}$ are taken from Table 9.1.

Table 9.1 Numbers (or Ranking Position) R

N^*	$R(q_{0,1})$	$R(q_{0,9})$
20	2	19
21	2	20
22	2	21
23	2	22
24	2	23
25	2	24
26	3	24
27	3	25
28	3	26
29	3	27
30	3	28
31	3	29
32	3	30
33	3	31
34	3	32
35	3	33
36	4	33
37	4	34
38	4	35
39	4	36
40	4	37
41	4	38
42	4	39
43	4	40
44	4	41
45	4	42
46	5	42
47	5	43
48	5	44
49	5	45
50	5	46

*Size of the sample (number of random sample values).

EXAMPLE 9.1

Quantify the span of 30 values of levels measured at the same location at different times and with different meteorological conditions. The values sorted in ascending order are shown in Table 9.2.

With a sample size of 30 one gets the relevant numbers $R(q_{0,1}) = 3$ and $R(q_{0,9}) = 28$ from Table 9.1. Based on the sorted values in Table 9.2 the quantiles $q_{0,1} = 56.8$ and $q_{0,9} = 62.8$ are determined. Therefore 80% of all level differences shall be expected to be in an interval between 57 dB and 63 dB.

Characterisation by mean value and standard deviation

Another possibility is to characterise the basic population of possible numbers represented by the sample of Table 9.2 by the mean value

$$m = \frac{1}{N} \sum_{R=1}^{N} dL_R \tag{9.1}$$

and the standard deviation

$$s = \sqrt{\frac{1}{N-1} \sum (dL_R - m)^2} \tag{9.2}$$

In many cases, the true distribution of values can be approximated by the well-known normal distribution, as it is shown in Figure 9.2.

In this example, the distribution is characterised by a mean value $m = 60$ dB and a standard deviation $s = 3$ dB. The area below the curve between two vertical lines at levels L_1 and L_2, or mathematically the integral from L_1 to

Table 9.2 Levels Sorted Ascending

R	LR dB	R	LR dB
1	55.7	16	59.4
2	56.1	17	60.1
3	56.8	18	60.4
4	56.8	19	60.8
5	57.2	20	60.8
6	57.3	21	61.3
7	57.6	22	61.4
8	57.9	23	61.8
9	58.0	24	61.9
10	58.2	25	62.1
11	58.5	26	62.2
12	58.6	27	62.3
13	58.7	28	62.8
14	58.8	29	63.2
15	59.3	30	63.5

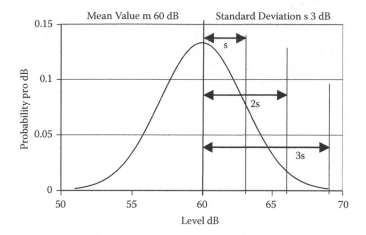

Figure 9.2 Normal probability distribution of levels with mean value m = 60 dB and standard deviation s = 3 dB.

L_2, is exactly the probability that one level determined will have a value between these two boundaries.

The assumption of a normal distributed basic population is helpful if one wants to describe the probability that a certain level, L_x, will not be exceeded.

Needed are the mean value and the standard deviation of all possible results. Then the coverage factor

$$k = \frac{L_x - m}{s} \tag{9.3}$$

and with Figure 9.3, the confidence level, w, that L_x will not be exceeded by all possible results can be determined.

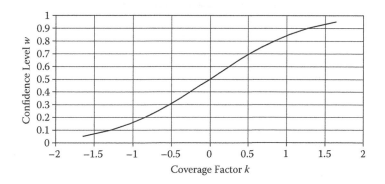

Figure 9.3 Confidence level, w, in dependence of the coverage factor, k.

EXAMPLE 9.2

From subsequent determinations a sample of levels with the mean value 60 dB and the standard deviation 3 dB have been determined. Wanted is the probability or confidence level that a certain limiting value of 63 dB will not be exceeded by the basic population of all possible results.

The coverage factor is

$$k = \frac{63 - 60}{3} = 1.0 \tag{9.4}$$

From the diagram one gets a confidence $w = 0.83$ with the coverage factor 1 and therefore the confidence level of not exceeding 63 dB is 83%.

For questions with respect to uncertainty, the adverse problem is more important. The results of a measurement may be the mean value L_m, and from literature or standards the uncertainty of this type of determination may be characterised by a standard deviation. Needed is the level L_x that will not be exceeded by the true value with a certain confidence level. The relevant coverage factors, k, for some important and often used confidence levels, w, are shown in Table 9.3.

EXAMPLE 9.3

Applied is the same situation as before: the level determined is 60 dB and the uncertainty of this procedure corresponds to a standard deviation of 3 dB. Wanted is the level L_x that will not be exceeded with a confidence of 90%.

Table 9.3 leads to the coverage factor of $k = 1.281$ and therefore

$$L_x = 60 + (1.281 \times 3) \; dB \approx 64 \; dB$$

Table 9.3 Relation of a Wanted Confidence Level (w) and the Necessary Coverage Factor (k)

w (%)	k
50	0
55	0.125
60	0.253
65	0.385
70	0.524
75	0.674
80	0.841
85	1.036
90	1.281
95	1.644

Figure 9.4 Basic element of noise map calculations.

INTERACTION OF SOURCE AND PROPAGATION UNCERTAINTIES

(Principles stated by Probst.[2])

The basic element or "atom" of a noise map calculation is the scenario point source → sound propagation → receiver, often abbreviated as P2P (point to point) (Figure 9.4). Even extended sources like roads, railway tracks (acoustically considered as line sources) or noisy areas like loading zones or parking areas (acoustically considered as area sources) are partitioned and at the end treated like arrangements of point sources. The emission of such a point source can be expressed as sound power level, L_W.

The receiver level, L, is calculated from

$$L = L_W - A$$

(9.5)

with A including all attenuations caused by air absorption, ground effects, screening, and others.

Equation (9.5) may be related to each single frequency band or the total A-weighted levels. But in all cases it should be taken into account that L_W is not absolutely certain. Then, like all input data, it is characterised by an uncertainty σ_{source}.

The sound power level of sources in industrial facilities (such as motors, pumps, or other machinery as well as forklifts or even trucks) can be measured according to, for example, one of the International Standards of the ISO 3740 series. Each of these standards contains information about the uncertainty or standard deviation that results from its application (Table 9.4).

Table 9.4 Some Examples and Recommendations to Quantify the Uncertainty, σ_{source}, of Emission Values for Industrial Sources

Source of Information	Standard Deviation
Information by colleagues (professionals)	5 dB
General literature (related to the machine family)	4 dB
Specific literature (based on machine type and parameters)	3 dB
Measurement, grade 3 (e.g., ISO 3746)	4 dB
Measurement, grade 2 (e.g., ISO 3744)	2 dB
Measurement, grade 1 (e.g., ISO 3741)	1 dB

If the level, L, at the receiver is calculated according to Equation (9.5), the uncertainty, σ, of the resulting, L, is influenced by the uncertainty of the source emission, σ_{source}, and by the uncertainty of the propagation calculation, $\sigma_{propagation}$. It can be calculated from

$$\sigma = \sqrt{\sigma^2_{source} + \sigma^2_{propagation}} \qquad (9.6)$$

The uncertainty of the propagation calculation depends on the method applied. For engineering methods like ISO 9613-2,[3] the following expression may be used to estimate it[4]:

$$\sigma_{propagation} = \begin{cases} 0 \; dB & for \; r \leq 10m \\ 2 \cdot \lg\left(\dfrac{r}{r_0}\right) dB & for \; r > 10m \end{cases} \qquad (9.7)$$

where $r_0 = 10$ m.

Another approach legally fixed in the German state Brandenburg for the calculation of wind turbine noise with ISO 9613-2 is

$$\sigma_{propagation} = \begin{cases} 1 \; dB & for \; r \leq 100m \\ 2\lg\left(\dfrac{r}{r_0}\right) - 3 \; dB & for \; r > 100m \end{cases} \qquad (9.8)$$

with $r_0 = 1$ m.

Generally the calculated noise level is influenced by many sources (Figure 9.5).

The partial levels, L_1, L_2, ..., L_N, are the contributions of the individual sources at the receiver and include all propagation effects summed up in A. The uncertainties σ_1, σ_2....σ_N are calculated with Equation (9.6) for

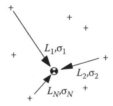

Figure 9.5 Many sources influencing the noise level at the receiver.

each source individually. The resulting noise level, L, at the receiver is then calculated by summing up the partial levels

$$L = 10 \lg \left(\sum_{n=1}^{N} 10^{L_n/10} \right) dB \tag{9.9}$$

The uncertainty of this receiver level—the final result—is

$$\sigma = \frac{\sqrt{\left(\sigma_1 \cdot 10^{0,1 \cdot L_1} \right)^2 + \left(\sigma_2 \cdot 10^{0,1 \cdot L_2} \right)^2 + \cdots + \left(\sigma_N \cdot 10^{0,1 \cdot L_N} \right)}}{10^{0,1 \cdot L_1} + 10^{0,1 \cdot L_2} + \cdots + 10^{0,1 \cdot L_N}} \, dB \tag{9.10a}$$

or abbreviated

$$\sigma = \frac{\sqrt{\sum_{n=1}^{N} \left(\sigma_n \cdot 10^{0,1 \cdot L_n} \right)^2}}{\sum_{n=1}^{N} 10^{0,1 \cdot L_n}} \, dB \tag{9.10b}$$

This opens the possibility to integrate in noise mapping calculations the aspect of uncertainty. Input parameters are the uncertainties of the emission levels of all sources. The noise map as it is shown for two sources in Figure 9.6 is the result of a calculation of the levels according to Equations (9.5) and (9.9) at each grid point. Then in a second step the uncertainty can be calculated at each grid point from Equations (9.6), (9.7), or Equations (9.8), and (9.10). The result is an "uncertainty map" as it is shown for this case in Figure 9.7.

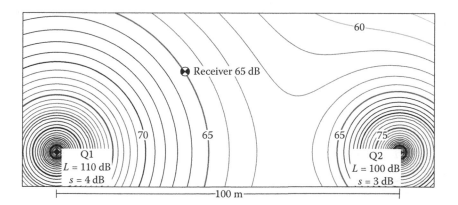

Figure 9.6 Noise map with two sources; sound power levels 110 dB and 100 dB and uncertainties 4 dB and 3 dB. **(See colour insert.)**

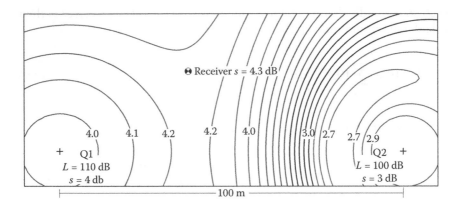

Figure 9.7 Uncertainty map with these two sources. **(See colour insert.)**

With a simple arithmetic operation the levels in Figure 9.6 are combined with the uncertainties s in Figure 9.7 to calculate the levels $L_{95\%}$ in Figure 9.8 that will not be exceeded with a confidence of 95%.

These examples based on point sources demonstrate the principle how the uncertainty of the source emissions and the uncertainty of the propagation calculation influence the uncertainty of the finally calculated receiver levels and the noise map.

Figure 9.9 shows the computer model of a car production factory where ventilation openings and other radiating parts have been simulated by 3500 point sources. The sound power levels of these sources have been determined

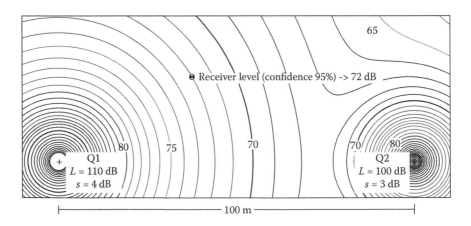

Figure 9.8 Noise map with levels that will not be exceeded with a confidence of 95%. **(See colour insert.)**

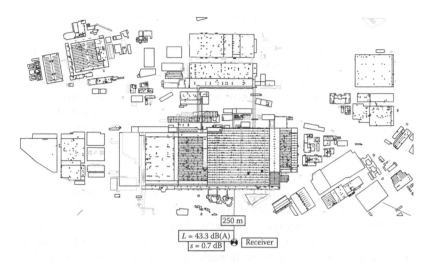

Figure 9.9 Computer model of a car factory where 3500 point sources are integrated. **(See colour insert.)**

by measuring the sound pressure level at one microphone position at each source in a defined distance.

Due to this simple determination of L_W an uncertainty of 5 dB was used as input parameter for each source. The level in a distance of 250 m, calculated with ISO 9613-2, is 43.3 dB(A). The uncertainty of this level is calculated with the procedures explained earlier to be 0.7 dB.

The diagram in Figure 9.10 shows the resulting uncertainty at the same receiver position in case of varying the uncertainty of the 3500 sources and repeating the calculation.

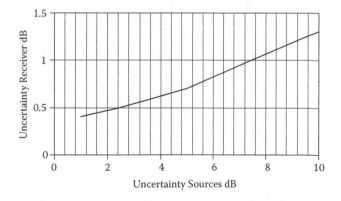

Figure 9.10 The uncertainty of the calculated noise level at the receiver in dependence of the assumed uncertainty of the emission of each of the 3500 sources.

This is a finding of general interest. Even in case of limited knowledge about the individual sources with quite uncertain input parameters, the resulting calculated noise levels are fairly accurate and characterised by an uncertainty that is much smaller.

This is true if the receiver is far away and many comparable partial levels determine the result: if one moves toward one of these sources and its contribution increases, the uncertainty will converge against the uncertainty value of this source dominating the result more and more.

The procedure demonstrated is widely used to check the uncertainty of noise levels calculated in the vicinity of wind parks, power plants, or any other industrial facilities.

But the method is not applicable if the sources taken into account in Equation (9.10a,b) are correlated. In case of subdividing a road segment or any other extended source characterised by one set of parameters, these parts are not uncorrelated, because an error in the estimation of an emission value is the same for all subparts. In such cases the covariance of the different contributions cannot be neglected and further measures are necessary.

The method is and can be integrated in software for noise prediction.

CONTRIBUTIONS TO THE FINAL UNCERTAINTY OF CALCULATED NOISE LEVELS

Uncertainties due to the methodology applied

As it was discussed earlier, all our calculation methods are not perfect and there are always physical phenomena influencing the noise levels that are not simulated or taken into account in the calculation method. A lot of approximations are necessary if the three-dimensional sound wave propagating through the layered atmosphere and sweeping over objects like buildings and walls shall be replaced by geometrically defined rays.

This should be taken into account if a certain phenomena of sound propagation shall be considered very detailed at the cost of calculation time and transparency of the procedure. An example is the inclusion of ground effects taking into account interference effects and partially coherent superposition of direct ray and the ray reflected at the ground surface. Mathematically correct results can only be obtained by performing the calculation separately with narrow frequency bands, but this makes it very tedious to trace back unexpected results. Such a detailed calculation may be advantageous in special "pure" cases with few dominating sources where such interference effects may influence the resulting frequency spectra. But it can be questioned if in situations with dense traffic, as shown in Figure 9.11, the accuracy of the finally determined A-weighted sound pressure level will be increased if such a detailed mathematical description of a detail is applied.

Figure 9.11 Dense traffic on a multilane road.

It is obvious that the heights of the sources above the road surface are different and diffraction of direct sound and ground reflection by cars in the other lanes will destroy any interference.

The imponderables of the acoustical description of the sources are one of the fundamental reasons why it is not possible to fall below a minimal remaining uncertainty. For road traffic on European road systems this is especially true for the distinction between light and heavy vehicles, because, independent from the number of different vehicle classes, there is no single parameter like length or weight or number of axles that correlates exactly with the sound emission. If the gross vehicle weight is used to distinguish light and heavy vehicles as it is the case with many national standards, the numbers of heavy vehicles determined with automatic counting stations are not very accurate. And as the portion of heavy vehicles is relevant for the noise emission, all our traffic emission data are of limited accuracy.

It must be accepted that models are always simplifications of the physical reality; a very detailed description of a certain phenomenon may reduce the performance of the calculation and reduce precision and transparency in cases where the influence of this phenomenon may even be negligible. Therefore a thorough balancing of all these aspects is needed.

Input parameters

If a model to calculate noise levels is developed, a lot of input parameters have to be determined. Especially in noise mapping projects, according to Directive 2002/49/EC, the scenarios are large and it is not easy—not to say impossible—to check geometries and other acoustically relevant parameters in all details. But it is helpful to have possible traps in mind to avoid them.

As shown in Figure 9.12, the model of a city or agglomeration is generally an assembly of ground or terrain, artificial objects like buildings and barriers, and sources like roads and railways.

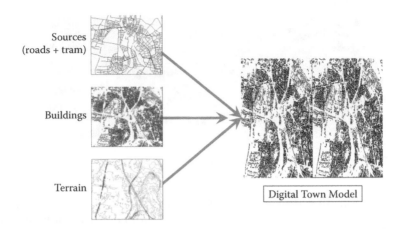

Figure 9.12 Integration of input data to a complete town model. **(See colour insert.)**

The terrain model

Height points and contour lines define the terrain model. One can define a "geometrical deviation" of the model by statistically analysing the difference of heights of the model and the real terrain at each point at a regular grid, and calculating mean value and standard deviation or the uncertainty interval $q_{0,1}$ to $q_{0,9}$. Repeating this with stepwise decreased grid spacing the geometrical uncertainty can be determined.

But the problem is to transform the geometrical uncertainty to an uncertainty of noise levels expressed in A-weighted decibels, or dB(A). From practical experience it is known that the influence of ground height variations or of object height remains small as long as the line of sight is above the ground and not grazed or even interrupted. But this influence increases dramatically if a diffracting edge comes into play.

Figure 9.13 shows a simple barrier in 7 m distance from a road that was erected to reduce the noise level at the façade behind. The receiver is in a height of 5 m above ground, and the height of the barrier is varied from 3 m up to 3.5 m in steps of 5 cm (Figure 9.14).

The detailed results are dependent from the calculation method, but the level jump with varying height in the small interval where grazing incidence occurs is generally true. Therefore considerable uncertainties of calculated levels are caused by terrain models that are too rough, and heights of barriers and other screening objects are defined relative to the ground.

Terrain models are often created by importing height points or contour lines from geographic information systems (GIS). These digital terrain models (DTMs) are often by far too detailed and cannot be processed without simplification. If software programs are used to simplify the geometry, the applied

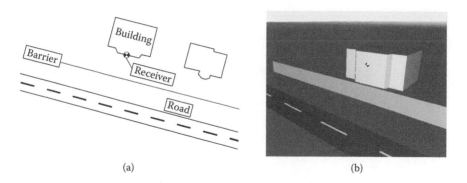

(a)   (b)

Figure 9.13 (a) Barrier between road and receiver, 2D. (b) Barrier between road and receiver, 3D.

strategy to reduce the number of points must allow a maximal and not have exceeded the deviation between original and simplified terrain model. If this is not the case no upper limit of the error of calculated levels can be given.

Figures 9.15a and 9.16a show the height point pattern before and after simplification with a maximal allowed deviation of 10 cm. This difference can hardly be seen in the resulting 3D-views, Figures 9.15b and 9.16b, but the number of points is halved.

Buildings and other objects

All artificial objects like buildings or barriers are generally modelled as a polygon line in ground view. If a receiver point is exposed to diffracted sound, the geometry of the diffracting edge may influence the result as it was shown earlier.

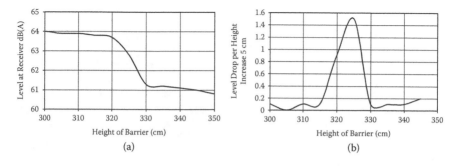

(a)   (b)

Figure 9.14 (a) Level in dependence of height. (b) Level change in dependence of height increase 5 cm.

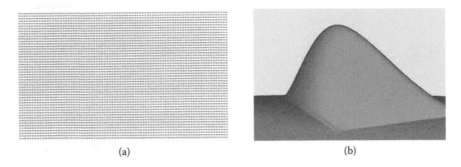

(a)   (b)

Figure 9.15 (a) Height points original (7500). (b) 3D view of original model.

But it should be mentioned that too detailed modelling may even increase the uncertainty of calculated levels, if receiver positions are screened from the source and if the resulting level is determined by reflections. If the reflecting façade in top view is not a straight line, but a polygon with many points to account for all little salients, alcoves, or bays the reflection will not be calculated in some methods (e.g., ISO 9613-2). Each line segment between two points is regarded separately as a reflector and if this is too short relative to the wavelength no reflection will be calculated. Due to this "fence-effect" geometrically detailed modelling may be advantageous to produce nice 3D views but will also decrease the accuracy in some cases. But this depends on the calculation method applied.

Special objects like bridges and galleries

There is no general rule to model geometries of such objects and most of the calculation methods contain no generally applicable method to simulate their acoustical behaviour. Therefore it is always necessary to know how the software

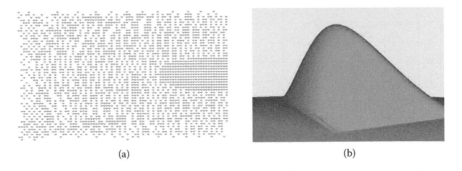

(a)   (b)

Figure 9.16 (a) Height points simplified (3200). (b) 3D view of simplified model.

supports such special objects. In noise mapping projects it is recommended to use such special and software-specific solutions only where it is absolutely necessary, because it is complicated to check the influence in all directions.

Absorption properties of ground, buildings, and barriers are often difficult to estimate. In most cases only the absorption of barriers and screens built for noise reduction are individually entered. For all other objects default absorption values are used after having created or imported the geometrical shape.

Source modelling and parameters to determine the acoustical emission

It depends on the method applied what parameters are needed and how accurate the emission will be determined. The most important input parameters for road traffic noise are the data describing the vehicle flow for passenger cars and heavy vehicles separately, their speed, and the acoustically relevant specification of the road surface. As was mentioned earlier, this is often one of the largest contributions to uncertainty, because in many cases flow data are not available and must be estimated. The influence of trucks and other heavy vehicles is large even in cases where the truck portion of the traffic is low. Unfortunately these numbers are often very roughly estimated and automatic counting systems are in most cases not able to distinguish exactly the vehicle groups as they are specified in the calculation standard.

Taking into account the uncertainties in the determination of traffic flows and truck portions by automatic counting devices and the different modelling approximations for roads in calculation methods, it is estimated that the variation of the predicted acoustical emission with the same road data is about ±2.5 dB for passenger cars and about ±5 dB for trucks. Based on this and other investigations it can be assumed that a basic uncertainty expressed as standard deviation cannot fall below 3 dB for the acoustical emission of predicted emission of road noise.

The distribution of driving vehicles on different lanes is another contribution to a base uncertainty that cannot be reduced. It has been shown that the simulation of a multilane road with two line sources above the axes of the outmost lanes is a good compromise.[5] The levels in typical distances of 10 m are quite similar to those calculated with a simulation of each lane separately.

Similar results were found for railway noise emissions according to the standards Schall 03[6] and SRMII[7] in a measurement campaign near a Slovakian railway line.[8] The differences of calculated and measured levels of passing trains were found to be distributed with a standard deviation of 3 dB with both calculation standards. This means that the prediction of the noise from one single pass-by event is characterised by this uncertainty; the prediction of mean levels for longer periods is more accurate.

Lack of input data like traffic flow is one of the main problems producing uncertainties in calculated noise maps. It is often recommended in such cases to install permanent monitoring stations and to calculate the emission values of sources from the measured levels.

It can certainly be advantageous to install noise monitoring stations and to measure noise levels additionally. This may increase the acceptance of the determined noise levels by the residents, allows one to check the validity of assumed input parameters and the calculation method, and in some cases may even be used to update the applied computer model. But it needs a thorough investigation of uncertainties if such measurements are used to avoid a detailed investigation of input parameters and to do a sort of "back calculation" of emission values from such measurements. Words like "reverse engineering," "inverse engineering," or "dynamic noise mapping" are used to label such techniques, but at the end it is all based on the same thing: a replacement of modelling with detailed input data by the measurement of sound levels. See Manvell et al.,[9] Reiter et al.,[10] Comeaga et al.,[11] and Stapelfeldt et al.[12] for examples.

The application of such methods may in many cases be justified, but it is recommended to thoroughly weigh pros and cons before deciding about a costly monitoring system and to check the uncertainties if measured sound levels are used as input parameters for a larger noise map. It can be questioned if the detailed data of flight paths and movements at an airport can be replaced by the input from one station and the railway traffic in a city from two stations as it is mentioned by Reiter et al.[10]

Probst[13] investigated the uncertainty aspect of this back-calculation and found that measurement errors of some tenth of a decibel can result in deviations of more than 5 dB for the back-calculated emission value. Doubling the traffic flow is equivalent to a correction of 3 dB, therefore little deviations in the measured levels can change the noise map equivalent to considerable changes in traffic flow.

The problems increase if monitoring positions are influenced by more than one dominating source each. If larger parts of the road network are taken as one source as it was done in some of the reported cases to reduce the necessary monitoring positions, then the accuracy of the resulting noise map must be questioned.

If monitoring stations are to be used to update noise maps, it is recommended to locate them always near the main noise sources. This minimises problems and reduces uncertainties.

Acceleration techniques

The calculation methods imply a lot of approximations to replace the real sound wave propagation by geometrically defined rays. The two most applied methods are ray tracing (RT) and angle scanning (AS).

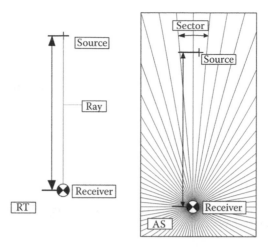

Figure 9.17 (Left) Ray tracing (RT), (right) angle scanning (AS). **(See colour insert.)**

Ray tracing in the frame of this context means that the possible ray paths between sources and receivers including direct rays and reflected rays are found by looping over all sources for each receiver, or vice versa, and are constructed geometrically, as shown in Figure 9.17: left side is for one source and one receiver.

Angle scanning (right side of Figure 9.17) works different. The 2D-angle of 360° around the receiver point where the level shall be calculated is partitioned in equal angle sectors (e.g., 100 sectors of 3.6°) and one search ray in the axis of each sector starting from the receiver is used to find the relevant sources. Extended sources like a road are logically subdivided in smaller parts and the contribution of each of these parts is calculated separately.

As shown in Figure 9.18 with the RT method, the number of calculation rays is increased if the distance from the receiver to the source is reduced. If objects like buildings are located between the extended source and the receiver, these objects are projected on the source as shown in Figure 9.19a. Such a fine sub-division gives a resolution good enough to take screened and unscreened parts of the road correctly into account, as it is shown in Figure 9.19b.

Such techniques are extremely time consuming and cannot be applied at each grid point for all roads of a city. If in a scenario, like the one shown in Figure 9.19b, the receiver is far away from the road and thousands of buildings are between them, then the road will be "atomised" and the number of necessary calculations as well; calculation time will increase dramatically.

Therefore the projection method is used in many cases only for exact and detailed calculations and it will be deactivated if large-scale noise maps are to be produced. A reduced accuracy for the noise map is the consequence.

It is also possible to apply the projection method only for receivers or grid points with a definable maximal distance from the source or to apply it for

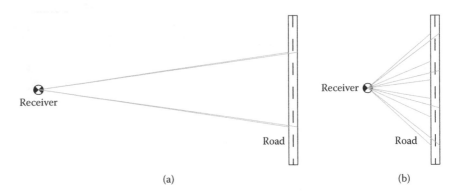

Figure 9.18 (a) RT with large distance source–receiver. (b) RT with small distance source–receiver.

objects with a defined maximal distance between the source and receiver. Making this upper limit smaller will increase the uncertainty introduced by this approximation.

With the AS method, the angle steps must be defined extremely small to "see" the gaps between the buildings in Figure 9.19b. This again may not be possible for large-scale noise mapping and this will lead to an increased uncertainty of the results.

Many such acceleration techniques can be applied to ensure acceptable calculation times. Depending on the software package used, the distance between the source and receiver, and the distances between the source and reflector and the receiver and reflector can be restricted.

Noise calculation software with the requirements according to standardised rules, for example, of DIN 456 87,[1] must allow the uncertainty introduced by such acceleration techniques to be checked.

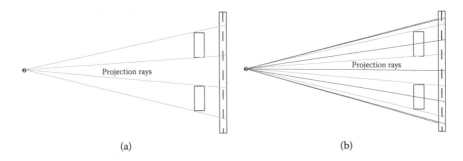

Figure 9.19 (a) Projecting diffracting objects to the source to produce screened and unscreened parts. (b) Calculation rays (red) to take screened and unscreened parts separately into account. **(See colour insert.)**

The procedure is the following:

- Depending on the size of the calculation area for noise mapping a number, N, (minimum 20) of receiver points is statistically distributed. The minimum distance to reflecting surfaces, diffracting edges, and sources of each receiver is 2 m.
- A reference calculation without application of any acceleration technique is performed. The results of this calculation are N reference levels.
- A second calculation is performed with the software configuration that is or will be used to calculate the complete noise map. This will generally include acceleration settings. The results of this calculation are N project levels.
- The N differences (Project level – Reference level) are statistically analysed and the quantiles $q_{0,1}$ and $q_{0,9}$ are used to describe the uncertainty of the noise map caused by the special configuration setting applied. Additionally, mean and standard deviation of these differences are determined.

Figure 9.20 shows an example. The noise map has been calculated without projection and other special configurations to speed up the process.

The receiver points shown are distributed across the calculation area automatically (if the software applied offers this feature) and the two calculations mentioned earlier—one with reference and one with project configuration—are performed. Table 9.5 through Table 9.7 are part of the report about this analysis according to DIN 45 687.

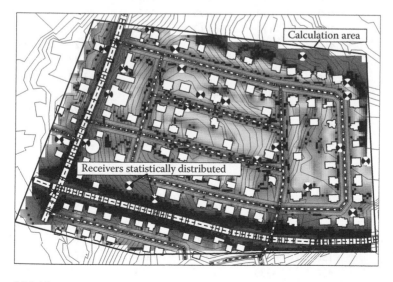

Figure 9.20 Noise map calculated with special configuration and receiver points to analyse uncertainty according to DIN 456 87. **(See colour insert.)**

Table 9.5 Information about the Two Configurations Applied

Configuration, Settings	Reference	Project
Maximum error (dB)	0	1
Maximum search radius (m)	3000	1000
Grid Interpolation	(none)	17 * 17
Proj. line sources	1	0
Proj. area sources	1	0
Maximum distance between source and receiver	3000	200
Search radius source	3000	100
Search radius receiver	3000	100

Notes: Reference—Calculation configuration with no acceleration techniques. Project—Calculation configuration with acceleration techniques to be applied for the noise map calculation.

Table 9.6 Detailed Report about Calculated Levels and Differences

	Coordinates			Level		
Number	X (m)	Y (m)	Z (m)	Reference (dB)	Project (dB)	Difference (dB)
1	270	−160	59.35	44.72	44.51	−0.21
2	110	−110	39.24	55.85	55.68	−0.17
3	400	−290	32.95	48.5	48.37	−0.13
4	200	−110	61.13	48.68	48.51	−0.17
5	460	−150	25.65	49.32	49.2	−0.12
6	230	−150	64.05	44.67	44.47	−0.2
7	160	−310	47.05	63.24	63.24	−0.01
8	350	−230	49.12	48.17	47.93	−0.24
9	390	−90	29.26	33.55	32.96	−0.59
10	460	−130	22.21	43.4	43.34	−0.06
11	470	−330	31.6	50.97	50.56	−0.42
12	210	−170	68.48	49.69	49.56	−0.12
13	10	−220	17.8	53.58	53.77	0.18
14	120	−290	42.84	59.82	59.82	0
15	190	−80	55.74	50.09	49.96	−0.13
16	0	−300	14.8	72.53	71.24	−1.29
17	490	−250	28.87	40.58	41.02	0.44
18	150	−50	40.93	40.43	50.06	−0.37
19	450	−230	29.03	53.39	53.29	−0.1
20	320	−250	54.97	50.59	50.41	−0.18

Table 9.7 Statistical Analysis of the Differences
at All the Receivers

Quantile $q_{0.1}$	−0.4
Quantile $q_{0.9}$	0.4
Mean	−0.2
Standard deviation	0.3
Minimum	−1.3
Maximum	0.4

In this case the uncertainty of the noise map (Figure 9.20) due to the selected settings and applied acceleration techniques is characterised by an interval ±0.4 dB.

It is recommended to add this information about the uncertainty due to the configuration applied to the other information in the legend of each noise map.

Interpolation of results

The bases of each noise map are levels that have been calculated at specified points. Between these points the information is nothing else but an interpolation of the values at neighbouring points. It is obvious that the uncertainty of the noise levels taken from a noise map increases with increasing grid spacing of the grid points where the complete calculation has been performed.

This has to be taken into account if the noise exposure of façades is interpolated from noise levels calculated for regular distributed grid points. The spacing of these points is in most cases 10 m and this is by far too distant to give accurate interpolations at façade points.

To check the uncertainty caused by interpolation from a regular 10 m × 10 m grid, the levels around the façades of all buildings in the example shown in Figure 9.21 have been calculated twice. The first time the calculation

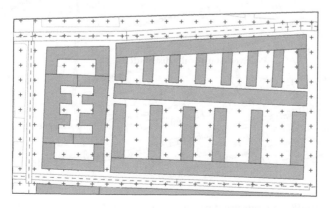

Figure 9.21 Grid points with 10 m raster; not a good basis to evaluate façade levels.

Table 9.8 Distribution of Residents on Intervals of L_{den}

L_{den}	Directly	From Grid
<50	711	471
50–55	543	725
55–60	368	191
60–65	456	615
65–70	54	130

was performed at all façade points (4 m height), the second time the façade levels have been interpolated from the points on the horizontal grid.

The distribution of the exposed residents on L_{den} values is shown in Table 9.8 for both cases. From this little exercise that interpolating façade levels from the levels obtained for a 10 m grid causes large uncertainties that are not acceptable.

Interpolation also comes into play if calculation results are presented as lines of equal level or iso-dB lines. The exact coordinates of these lines are derived with interpolation techniques from the position and level at grid points.

According to DIN 45 687[1] a similar method can be used to check the uncertainty of iso-dB lines caused by interpolation and by the aforementioned acceleration techniques. This is shown with the same little project shown in Figure 9.20. But now the results are presented as iso-dB lines and the uncertainty of the 60 dB line as shown in Figure 9.22 shall be determined.

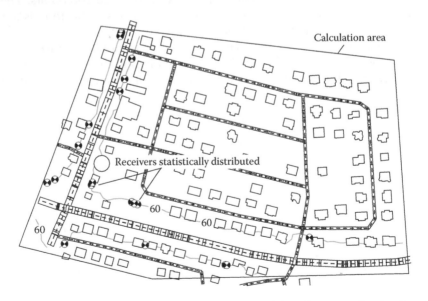

Figure 9.22 Iso-dB line representing the level 60 dB(A) and statistically distributed receiver points to analyse uncertainty according to DIN 45 687.

Again a sample of receiver points (minimum 20) is statistically distributed along this line under question. The level difference to be evaluated is the difference of the level represented by the iso-dB line (60 dB in this example) and the level calculated at these points with the reference settings. The resulting uncertainty interval can again be presented by the quantiles $q_{0,1}$ and $q_{0,9}$ in the same way as it was shown in Table 9.7.

These techniques should preferably be integrated in the software so that the distribution of receiver points can be performed automatically.

Deviation of calculated and measured noise levels: An example

It is a difficult task to estimate the accuracy of a noise map, if one defines it as an accurate result in analogy to determine it by performing an "ideal" measurement. Ideal means in that case that the measuring process shall be characterised by negligible errors or deviations. The problem is further that an undisturbed real measurement over months to get the yearly average (this is the target value of the Directive 2002/49/EC) is obviously impossible.

Comparing short time measured levels with calculated levels it should be taken into account that the input parameters like traffic flow data during the measuring time interval may differ from its mean value related to a full year. A real check of the uncertainty of the calculation method would need to determine or to estimate these traffic flow parameters related to the measuring time period and to perform the calculation of noise levels with these parameters.

But this is a difficult if not impossible task. Therefore most comparisons of that type are based on measured levels without parallel determination of traffic flow parameters. An example is the comparison of calculated and measured levels performed at 60 microphone positions in the Fildern Area in Germany (Figure 9.23).

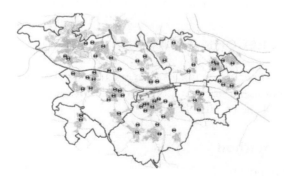

Figure 9.23 Fildern area around Stuttgart airport—60 locations where levels have been measured 2 weeks each.

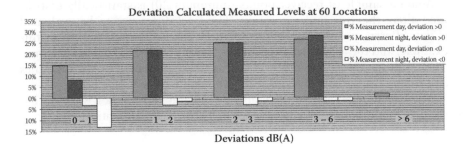

Figure 9.24 Differences calculated–measured levels not corrected for traffic flows and movements.

The noise map in the area around the Stuttgart airport was calculated and afterward measurements were performed at 60 different positions. Calculated levels are yearly averages and measured levels were not corrected with parameters related to the measuring time interval (Figure 9.24). Here only the statistical analysis of differences of calculated levels versus measured levels shall be presented—the result is only an example and cannot be taken as representative.

The horizontal axis shows the deviation of calculated from measured levels. As it can be seen from the first bar, 15% of all day results showed a difference between calculation and measurement of less than 1 dB.

These results show that in most cases the calculation is conservative and the deviations are on the safe side with respect to the population exposed. It should also be mentioned that there is a correlation between mean deviation and absolute levels; in more noisy positions the deviations are smaller. In about 25% the deviations are larger than 3 dB but these are mainly positions far from the sources with low levels.

This seems to be a general result; most calculation methods used in different countries overestimate a little bit to cover the unavoidable uncertainties.

SOME FINAL RECOMMENDATIONS

From all these influences on uncertainty it can be concluded that there is no simple way to design a noise mapping project without taking into account the confidence of the results obtained.

Calculation method

The first step important for the final uncertainty of obtained results is the selection of the calculation method to be used. In the case of calculations for European Directive 49/2002/EC this is not really a choice. The directive requires applying the harmonized method as soon as it will be published and in the meantime the Interim Methods NMPB,[14] SRMII,[7] ISO 9613-2,[3] and ECAC Doc.29[15] have to be used.

Generally a calculation method should reflect and take into account all the acoustical phenomena that have noticeable influence on the noise levels in reality and that may be important for the decisions to be taken. But on the other side very complex methods offer many screws to adjust the calculated results and this may reduce the precision of the method more as it can be justified by the improved accuracy relative to measured results. The calculation method should not require input data that are not available in most cases, because this leads to additional uncertainties due to the necessary estimation of these parameters.

It is a good and well-proven rule to apply a calculation method accurate enough to support the acoustical optimization of a scenario, but as simple and transparent as possible. Calculation times are also an issue. Even with best possible and powerful computer networks it is not possible to calculate a noise map of an extended agglomeration taking into account all possible reflections and diffractions for each partial source at all buildings and other objects. Therefore acceleration techniques must be used and many contributions must be neglected. This again increases the uncertainty and therefore the possibly better accuracy of more complex methods will be destroyed by the need to speed up the calculation by such approximations.

Modelling aspects

Many different input data are combined to produce a 3D model of the environment. This data acquisition is extremely time and cost relevant, and therefore the detailing of the model should be oriented at the acoustical needs. The wish to create good-looking 3D views is misleading. It is by far better to use only those properties of ground and built-up areas that are taken into account in the calculation instead of modelling details of buildings with jutties, balconies, and complex roofs.

In many cases terrain data are available as contour lines or as height points with an unacceptable high density of points. Such models should be simplified according to acoustical needs, because each line segment between two points may be checked if it could be a diffracting edge and this will increase calculation times enormously. Modern software products are able to perform this simplification of the ground model automatically if the software user defines an acceptable maximal deviation of the simplified ground model from the original ground model.

Calculation configuration and interpolation techniques

As it was mentioned earlier, it is always necessary to speed up the calculation by neglecting contributions that are not relevant and by other approximations, because it is not possible to calculate the contribution of all partial sources at all receivers by taking into account all possible—even screened—reflections at all surface elements. The software products available offer different approximation techniques, but it is not possible to estimate the additional uncertainty that is introduced by such methods. The same is true for noise level information shown by coloured maps between the grid points where the calculation was performed. These levels are interpolated from the values calculated at the grid points and are characterised by an additional uncertainty.

Therefore it is recommended to apply techniques similar to those described in DIN 45 687 and presented earlier to calculate at some statistically distributed receiver points with the intended calculation configuration, to repeat this calculation with the best possible accuracy without applying the acceleration techniques, and to analyse the level differences obtained to get a measure for the uncertainty caused by the applied calculation configuration.

REFERENCES

1. DIN 45 687, Acoustics—Software products for the calculation of sound propagation outdoors—Quality requirements and test methods, Beuth Verlag Berlin, 2006-05.
2. Probst, W., Uncertainties in the prediction of environmental noise and in noise mapping, Proceedings of "Managing Uncertainty in Noise Measurement and Prediction," Symposium INCE–Europe, Le Mans, France, June 2005.
3. ISO 9613-2, Acoustics—Attenuation of sound during propagation outdoors—Part 2: General method of calculation, International Organization for Standardization, Geneva, Switzerland, 1996.
4. Probst, W., Donner, U., The uncertainty of predicted sound exposure levels (German: "Die Unsicherheit des Beurteilungspegels bei der Immissionsprognose"), Zeitschrift für Lärmbekämpfung, 2002.
5. Comparison of the German calculation model for road noise with other European models (in German), Heft 1030, BAST (Bundesanstalt für Straßenwesen), 2010.
6. Richtlinie zur Berechnung der Schallimmissionen von Schienenwegen—Schall 03, Deutsche Bundesbahn, 1990.
7. Reken- en Meetvoorschrift Railverkeerslawaai 1996, Ministerie Volkshuisvesting, Ruimtelijke Ordening en Milieubeheer, 20 November 1996.
8. The assessment and management of environmental noise, Final report of project Czueozova31-Phare-SR-PAO/CFCU, Framework Contract IB/AMS/451, February 2005.

9. Manvell, D., Marco, L.B., Stapelfeldt, H., Sanz, R., SADMAM—Combining measurements and calculations to map noise in Madrid, InterNoise 2004, Prague, 2004.

10. Reiter, M., Kotus, J., Czyzewski, A., Optimizing localization of noise monitoring stations for the purpose of inverse engineering applications, Acoustics '08, Paris, 2008.

11. Comeaga, D., Lazarovici, B., Tache, G., Reverse engineering for noise maps with application for Bucharest city, InterNoise 2007, Istanbul, 2007.

12. Stapelfeldt, H., Vukadin, P., Manvell, D., Reverse engineering—Improving noise prediction in industrial noise impact studies, EuroNoise 2009, Edinburgh, 2009.

13. Probst, W., Noise Prediction based on measurements, DAGA 2010, Berlin, 2010.

14. NMPB-Routes-96, French standard XPS 31-133, Acoustique—Bruit dans l'environnement—Calcul de niveaux sonores (Acoustics—Outdoor noise—Calculation of sound levels).

15. ECAC.CEAC Doc. 29, Report on standard method of computing noise contours around civil airports, 1997.

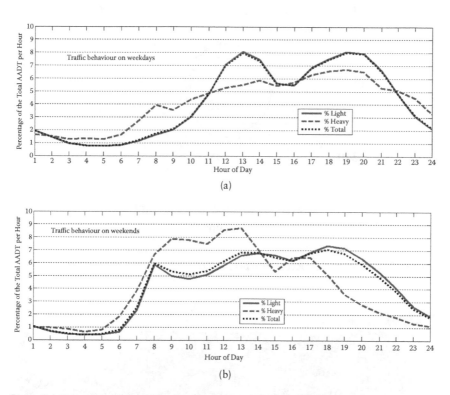

Figure 3.4 Percentage of the total average annual daily traffic (AADT) per hour during a 24-hour period. (a) Statistics for working days are presented, and (b) only for weekends. Distinctions between percentages of heavy and light vehicles are considered, too.

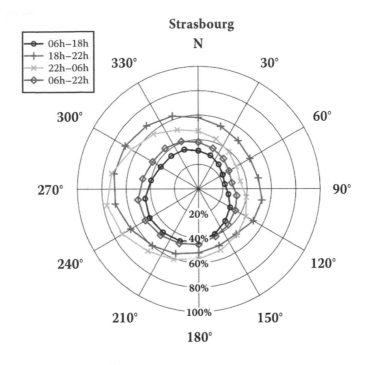

Figure 4.10 Sample rose of occurrence probabilities.

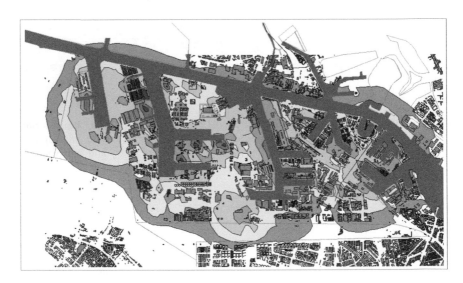

Figure 6.6 Contours of the industrial area Westport near Amsterdam, Netherlands. Contours of 50 dB are situated at 500 to 900 m from nearest industry.

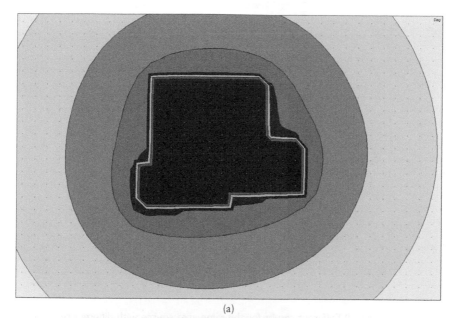

(a)

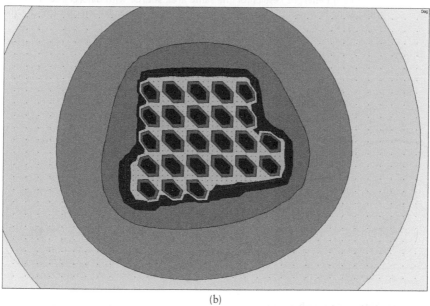

(b)

Figure 6.8 Different results for the same area source with different selection of grid spacing. From left 50´50 m, 40´40 m, 41´41 m while calculation grid is 100´100 m. (*Continued*)

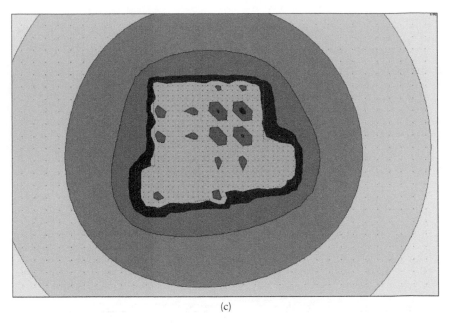

(c)

Figure 6.8 (Continued) Different results for the same area source with different selection of grid spacing. From left 50´50 m, 40´40 m, 41´41 m while calculation grid is 100´100 m.

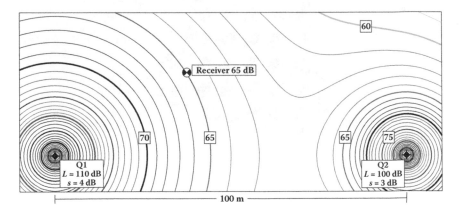

Figure 9.6 Noise map with two sources; sound power levels 110 dB and 100 dB and uncertainties 4 dB and 3 dB.

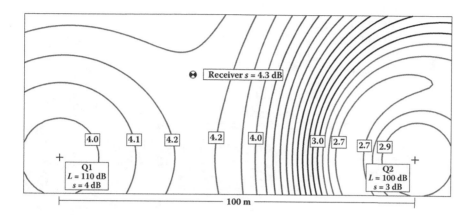

Figure 9.7 Uncertainty map with these two sources.

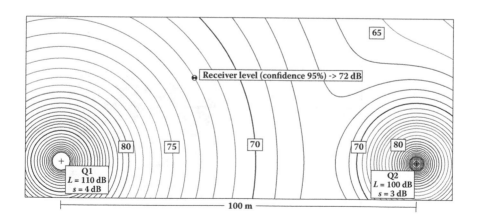

Figure 9.8 Noise map with levels that will not be exceeded with a confidence of 95%.

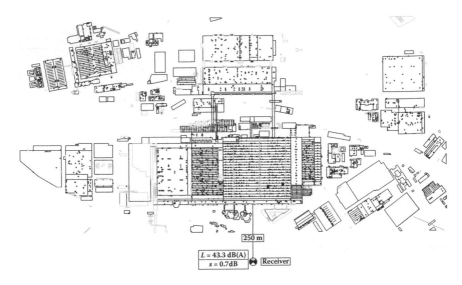

Figure 9.9 Computer model of a car factory where 3500 point sources are integrated.

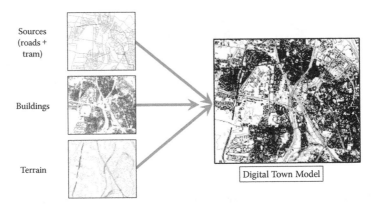

Figure 9.12 Integration of input data to a complete town model.

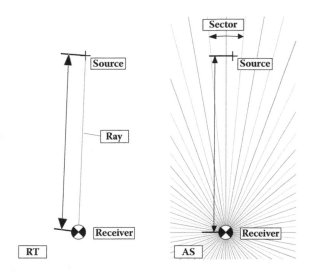

Figure 9.17 (Left) Ray tracing (RT), (right) angle scanning (AS).

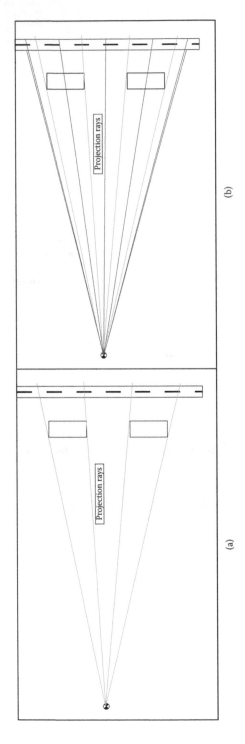

(b)

(a)

Figure 9.19 (a) Projecting diffracting objects to the source to produce screened and unscreened parts. (b) Calculation rays (red) to take screened and unscreened parts separately into account.

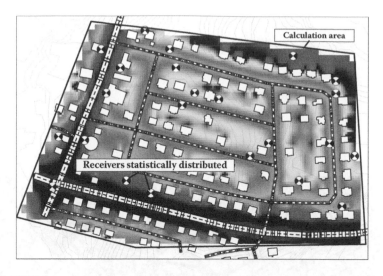

Figure 9.20 Noise map calculated with special configuration and receiver points to analyse uncertainty according to DIN 456 87.

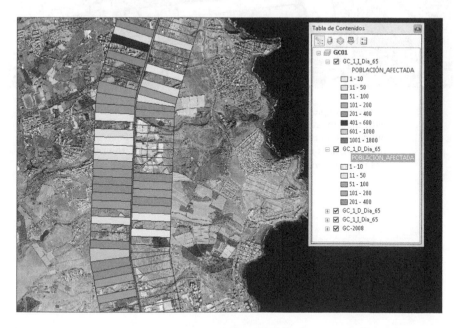

Figure 10.7 A choropleth map created via buffering the highway, highlighting the density of exposed people in residential buildings.

Figure 10.8 In the example of the figure the residential buildings highlighted are those with their most exposed façades affected by more than 65 dB by night. Some buildings in between with the same noise levels have been excluded because they have uses other than residential, and in one case, the residential building has no windows in its most exposed façade.

Figure 11.2 Some of Hong Kong's breakthrough 3D noise maps. (Courtesy of Hong Kong Environmental Protection Department.)

Figure 11.3 Examples of 3D maps for new developments in Amsterdam.

Figure 11.4 3D noise maps are now possible in Google Earth. (From Google Earth. With permission.)

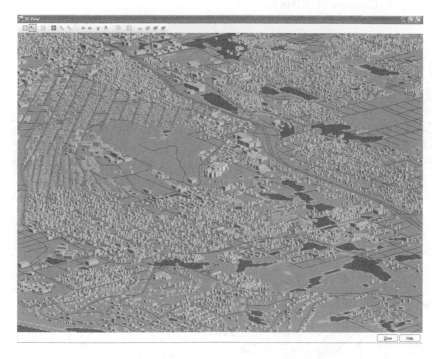

Figure 11.5 Sample data showing that correctly imported GIS data needs evaluation prior to use.

Figure 11.9 An example of a noise map with a categorization of noise levels at the façade. (Courtesy of DGMR.)

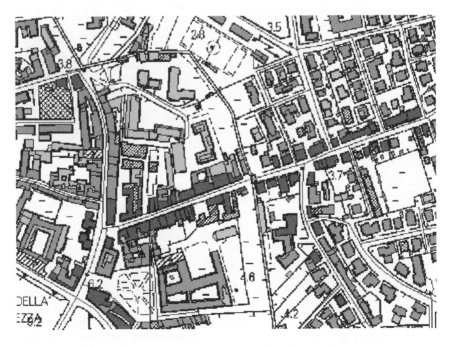

Figure 13.19 Coloured map showing most exposed façades.

Figure 17.5 Map of an area-related noise score in a 3D presentation showing the hot spots in red.

Part 2

Noise mapping and geographic information systems (GIS)

Chapter 10

Geographic information system tools for noise mapping

J.L. Cueto and G. Licitra

CONTENTS

Noise pollution is a spatial-dependent phenomenon strongly correlated to the morphology of terrain, both in open country and in urban environments. A geographic information system (GIS) is a suitable instrument to tackle this problem from all angels, from the assessment of noise levels to the definition of action plans and the presentation to the public as required by Directive 2002/49/EC (Environmental Noise Directive).[1]

To describe GIS is not easy, as there are as many definitions as users of such systems. GIS could be defined, in this context, as an information system capable of presenting and solving geographical problems.[2] GIS instruments can analyse the "pictures" of noise levels, both in space and in time, illustrating the effects of measures to reduce noise exposure or the possible effects of different planning policies. So, GIS can be not only a database of information but, in combination with expert knowledge, technical procedures and mathematical algorithms, it is also part of a wider system known as a decision support system or DSS.[3,4]

Although most noise prediction software offer many GIS features, usually it can be useful when managing noise propagation by dedicated programs, and spatial analysis and presentation by a GIS one. Once noise maps have been made, GIS allows evaluating relevant data to be collected for the EU Commission for the number of people exposed to different noise levels.

After the noise mapping procedure, action plans (Figure 10.1) are the next step focused in reducing noise exposure. In order to define hot spots for noise remediation and the temporal (economic) prioritization, it is of paramount importance to have a reliable instrument to analyse noise and

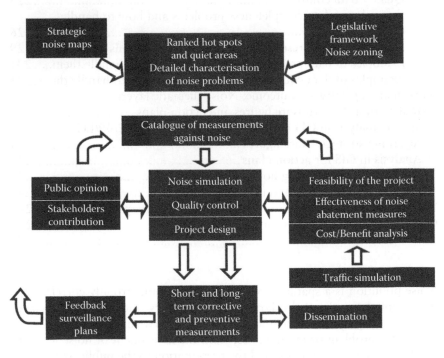

Figure 10.1 Flow diagram representing the processes involved in the design and implementation of action plans in agglomerations.

spatial data, to enlighten the different measures, and to present the effects to stakeholders and the general public. GIS represents, again, a powerful tool.

In defining action plans from noise mapping, both phases can be viewed as a dynamic system that requires a control mechanism over the outcomes that maximises the efficiency and profitability of the measures adopted. Finally, GIS provides a solution for import/export data from/to GIS format as a best option for geographic data exchange. Let's not forget that the adoption of Directive 2002/49/EC generates more responsibilities for European governments than carrying out plans to control environmental noise. One of them is to guarantee the distribution of the relevant information to the public, and the other is pertinent to the data being submitted to the European Commission; both of them are related to the capability of GIS to represent noise maps.

GEOGRAPHIC INFORMATION SYSTEMS (GIS) AND NOISE MAPS

The development of strategic noise maps and subsequent action plans fundamentally comprises many complex processes, which need a multidisciplinary approach. From acoustic experts to civil engineers, urban planners, and GIS specialists, all of them can be involved in one or more steps in these processes. Here, we will focus on describing how GIS can contribute to the development of some of them. But first, we need to understand the connections between GIS and noise mapping, and then set what is inside and outside the interest and scope of this work.

The general structure of all noise prediction software consists of

- a noise source model and
- a propagation model.

To better understand the possibilities of GIS software it has to be stressed that GIS is not only designed to display and build maps. Its applications go beyond mere preparation, presentation, and export of data to noise prediction software. Actually, one of the most important advantages of GIS for its users is the opportunity to develop and program their own tools of spatial analysis. Today's GIS packages provide a wide range of these analysis tools and even links to other programs such as statistical packages. Some software developers have also produced sound propagation tools that work embedded in GIS platform, while the calculation core works independently.

Nowadays, noise prediction software incorporates procedures that automatically generate noise sources to avoid having to create them from scratch. Generally, software packages could include a catalogue of noise sources that help us to find the most suitable one for a particular application. For example, a polyline represents a road. Immediately, the software puts the linear noise source to a relative height above the road surface and models the entire width of the road platform as reflexive. The noise power of the source is estimated indirectly from "easy to gather" traffic data, such as traffic flow, percentage of heavy vehicles, and speed of the fleet. As you can see, the attribute databases for noise mapping purposes form an essential part of work in GIS.

Some guides have been published to help noise consultants manage the input data necessary for noise maps.[5,6] The WG-AEN Good Practice Guide[5] (GPG) provides a set of toolkits to face what to do in case of total or partial lack of data. The IMAGINE document[6] provides a list of input data requirements for noise mapping together with a classification of availability, cost, and benefit for each topic. Based on this classification a checklist has been prepared to divide the level of exigency on the input data in two categories: compulsory and advanced information. Each topic covered by the guide—terrain model, ground impedance, buildings, transport infrastructures, industrial facilities, and population data—contains a discussion of the problem.

This review on GIS and noise mapping and action plans development covers essentially three areas of interest (Figure 10.2):

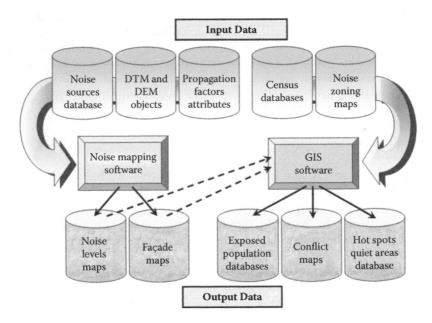

Figure 10.2 Data structure (input–output) in noise prediction software and GIS. GIS is also playing as a central repository platform for the information needed for noise mapping and action plans. Layers should be compiled in a format compatible with noise software.

1. Data collection and preparation.
   - Collecting the spatial input data necessary to run the noise model software.
   - Controlling the quality of the input spatial data.
   - Editing and improving these data, to the construction of a digital representation of propagation scenario, with the level of approximation required to obtain the desired results.
   - Controlling the quality of the output after correction and adjustment. Taking into account the particulars of the requirements and formats of data exchange.
2. Analysis of noise levels distribution.
   - Spatial analysis of noise mapping outcomes in GIS. The output of the noise prediction software could be another spatial variable composed of isolated noise immission points or usually by a network of these points that draw a surface. As noise receivers are actually in every point of the grid cell, the human being as a target of that noise does not appear explicitly in noise models. However, it really does not matter because starting with these noise spatial information, GIS can estimate the figures related to people exposed at their homes and other information concerning action plans on a routine basis.
   - To yield valuable results at this level, the quality of census databases and accuracy in the assignment of people to buildings and dwellings are key issues.
3. Dissemination.
   - Displayed information and public dissemination of noise mapping and action plans' main results. This is achieved showing maps that explain how noise levels are distributed all over the city, figures for the exposed populations, total surface affected by noise, number of sensitive buildings, and so on.

GIS FRAMEWORK

We are going to start with the concepts and theoretical basis of modern GIS models as described in many guides and handbooks.[2,6–10]

Briefly, GIS manages two categories of data required for noise mapping:

1. Geometric data describing the real physical scenario using objects classified in accordance with their topology, such as points (nodes), polylines (arcs), and polygons (areas). These objects are the digitalised representation of the geographic features positioned in the real world. This classification of object types is scale dependent and only embeddable in a coordinate space system of the same or

higher dimension. A vector model or a raster model can represent geometric data. Vector GIS representation builds reality from polylines and polygons as a composition of a finite number of control points defined by its coordinates system and connected by straight lines. Instead, raster GIS representation divides the space in a regular grid of cells and assembles these discrete entities until polylines and polygons have been completed. Instead of scale, maybe it is more appropriate to talk about resolution of raster model, conditioned by the size of the grid cell or positional accuracy of the control points describing objects when applying the concept to vector maps.[8]

2. Attribute data, which defines the properties and description of the objects (as a whole or as a part). Object ID is the simplest type of attribute data that identifies univocally every object. Normally, attribute information is completed with a set of values (quantitative or qualitative) related to different variables that describe the object. Every variable is only defined inside a domain of applicable values. Attribute information is codified in a database in two-dimensional tables (like spreadsheets, matrix in ASCII files, etc.) and attached to locations that remain in the spatial information, like the centroids of a set of polygons or control points of polylines.

The composition of geometric or attribute data logically related can be included as a thematic layer in the same thematic model (Figure 10.3). This thematic model is created with the features of the real world recognised as relevant for noise mapping and estimation of exposed people:

- Area to be mapped.
- Terrain and relief.
  - Equal height ground contour.
  - Embankment and cutting edges.
- Infrastructure.
  - Road infrastructure.
  - Traffic models.
  - Railway infrastructure.
  - Spatial attribute layer with bridges, viaducts, and tunnels.
  - Industrial facilities.
- Noise zoning in accordance with regulations. In case of lack of legislation, the land use areas are classified according to noise limits. Quiet areas inside agglomeration could be included in this layer.
- Land cover extracted from land uses.
- Population geodatabases extracted from the census, polling data, cadastre, and so on.

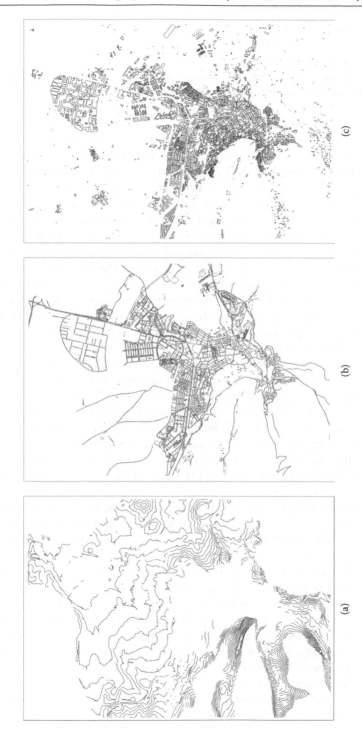

Figure 10.3 Thematic layers ready to feed noise mapping software: (a) Topographic contour lines, (b) building footprints, and (c) road network.

- Buildings and barriers.
  - Residential buildings.
  - Sensitive buildings.
  - Buildings with other uses: commercial, industrial, mixture, and so on.
- Constructions, barriers, and other screening objects.

One of the key characteristics of GIS is that every database is georeferenced, that is, located in space by the coordinate referencing system (CRS). The Universal Transverse Mercator (UTM) coordinate system is the most frequently CRS used in GIS. Under this constraint it is easy to merge layers with new ones. If two features have different CRS in GIS, then one can be converted in the second one that is in use. These thematic layers are, at the same time, the best way to exchange data between noise mapping software and GIS.[8] GIS includes a metadata tool to provide a tracing over the origins of the raw data and modifications suffered during all the stages covering the noise mapping process. The metadata describes the data, the type, and the properties of the data, both of attributes and geospatial databases; this way, you can decide whether these databases are appropriate for the study purposes and comply with the requirements. For example, in a noise mapping context, a simple and important factor to check by metadata is the date the data was obtained.

The relationship between the spatial data structure and attribute data varies from one GIS software package to another. It is not uncommon that attribute data appears embedded into commercial relational database management systems (RDBMS) software attached to every object. These relational DBMS work with multiple tables related by the same object ID and used to be managed by SQL, a query language tool that manages the access to data and filters these data until meeting the desired specifications. DBMS can manage spatial characteristics of topologic objects in the same geodatabase than the attributes.

Unfortunately, the previous definitions do not constitute a standard and the suitability of all kinds of data for noise prediction software must be evaluated considering formal and quality specification of data exchange.[8] A clear case of fault in harmonization is the semantics used in databases, which is not really an integral part of GIS software. We have many examples of attributes not fitting because they are codified in very different ways, like building heights, which, depending on the domain of application, could be expressed in metres, floors, high/middle/low, and so forth. In that sense, INSPIRE[11] is an initiative from the European Union that is looking to share the same standards and protocols in all countries inside the European Union. INSPIRE specifically seeks to guarantee compatibility of data exchange and visualisation.

Public access to information is a crucial issue in the EU noise policy, which is why EIONET is now providing a structure to join all EU noise maps and exposure data in order to provide a harmonized visualisation platform of noise problem at the European level.[12] Various recommended descriptions of how to create standardised maps and metadata—format, projection, extension, type of layout, the type, the order, and the colour of different features—are exactly indicated.

Geometric basis: Digital terrain model, digital elevation model, and triangulated irregular network

The propagation model is built using geometric data that creates a terrain model DTM (digital terrain model) and DEM (digital elevation model),[8] which is defined here as a geometric model that includes the rest of the man-made objects that influence propagation. Buildings (whether residential or industrial facilities), barriers, roads and railway networks, and other sound screening objects are part of DEM. On the other hand, DTM is integrated by elements that describe the shape of the ground surface, such as plateaus, slopes, cliffs, cutting edges, and embankments. In vector models these elements are contour height isolines; raster models use an interpolated surface consisting of a regular grid of height points or may be based on a triangulated irregular network (TIN). TIN is widely used by noise software packages as a preferred media to create the terrain skeleton. TIN control points are connected by edges to form a set of irregular and nonoverlapping triangles used to represent surfaces. TIN allows dense information in complex areas and sparse information in the simplest. The TIN data set includes topological links between points and their neighbouring triangles. Each sample point has an X, Y coordinate and a surface, or Z value. In fact, the generation of TIN was based upon the Delaunay triangulation.

The most familiar objects used in DEM for noise mapping applications are buildings and infrastructure. Buildings and its shapes are defined by its footprints. The footprint is the projection over the XY plane of the polygon intersecting the DTM with building walls (or other objects from the DEM). Using a polygon guarantees that vertical surfaces are included in the propagation schema. Usually, every control point keeps the relative height over DTM. Noise source layers include, as well, a geometric model for the geographical description of the infrastructure itself. The road network, for example, is normally described geometrically as "arc and nodes" in a vector model.

The composition of DTM and DEM could be seen as an oriented surface that can be matched to a 2D topological complex embedded over a 3D coordinate space (UTM X, Y, Z).[8] In other words, the geometric model is defined as a 2.5-dimensional model, which describes a continuous field where the height (z coordinate) is encoded as an attribute attached to control points (defined in X, Y coordinates) of every topological form. Buildings, barriers, and ground look like a continuous 2D surface folded in a 3D space.

Creating a 3D continuous surface in that way has some inconveniences and puts some limitations on the Harmonoise model. These restrictions on the geometric model could separate noise models output from the real outdoor propagation schema.[8] Therefore, software developers introduce in their products some 3D tools to solve some of these limitations. Thus, the thematic model is enriched with the experience of developers until fitting a most appropriate physical scenario for the purpose of noise mapping. And next, a list of constraints and solutions (when possible) are considered:

- The most striking one is the impossibility to get a 2D point with more than one height. That is why special care must be taken with complex solid structures that share vertical surfaces and vertical edges. Hence, overlapping between 3D objects is not allowed. To overcome the problem, these volumes are built by assigning to each object a constant height attribute.[8]
- The noise sources and receivers must be placed over the surface field.
- The boundary of the working area to be mapped is shaped in such a way that every possible emission–receiver path cannot cross over an empty space. Thereby, no holes are tolerated inside this area.
- Topological objects define an orientation. Actually, the continuous field consists of discrete 2D planes, which must be oriented in such a way that no normal vectors are pointed downward. Negative slopes, overhanging, and floating obstacles are not allowed. If we visualise a vertical section along a propagation plane transversally cutting an overhanging or a bridge, what we see in the segment is a floating obstacle. But bridges, viaducts, overhangs, tunnels, and so forth can be considered in some occasions an essential part of the physical scenario. Some solutions have been developed using sound-transparent objects with an attached attribute that adds an insertion loss to the propagation path. These virtual volumes could be used to model scattering areas, like forests and low-density residential areas.
- The infinitely thin courtyards must be avoided.

GIS works before noise mapping calculation

The first steps in a noise mapping project cover a series of works focused on the data management covering the following steps:

- Searching for data. Raw data collection, storage, and organisation.
- Converting raw data in valuable information. Checking the quality of data for noise mapping applications. Quality control and data treatment until corresponding with the project team requirements.
- Checking the final physical model. Quality control of the DTM/DEM final scenario, before noise mapping calculation phase.

Data acquisition

The Good Practice Guide[5] faces obstacles arising in this crucial phase of the project and releases a set of toolkits to cope with missing (or incomplete) geodata. Fortunately, in recent years the availability of digital maps has been constantly increasing, probably due to relatively low cost remote system technologies that provide quality geographic data sets like[13]:

- Photogrammetric aerial survey of 1 m resolution
- Lidar technology providing 3D vector mapping
- Satellite raster photo images

From one end to the other, the worst case to deal with is, possibly, when a printed map has to be scanned, georeferenced, and finally every entity manually digitised.

But more often than not the available geodata is gathered by urban departments of municipalities and then processed and stored with no concern about the acoustic software's specifications. Being spatial data is one of the foundations of successful strategic noise maps, dealing with problems arising in the physical scenario is one of the most resource-demanding phases in the development of this kind of project. In general, the Z coordinate of the DTM model should be defined more accurately than XY, but is less available and the accuracy more difficult and expensive to improve. Nowadays, fortunately, the dispersion pollutant models require more or less the same GIS approximation.

Quality data control

Quality is a concept describing how suitable data is for a specific purpose and if these data are error free. Geographical data quality can be focused on space, time, and thematic dimensions (recognizing these dimensions are not independent variables). Error is defined as the discrepancies between the encoded and actual value (recognizing the uncertainty to measure the real value) of an entity in at least one of these three dimensions. Evidently, data requirements must reach a balance between quality (approximation to benchmark) and other parameters as cost, time of acquisition and availability, size and manageability of files, and that the calculations become computationally burdensome. So, it is necessary to realize when to simplify, when to update, and when to improve data. Some studies have been carried out to understand the model's sensitivity to different data inaccuracies,[5] but the reader should take into account the different expectations when using different noise prediction models as P2P (Point-to-Point), Nord-2000, or NMPB (Harmonoise versus interim methods).

The fact is Harmonoise is more sensitive to geometrical input data than interim models, which do not consider all features affecting propagation and tend to provide unbiased results on average over large areas.[6] Then, simplification of DTM and DEM looks more appropriate for applying to interim models than to P2P.

The Harmonoise method[14] uses a 2.5D geospatial schema composed by objects, and these objects have to be created in such a way that their geometric and attribute characteristics have to be relevant to noise propagation calculation. *It is good for nothing using the best geometric model without any consideration whether the propagation model can exploit these accurate structures.* First, propagation paths are found in a 2D zenith view. These paths can connect the receiver with the noise source by means of a direct ray, by lateral reflections in vertical obstacles, and by lateral diffractions over vertical edges. Finally, every propagation path is defined by its own 2D vertical plane and intersecting the 2D horizontal objects built-in DTM and DEM. The geometric model must be created in such a way that every segment defined in this intersection has information about the impedance. Then, all the propagation effects along these paths have to be calculated with the height attribute provided by the objects that are part of the DTM plus DEM scenario. This time, the total paths are a composition of direct shots together with vertical reflections on the ground and vertical walls and diffractions by horizontal edges. A good definition of control points of DTM and DEM objects, like footprints of buildings and other screening obstacles and contour terrain lines, it is important to incorporate reflections in the propagation schema. But a realistic incorporation of cutting edges (intersection between two surfaces) vertical and horizontal, from DTM and DEM (like discontinuities of terrain describing a breaking line) is also crucial when incorporating diffractions in the propagation schema. Meteorological refraction is not considered here.

Quality control (QC) checks the data set's quality, measuring if it complies with the minimum requirements for a noise mapping project. QC consists of a combination of manual and automatic procedures. Probably, the first step is checking the lineages of geodatabases via metadata. Once the geodatabases are approved, and if some problems are detected, corrections, improvements, and updating of data must be carried out. Finally, a visual inspection of features and their attributes are programmed before approving the suitability for noise mapping uses. For that reason, it is important to know the threshold of sensitivity of Harmonoise method regarding spatial data and their attributes.[6] The functional reliability of the spatial system depends completely on the following categories of quality factors in GIS databases: data completeness, data resolution, data precision, and data consistency. In the following paragraphs, some problems that noise engineers can run into are described and subsequent decisions via data processing are briefly outlined.

Examples of data completeness problems and how to handle them

Data completeness must be understood as a thematic universe of definition covering all aspects of features' suitability. These features must be adequately represented by objects and their attributes. The first constraint for the geometrical model can be synthesized by the following question: Is this object relevant and should it be included in the geometrical model? The data set could be qualified as complete and comprehensive because it includes, inside the thematic layer, all relevant features from the space–time domain, its characteristics, and relationships. The notion of completeness includes entity, formal (semantics), and attributes completeness.

One of the most common examples of data shortage for noise mapping is the road gradient. But this quantity can be estimated directly in GIS by means of DTM or the road "z-coordinate" profile.

Deciding the number of road lanes to be digitised is another issue influencing road model quality. To establish when it is important to distinguish between forward and backward carriages (even between lanes inside carriages) in the creation of strategic noise maps and action plans, two things have to be revised:

1. Source factors, like the distribution of traffic variables over the lanes and carriages.
2. Geometrical factors, like distances and relative positions between carriages and from the carriages to close buildings and barriers.

Some things can be done, for example:

- Checking the number of carriageways on a road and the total width by aerial photograph. Width of road platform is a key issue in the determination of noise levels in the neighbourhood. Dual carriageways are considered as two different line sources when the reservation width is more than 10 m or those carriages differ in height by more than 2 m.
- Taking into account special lanes designed for bus transit, taxi lanes, reversible, and heavy-traffic-only lanes, slow vehicles lanes, maintenance passageways, and entrance/exit lanes.
- Considering the possibility of different traffic figures between the two carriages during day, evening, and night, including the consequences of slopes (uphill/downhill traffic of heavy vehicles).
- In the vicinity of barriers and close buildings, the correct definition of, at least, the forward and backward carriages, becomes a requirement.
- Roads must be split in stretches as a consequence of changes in the attributes per stretch of such magnitude that deal with detectable changes in noise power emissions.

Usually, noise technicians decide to overlay DTM and DEM models from different origins for several reasons: for comparison purposes (to check the fit), to compound the best map selecting the best features and attributes from every model, to fill the total area to be mapped, to introduce accuracy contour lines near roads (embankments and cuttings), to introduce a new infrastructure, to update residential developments, and so forth. All these reasons try to avoid the incompleteness of the data. If one of the layers is satellite photography, the new information can be digitised and then completed by others means, like on-site visits. There are many other practical applications of this method; for instance, to complete the height of buildings, and to confirm the presence of bridges, viaducts, and tunnels, which appear in the official cartography as attributes of different layers. One of the problems when rearranging the thematic layer is the possibility of geometrical and semantic interoperability confusion between different geodatabases, as we will observe later.

The buildings' layer usually suffers its own completeness problems:

- Buildings are represented as its footprint with the information of its height or number of floors in a text label. Sometimes it is possible to prepare macros in GIS that could recognize the figures and transform them into heights of buildings.
- Buildings' total heights is not explicitly supplied, only the number of floors. In noise maps, the number of storeys is more important for the estimation of the exposed people than the height of buildings.[5] Special care must be taken with the height of buildings in front of major infrastructures or when fine analysis for action plans is undertaken. The real height of the commercial ground floor and the shape of the roof are considered for a better estimation. Both data, when generated from scratch, very often demand a lot of time and cost from the engineering; for example, measuring heights and taking the number of storeys on site or from development plans or aerial photographs.
- Courtyards of buildings have to be included only if the estimation of quiet façades is covered by the study. One should be careful with the construction of buildings with courtyards in order to prevent the overlap of receptors.
- Some buildings (e.g., garages, garden huts) could be excluded from noise maps intended for strategic purposes.[15] However, depending on the situation these buildings should be included in DEM as barriers, considering their implication in propagation.

Once the geometric model for every thematic layer is completed, it is necessary to add all relevant attributes intended to run the noise prediction model. The toolkits included in the GPG[5] describe how to get these data sets, even *ex nihilo*. The complexity of the tool depends on the starting data available, the cost and time we can assume, and the accuracy we need.

To put some of the tools into practise, GIS could play a useful role, even though it only serves as a simple storage platform. Some examples in which GIS can help to achieve the completeness of the attribute of different layers are:

- Ground impedance can be established easily, but roughly, from land usage maps using a twofold distinction between hard and soft ground.
- Identification of the boundaries of the area to be mapped. It could be done using distance criteria to the significant sources or maybe population density criteria. In any case, GIS will identify these areas in an easy manner.

Examples of data resolution problems and solutions

The concept of data resolution was covered previously and defines the scale or, in other words, the spatial positional accuracy answering the question: Do the control points define the object correctly positioned in the space and do they accurately match the requirements for strategic noise mapping? Regarding attributes, these can be termed as the discrepancy between the actual attributes value and coded attributes value (see Table 10.1).

In previous examples, when a noise engineer decided to overlay models from different administrations and tried to complete the information to input noise map software, some problems emerged. The first one was the geometrical interoperability found when geometric models do not match each other spatially and cause inaccuracy in position, for example, geographical errors in XY road alignment, road passing through buildings, and so forth. These inaccuracies cause errors in noise mapping calculations and have to

Table 10.1 P2P Positional Accuracy Requirements for Geometric Model That Refer to Control Points

Position Accuracy for Geometrical Items	Level of Detail		
	High	Medium	Low
Horizontal position	1 m	2 m	5 m
Vertical position in general	0.5 m	1 m	2.5 m
Height of obstacles near low sources	0.25 m	0.5 m	1 m
Height of obstacles above ground	10%	20%	50%
Typical accuracy	<0.5 dB	<1 dB	<3 dB

Source: IMAGINE, 2007, Specifications for GIS-Noise Databases, Work Package 1, Deliverable 4, Document IMA10-TR250506-CSTB05, http://www.imagine-project.org/.

be homogenized. If noise sources are not well located (horizontal definition); technicians have to decide what to do to break the ambiguity. Exporting the best attribute model and attaching it to the best geographic model is an option. Possible causes of that mismatch is found in georeferenced errors induced by the use of different projection systems, different datum, maps collected with different spatial resolution (scale) and within different coordinates systems, or maybe the problem could be in the UTM zone.

Semantic interoperability takes place when features from the same thematic layer are described differently in different data models. Consequently, the integration of geometric and thematic geodata causes errors and the geodata has to be homogenized.

Road and railway infrastructure are not geometrically well defined with regard to the surrounding DTM and DEM. Establishing the noise line source elevation regarding the closed terrain is one of the most important issues on the propagation model results, especially under P2P specifications. The presence and accurate contour heights definition of viaducts, tunnels, embankments, cuttings, and trenches as well as a proper determination of the road platform width can affect the expected noise level in the far field receivers. Barriers, buildings, and screening features close to the source line must be accurately drawn. By default and without precise information available, it is recommended to elevate the road 0.25 to 0.5 m, and railways 1 to 2 m outside agglomerations.[6] It is preferable to overestimate the noise space range affected.[6]

The DTM and DEM close to buildings are not precisely defined. Secondary but important too is the correct characterisation of ground relief and other screening objects close to the receiver points or buildings, especially in the presence of cliffs.

The road layout must fit well with the terrain model (vertical definition). It is particularly difficult to obtain an appropriate layout of road profile that fits well with the surface of the DTM. In fact, it is quite common that a part of the polylines that represent highways are underground or floating. Obviously, this causes problems in the estimation of noise levels and therefore it is necessary to pay attention to adjusting the heights of both geometric layers.

The relief model is a vector or TIN model restored from a raster model. In these cases the grid height points of the terrain must be transformed into terrain contour lines. There are advantages to doing so. First, the size of the files decreases, especially when almost flat terrain is oversampled by the raster model that manages too much redundant land information. Another advantage is that vector and TIN models provide better diffraction predictions. But there are some disadvantages. For example, the slopes are smoothed and important relief information is lost. This information is based on the correct definition of contour heights of embankments, cuttings, trenches, platforms, and cliffs close to the roads and railways noise sources and close to buildings.

Examples of data precision problems and how to handle them

Data precision answers the following question: Is the shape of the object well defined and with enough sampling control points for the specific application suitability? Precision can be termed as the degree of details displayed on a uniform space. Generalisation is the process of simplifying in order to produce manageable maps. It includes deletion and merging of entities, decrease in details of objects, smoothing, thinning, and aggregation of classes. When there are no significant improvements using too many nodes in the construction of polylines, polygons, and fields, it is necessary to simplify the propagation scenario. Simplified models may be desirable in certain situations when examining noise spatial patterns at regional level. A clear example is a flat ground accurately defined in a fine raster model.

Buildings are one of the most prominent targets in simplified scenarios designed with noise mapping intention.

- A building has a complex roof with different heights. Put only one height, while keeping the total volume of the building.
- A group of buildings in the neighbourhood with similar heights. Merge them until forming a block with the lower height of the buildings.
- Small buildings and other DEM objects with a footprint area less than 10 m² can be ignored within the purpose of strategic noise maps.
- Must check if there could be an overlap between buildings or parts thereof.
- Traffic macrosimulation models can be employed in the development of strategic noise maps. But sometimes the geometric baseline of that traffic model is not accurate and precise enough for noise mapping purposes. In particular, the Z coordinate is not taken into account by traffic models. GIS can act as an intermediate platform to merge the attributes (traffic flow characteristics) of the traffic model with the more georealistic road network DEM.[16] (See Figure 10.4.)

Examples of data consistency problems and how to handle them

Data consistency can be termed as the absence of conflicts and contradictions in a particular database. It describes the compatibility of the data sets. To ensure compatibility for the exchanging of information, data must be captured, stored, and edited following the same processes.

Logical consistency measures the interaction between the values of two or more functionally related thematic attributes.

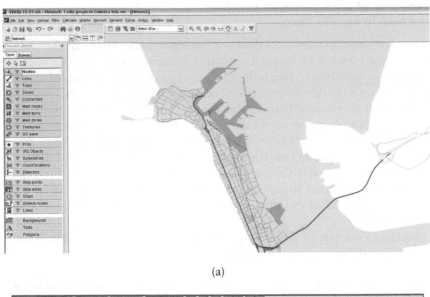

(a)

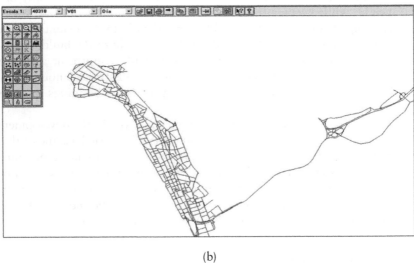

(b)

Figure 10.4 GIS performs as an intermediate platform to export and import roads between VISUM® and CadnaA®. (From VISUM and CadnaA. With permission.)

Physical consistency measures the topological correctness (in accordance with certain topological rules) and geographic characteristics of the geospatial database. A wide variety of examples can be extracted: only one point may exist at a given location, network lines must be intersected at nodes, and polygons are bordered by perimeter polylines. Other spatial inconsistencies can also be identified through redundancies in spatial attributes.

- Errors in feature classification codes hamper the automatic identification and edition of objects.
- A vector terrain contour map has been split in different segments, which must be merged.
- Difficult identification of the object that represents the target feature inside spaghetti polylines. For example, the polyline representing the road or the carriage, in a soup of sidewalks, gas pipelines, telephonic network, and so on that have to be deleted. Another example could be that terrain level lines, crossing one another, must be redrawn manually.
- Building contours appear in the map as a polyline, not a polygon; therefore, the model is unable to recognize it as a building. So, buildings and other DEM features drawn like polylines must be closed (transformed into polygons). A GIS has the ability to construct the topology of disjoint polylines. The nodes are tested in the proximity to another and connected together.
- Checking for inconsistencies and discontinuities in the final scenario. GIS provides tools for 3D visualisation (or XZ and YZ perspectives) in order to check the shape of the continuous field and outside objects. We can check all kinds of geometric integrity. Duplicate objects, like infrastructure digitised as forward/backward double lines or buildings considered twice, could overestimate the noise levels and the population exposed. The rule of thumb is that errors automatically detected can also be automatically corrected. Thus, many faults of topology can be corrected, such as unended lines and slivers; the rest might be visually identified as spikes and missing features, for example.

NOISE MAPPING SOFTWARE OUTCOME; NOISE THEMATIC LAYERS

The first results from the development process of noise mapping are the noise level maps saved in layers related to roads, railways, aircrafts, industrial facilities, or a composition of them. Noise indexes, such as L_{den} or L_{night}, are attached to a virtual receiver that characterises nonspatial information linked to a place. Noise level maps layers are composed by a set of these receivers forming a grid cell that describes spatial distribution of noise. Geometrically speaking, this network draws a surface enveloping the DTM model at the same height, usually at 4 m high with the exception of building roofs and the rest of DEM objects, if they are higher. In some cases, the noise level at 4 m is not representative of the actual noise exposure of the receptors and evaluations are necessary at other heights that will allow developing appropriate action plans.

- The maps of road infrastructure network inside agglomerations must also be split regarding to its owners and density flow of vehicles. In doing so, the levels of responsibility for noise pollution can be determined conveniently. Accordingly, the classification could be arranged in: national and regional major roads, in conformity with directive 2002/49/EC criteria;[1] the primary urban network, whose main function is to carry intraurban traffic and exceeds the flow vehicles requirements from major roads; and the rest of the network, including low-density traffic roads.
- The addition of layers of noise coming from different infrastructures to express the cumulative level of noise is an easy task when every layer is constructed with the points belonging to the grid with the same step located exactly in the same position xyz.
- Noticeable errors should be expected in noise level maps carried out with a low-density grid when complex DTM and DEM are contemplated. Moreover, in open country and in smooth relief landscape, or in aircraft noise maps, a high-definition picture of noise distribution in space is not justified at the expense of increasing the computational burden. Several recommendations point out the need to place the normal grid density from 30×30 metres in open country to 10×10 m in urban areas.[5] However, exceptional situations require different treatments, and aircraft noise maps can be developed with a 100×100 m grid cell, meanwhile narrow streets in city centres may require a detailed 2×2 m analysis.

The second outcome of the noise mapping process is the façade noise map. This information is fundamental to calculate the people exposed to noise. The façade noise map is composed of a belt of immission points surrounding the façade of the buildings usually at 4 m height relative to DTM. The peculiarity of the configuration of these noise receivers is that they have to be calculated without introducing into the propagation model any reflection coming from the building itself. The estimation of the most exposed façade of buildings based on noise at 4 m high has some inconveniences, which should be considered in action plans.

- Single-floor houses usually lower than 4 m high pass unnoticed in the estimation of exposure data.
- The noise assigned to the most exposed façade of large multistorey buildings, shielded by other lower buildings lined up in front of the major infrastructures, is generally underestimated.

With action plans, a finer approach consisting of the calculation of vertical surface of noise that envelops all building façades is best. This surface is constructed using a set of receivers shaped as belts that encircle floor by floor to cover the perimeter of the building (e.g., at 2 m, 5 m, 8 m, 11 m, etc.). With the aid of GIS, it is possible to define the distribution of immission levels (maximum, minimum, arithmetic mean, power mean, etc.) linked to every building. With this procedure, it is quite simple to define the quiet façades and the most exposed ones. The next step involves GIS assignation of noise levels with people living in these buildings, by means of different procedures.

SPATIAL ANALYSIS FOR EXPOSURE FIGURES AND ACTION PLANS

GIS technology provides an appropriate environment for spatial analysis and identifies potential spatial relationships between different phenomena displayed on maps. These geographic tools enable technicians to find a deeper understanding of noise problems inside the city and choose the best among all possible solutions. The objective of this part of the text is to give an overview of the most helpful data analysis tools available for noise mapping and action plans and to provide some examples of application.

Spatial analysis in GIS to comply with Directive 2002/49/EC regarding strategic noise figures

As we saw in previous sections, the propagation model's output yields noise surfaces in a raster format. But this is only a part of the job. To fulfil the requirements of Directive 2002/49/EC some additional information has to be sent to the commission. The requested data can be used to understand the extent and consequences of noise spreading over territory and affecting people. These figures are calculated from the two types of noise maps.

- From the façade noise maps. We need to calculate the distribution of the people exposed to different noise level bands in relation with noise at the façade of their buildings. For example, in the case of L_{den} the interest focuses on the total number of inhabitants living inside the following bands: 55–59, 60–64, 65–69, 70–74, >75 dB. These figures have to be estimated from most exposed façades to every type of noise source (separately and then totally).
- From the noise level map. The data asked by the commission is related to the areas that exceed certain noise levels. For example, if we focus on L_{den} the figures are calculated from areas covered by the following contours: 55, 65, and 75 dB. Specifically, these figures refer to the
  - Total area surfaces outside agglomerations.
  - Total number of inhabitants affected in their living areas.
- Total number of schools and hospitals located inside noise contours.

Before implementing spatial functions to extract these data, the raster map (grid cell) of noise levels must be converted into a contour vector map. In order to prevent some problems in the vectorization process, a previous step is needed. Interpolation is a process of resampling a 2D point matrix and then reassembling it in a continuous surface. Interpolating the noise network is a delicate operation because the new noise level points have not been estimated taking into account the acoustic propagation laws.[17] But interpolation is a basic step prior to the definition of smooth and error-free isophones. It appears that the TIN interpolation method shows a low RMSE (root mean square error), meaning this method is more accurate than others for building 2.5D noise models.[17] And obviously, the lower the density of the grid, the poorer its accuracy.

Under the term geoprocessing operations there is a group of tools covering a wide range of spatial data operations under the GIS environment. Overlay operations consist of the extraction of new information in relation between different layers. Suppose we have a collection of polylines of L_{den} 55, 60, 65, 70, and 75 dB running parallel to a major road in the noise level map layer; and a polygon layer showing residential buildings with figures of population assigned to every building. The composed overlay of both layers gives a new perspective that indicates the total number of buildings and then a total population inside the 65 dB noise isophone. Evidently, the total surface and number of sensitive buildings can be estimated as well.

But to do it we need another GIS tool, probably the most used geoprocessing tool. We are referring to a built-in command data search called query. This GIS tool implies, first, a spatial search and a retrieval of target features matching with certain attribute or geometric[13] constraints and, as a second criterion, these features must be located in a specific searching area. The reader has probably already guessed this tool's potential applied to noise mapping and action planning. Query can filter the noise status of façades exposed to L_{night} bands between 60 and 65 dB and can provide the number of people exposed to every noise band. There are also nonspatial queries that apply only to databases. A sequence of queries can be coupled in such a manner that identifies entities that satisfy the combination of a set of spatial and nonspatial criteria.

Flexibility in rearranging information is achieved by the possibility of integrating different types of independent data using simple spatial relations. This way, GIS can be used to try to overcome the data's lack of completeness regarding building usage, in general, and residential buildings, in particular. One of the principal problems in this field includes the identification of the residential and mixed uses buildings from the polygons mixture. In European cities, a mixture of residential and shop/office spaces in the same building is common. These buildings can be recognised from the rest of polygons in the building layer through the use of the postal delivery point geodataset.[15] The two layers' overlay clarifies the use of buildings

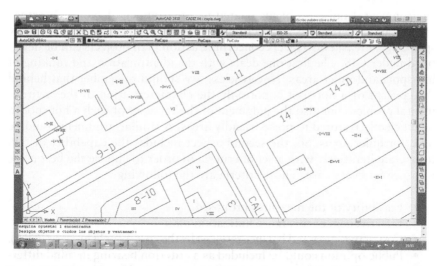

Figure 10.5 The address database georeferenced at the entrance of buildings is overlaying with the building layer. It helps us to distinguish each particular building from the block. Depending on the information available at the address database, this process can provide information about usage of buildings. The number of floors in buildings is codified in Roman numbers.

considering the address point nearest to them (Figure 10.5). We will see later that building usage is a decisive variable to devise the noise zoning layer.

Another problem in relation to receivers is the lack of information about the number and distribution of dwellings inside those residential buildings and the actual population that lives in each dwelling unit. In most occasions, municipalities manage census data covering population per district as the best resolution available. So, it is necessary to use statistical analysis to assign population to every building.[5,6] For example, one of the most exploited GIS procedures is to first estimate the total residential surface area of a district and then calculate the average square metres per resident. This way, we can easily estimate people living in detached and multistorey buildings.

This flexibility that characterises GIS is also essential for the implementation of informative maps for the general public and stakeholders.

Analysis in GIS for action plans

If the first steps to comply with the Environmental Noise Directive have consisted on the diagnosis of the current noise environmental situation, the second part corresponds to the design and development of action plans against noise. There is no standard procedure on how to handle environmental noise problems related to Environmental Noise Directive, even though, there are as many criteria as studies undertaken.[18] But roughly speaking, at least two aspects concerning action plan projects can be developed under GIS. The first one deals with the identification and ranking of hot spots and quiet areas. Nevertheless, we should remember that behind the concept of a hot spot is not only the characterisation of the environmental problem but also the judgment of the capacity of such a hot spot to be efficiently restored to noise-friendly areas. A similar argument could be used for quiet areas. So, the second aspect includes the capability of GIS to operate iteratively with a set of criteria in order to choose the best noise measure for every case. Those criteria are the following:

- Feasibility of the measures proposed.
- Effectiveness of this noise mitigation plan.
- Cost of measures and balance revealed by the cost–benefit analysis (CBA).
- Public opinion could be included as a criterion bearing in mind different approaches, for example, GIS can estimate the probable resistance of the community against certain measures (barriers cutting the landscapes). GIS can also look for spatial correlations between hot spots and the database of complaints against noise in order to prioritize intervention in some areas.

However, the appropriate way to guarantee the success of the whole process is not straightforward. Among other things, a DSS (decision support system) supported by GIS can be created to help environmental consultants to estimate the best way to achieve these tasks successfully.[3,4,13,18–20] In this review we will focus on a set of GIS tools to assist in the identification and classification of hot spots and quiet areas on the basis of relevant data.

Modelling is another GIS function that manages a complex of information kept on different thematic layers in order to elaborate a model.[2,13] This model includes overlay operations using Boolean logic to perform a set of queries whose outcomes are the optimal locations complying with multicriteria requirements. The queries could include statistical analysis and customized equations. The model will finally be defined by a flow diagram that can be implemented by scripts. The algorithm is a compromise of expert knowledge and stakeholder specifications codified in rules that guide the reasoning via decision trees comprising an IF–THEN–ELSE loops. The main Boolean operators are AND, OR, NOT and XOR (OR exclusive). These logic operators can be linked with some geoprocessing functions; for example, OR is equal to the union function, AND is equal to the intersect function, and the rest of the logic combinations between two layers can be performed using overlay operation.

With the same geoprocessing operation tools—overlay and query, which we used previously—we can design a simple procedure to identify quiet areas in agglomerations. We need to create a new layer with urban spaces that could be designated as quiet areas. It could be carried out by over-laying the noise contour of 55 dBs L_{day} (just the minimum level displayed on strategic level noise maps) with the layer of polygons that compound the catalogue of candidates for quiet areas inside agglomeration.[21] Since one objective of the Environmental Noise Directive is preserving environ-mental noise quality where it is good; for example, all the surface of the parks affected by noise levels less than 55 dB could be considered as can-didates for quiet spaces inside the city. GIS can assist even the final choice of consolidated quiet areas as well. We select several decision criteria that illustrate this claim, for example, area dimension, distance to major roads, density of population living and working around the area, and the distance to the nearest candidates of quiet areas.

To carry out this multicriteria model for decision making about quiet areas (Figure 10.6), it is necessary to put into action new GIS tools. Buffering is

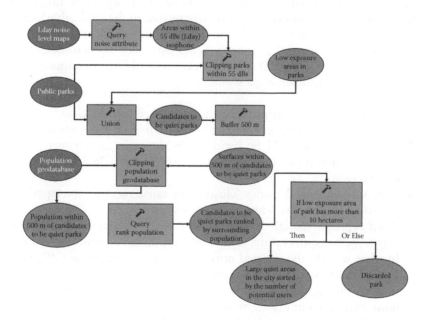

Figure 10.6 A flowchart designed to find the city's public parks that can be identified as quiet areas. This simplified ArcGIS® model represents a geoprocessing workflow covering different kinds of GIS tasks. In this case, the model con-sists of only three criteria (noise level, number of possible users of parks, and size of low-exposure area within park) for an optimal identification of quiet areas. The oval symbols represent the data; the blue ones are for input data and the green for output data. The orange rectangles represent the GIS tools and other operations. All symbols are linked by arrows showing the sequence of processes involved in the analysis of quiet areas.

the operation that enables the analysis of spatial and attributes data within a radius over a feature. As a tool that operates with distances, buffering builds a new virtual polygon around the topological object representing the feature. This operation yields a new layer that allows reexamining the situation inside the buffer polygon and undertaking an overlay analysis with other thematic layers. GIS software packages also include a variety of geoprocessing functions, which reorganise and reclassify the data sets until new practical information is emphasized. Clipping reduces the total area covered by a thematic layer to a smaller target for analysis via a cutting polygon. A buffer of one thematic layer can act as a clipping polygon for another layer, allowing the overlay to analyze only that buffer zone. The role of the union function is more or less the opposite of clipping. It remakes two or more different maps into only one that spatially covers the whole, and thematically includes all features with their respective attributes. Starting with the same premises as previous cases, the intersect function builds a new layer with no more than the features falling in the common area shared.

Identifying and ranking hot spots is a little bit tricky. Most tasks programmed to identify and rank hot spots are related to the applicable legislation. But in most countries, legislation only sets the noise limits threshold. This allows for finding different approaches using GIS tools to detect and classify hot spots.[22–30] But all of them have something in common; they have to answer the following questions.

1. What object will be identified as a "noise receiver" by GIS?
2. How is the noise impact going to be quantified on the noise receiver?
3. Which variables could be included in the noise annoyance analysis? How should this information be used to rank the importance of noise receivers?
4. How should high densities (clusters) of noise receivers be (spatially) identified in order to define them as hot spots?
5. What types of spatial operations have to be programmed in GIS to rearrange hot spots taking into account sources responsible for noise?
6. In what way do the selected noise measures influence the GIS procedures for the spatial identification of hot spots? Under what conditions can hot spots (and scattered noise receivers) be joined to obtain noise management macro-areas?
7. Are there other sources of georeferenced information (not necessarily required for strategic noise maps) that can be correlated in GIS to confirm the existence of a hot spot?
8. And finally, if sensitive buildings and areas require special GIS treatment or not.

Questions 1 and 2 have already been answered, but there are some alternatives for the definition and identification of noise receivers that depend on the available data, local legislation requirements, and possibilities of the approach used in GIS.[22] Moreover, ranking annoyance (question 3) tries to overcome doubts about assessing which is the worst-case scenario, for example, a large number of people exposed to 55 dB at night or a few exposed to more than 65 dB at the same hours. The most promising option is to weight people exposed to noise in the database with an equation that quantifies the level of annoyance.[27-33] This operation is easy to implement in GIS. Annoyance depends on noise level but in other variables. Here is a list of some of these variables to be included in the final equation: the nature of noise (noise emitted by traffic, airplanes, and industries do not have the same effects), the shape and distribution of pieces within dwellings and the quality of construction (the presence of quiet façades, good window insulation, bedrooms not placed in most exposed façades, all of these factors reduce the nuisance), the usage of the building, other local factors may cause sleep disturbances (e.g., detected high levels of L_{max} during nighttime). The equation should be contrasted in appropriateness with the result of surveys, information on complaints, and so forth. Finally, what we have is an index that applies to the area or noise score,[33] or maybe a number of "weighted exposed inhabitants" per building or area. A weighted person is equal to a real person when this citizen lives in a building whose most exposed façade is right at the limit for residential areas, without considering other factors mentioned before. To simplify, from now on, when we refer to exposed population, we are really talking about the weighted exposed population. In the case of sensitive buildings (schools, hospitals, etc.) a weight factor can be used to calculate the users "equivalent" (this is just one more of the possible answers to question 8).

Although to assess the noise impact we focus on analysis of receivers, we cannot forget that noise measures must be applied in the noise source or in the propagation channel. The other option (acting on the receivers) means letting the environmental noise remain high, and then insulate buildings. These ideas can help to better understand the replies to questions 4 and 5. If the cost/benefit is a challenge in developing action plans,[8] then density of noise receivers is an important issue and suggests the first definition of a hot spot. GIS can be programmed in many different ways to search clusters of noise receivers. But, a great density of noise receivers, within a hectare of built-up area, does not guarantee that all receivers are actually affected by the same responsible source. Therefore, simply using a GIS proximity analysis between noise receiver and the main road network of the agglomeration can create more consistent hot spots for noise management.

Different measures against environmental noise might require different methods when evaluating the optimal solution for noise mitigation. Therefore, and answering question 6, it is required to choose different GIS strategies to analyse the relationship between source and receiver depending on each chosen measure. As an example, measures relating to traffic management usually have to be executed throughout the entire section of the road (for example, between two main road junctions). Therefore, in reality, more people (macro-area) will benefit from this measure once adopted, than those (hot spots) for which the action plan was initially designed. Another clear example can be found in the differences between GIS search procedures of hot spots, when the measure solution is a barrier or repavement, as we see next in a case study of major roads.

A geodatabase of complaints can be correlated with hot spots in order to confirm (or not) these areas and rearrange the urgency of the intervention. This is just one more of the possible answers to question 7.

Examples of detecting hot spots using GIS

Next, we are going to outline two different methods of identifying hot spots using GIS functions. These two examples do not profess to give a comprehensive view, or a unique or best approach to the problem; they are only two case studies.

IMAGINE a town district affected by a ring highway. The GIS query tool can pinpoint the façade of buildings with more than 50 dB L_{night}. With this specific spatial information extracted from queries, the noise engineer can create a new thematic layer. Let's consider 50 dB as the maximum value judged acceptable by night in residential areas. Due to the surrounding topography, the cross-section shape of the highway, and the relative position between travel lanes and noise receivers located on the buildings' façades, barriers are expected to be the most probable measure against noise. This judgment includes a preliminary analysis of feasibility and effectiveness of project based, *inter alia*, in the experience of noise experts.

But the barrier project must integrate a new decision criterion: the cost-effectiveness analysis. There is a threshold in the relation: number of beneficiaries–cost of the project, below which other measures against noise should be taken into account and can be judged as more attractive. While the number of people benefited by such barriers increases at a fixed cost, the appeal of installing barriers increases at the same time. Therefore, the population density per unit length of road should be the main aspect in the decision process. That is why we can buffer our highway with a fixed width containing the total population exposed (alternatively, the overlay of an isophone is also possible), and then divide the area transversally to the right and to the left of the highway every 100 m longitudinally. Now, using query function we can extract the number of citizens exposed per road segment to the right and to the left and right to obtain a choropleth map (Figure 10.7). These maps provide an easy way to visualise how the density of exposed inhabitants varies from left to right along the road.

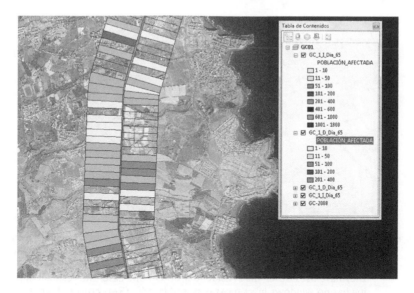

Figure 10.7 A choropleth map created via buffering the highway, highlighting the density of exposed people in residential buildings. **(See colour insert.)**

The described operations highlight the areas where the situation is more conflictive in the sense of number of people affected, not from the level of noise emissions point of view. In doing so, the owner of the infrastructure can plan the order of the intervention in the short and long term related to total investment and the urgency of the environmental situation. Dividing the road into segments of 100 m does not intend to argue that the barriers to reduce exposition levels must be 100 m. Neither should it be implied that each road segment by itself explains the exposure of people associated with areas. It is just a simple way to prioritize those areas of intervention based on the population-exposed density. Much care must be taken with this method when applied to sharp curved segments of roads. Of course, other types of evaluation are possible, for example, on the basis of the identification of homogeneous acoustic treatment (speed, asphalt, traffic, etc.) and processable through GIS.

The second case study takes place in agglomeration. The detection of hot spots will consider factors associated with sleep disturbance. Urban areas are characterised by multiple noise sources that can affect citizens at homes. But now, we are going to consider only the influence of primary and secondary road networks and railways during the night. The first step is to isolate buildings with façades exposed for more than 50 dB by night in a separate layer (Figure 10.8). On this layer will be processed the database of people living there with a ranking algorithm capable of providing an

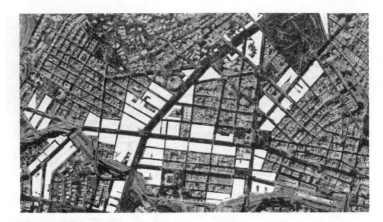

Figure 10.8 In the example of the figure the residential buildings highlighted are those with their most exposed façades affected by more than 65 dB by night. Some buildings in between with the same noise levels have been excluded because they have uses other than residential, and in one case, the residential building has no windows in its most exposed façade. **(See colour insert.)**

indicator that classifies all areas by order and importance of intervention, in a systematic way.

What distinguishes this case study is the spatial linkage of each acoustic problem with its responsible noise source. For this purpose we search a pattern of spatial distribution of exposed residential buildings. Buildings can be considered and then buffered as polygons or as polygon centroids, creating in the latter a circular area drawn from this point. In this case we prefer to buffer the most exposed façades as polylines (Figure 10.9). Basically, the reason for doing so is a greater ease in identifying the responsible line source in case the building block is exposed to multiple concurrent noise sources. The search buffer distance depends strongly on the design of the city analysed: the width of streets, the sidewalks and promenades, the building plants, and so forth. Now we use the dissolve function that deletes the shared boundaries of adjacent polygons that have in common a certain attribute. We refer to common attributes like usages of buildings, the noise control zoning, and noise bands at the façade. If there are no restrictions to consider on the joining, merge will be the right choice.

The merge (or append) function recombines a group of the same thematic layers (same feature class) that are spatially adjacent, in a single layer. The only condition is that the new layer can only contain the topology of the initial layer; the rest disappears. We uniquely introduce the prime road network and the major railway routes, and then a new buffer area is created around them. But we have to take into account that potential noise measures often cannot be implemented over the total length of avenues or streets

Figure 10.9 The polylines defining the most exposed façades of buildings considered in Figure 10.8 have been buffered and then have been dissolved. The next step is overlaying the current layer with the main road network in order to review the extent and importance of the areas where to introduce measures against noise.

in the same manner. Larger nuisance can be expected from traffic lights, roundabouts, junctions, and, in general, any sector of road that involves both accelerations and decelerations. Often, specific noise measures based on traffic management can only be applied to street sections. So, only those sections should overlap on the processed exposed façade layer to merge the buffered zones and identify areas to work on. As usual, what are identified are not the noisiest streets but the streets that affect the most people.

Noise conflict maps and urban planning

Noise control zoning (also known as environmental acoustic zoning or simply noise zoning) is the reclassification of land-use areas in the city, according to the environmental noise quality objectives for each use. Zoning should be considered by municipalities as one of the tools available to reduce the environmental pollution as part of the more ambitious urban environmental planning. Environmental plans are designed and integrated into the complete urban planning under the requirements of the sustainable development model for urban areas.

Responsibility for urban noise zoning lies with urban planning departments, which are responsible for releasing noise-zoning maps showing the boundaries between areas of different assigned noise quality objectives. Common examples of urban noise zoning are residential, touristic, industrial, commercial, leisure, sensitive areas like educational sites, hospitals, and quiet zones; of course everything depends on the national and local legislation. Each noise area has different regulations for different time

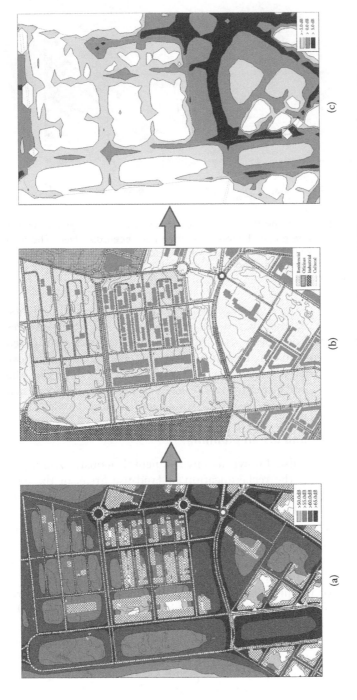

Figure 10.10 Example covering the sequence from a noise level map (a) to a conflict map (c). The second image represents the land-use areas layer. Every noise zone constraint polygon includes its own limit. The overlay of the level map and noise zoning map shows the conflict. In the conflict map, a colour code is used to identify situations in which the presented noise exceeds the limits and also the capacity figures (surface available inside certain isophone) for new residential developments in vacant plots.

periods of day. Noise zoning helps to regulate and manage the current uses and the future developments of areas and to ensure that adjacent land uses are compatible from a noise objectives point of view. Furthermore, the differences in acoustic classification intervals among adjacent areas should not exceed more than one class level. Urban noise zoning includes the acoustic easement areas surrounding countrywide infrastructures, where responsibility lies with the owner of the infrastructure.

Zoning changes would be evaluated using noise environmental impact assessment studies that have to confirm that new land uses are compatible with their surroundings and fit in with the rest of city planning (like a mobility urban traffic plan). At the same time, noise zoning sets the basis to environmentally evaluate the city's growth. Infrastructure projects, new residential developments, hospitals, schools, and so forth could be designed under certain noise requirements and the success of the project assessed by an environmental impact study.

Conflict maps are a decisive tool to disclose the difference between the predicted noise level and the noise limits in every land uses area.[34] Following these principles of action, with time, municipality departments will invest increasingly more funds in the prevention of noise pollution and gradually less in corrective measurements against noise. Taking into account that prevention is always more effective and cheaper than correction the objectives become clear. At the same time, conflict maps are mandatory in the Environmental Noise Directive to inform the general public where national or local limit values are exceeded.

Under a GIS environment, the development of conflict maps is possibly one of the most paradigmatic examples of overlay analysis inside a noise-mapping field. They cross the raster noise level map with virtual polygons describing the acoustic zoning map in order to spatially describe the capacity or the excess of noise in an area with regard to the objective limits. To overlay two different topological layers, the point-in-polygon GIS function is needed. Point-in-polygon connects two layers that have different kinds of topological objects; one of them composed by points (or polylines) and the other necessarily has to be a polygon layer. This operation overlays the two thematic layers in order to assign the attributes of one to the other. Graphically the steps involved in this operation are shown in Figure 10.10.

DATA PRESENTATION FACILITIES OF GIS AND MULTIMEDIA MAPS

GIS easily integrates data and graphs in maps, becoming the adequate platform for the public dissemination of strategic noise maps conclusions. An optimal presentation of the maps can show citizens the noise pollution in their cities and highlight what is relevant for a better understanding of noise problems. GIS manages the visualisation of data, offering a large diversity of point of views. There are many possibilities available for presenting information to the public, for example: classification tables of square kilometres of quiet areas per person per borough, statistics pie graphs showing the distribution of people exposed per district, and special icons or colours symbolizing the level of urgency of intervention in certain delimited areas. Another variable to contend with is time evolution. It is of great interest to the public to compare a noise environmental situation before and after certain measures against noise have been implemented.

There are more GIS resources helping people to understand noise in cities. 3D visual options allow citizens to virtually fly over a city's noise problems. Excess of noise in residential areas can be codified as colour painted on the façade of 3D buildings. These possibilities can generate a mostly accurate view over the town's noise distribution and pattern. Audio/video files could be attached to conflict areas and thus help people understand the actual situation. The use of DBMS is a good strategy for managing these multimedia files. Besides, we probably need to transform all spatial and nonspatial information into a SQL database, because shape files are not compatible with the Web application. And, of course, the Internet and World Wide Web are powerful tools to create feedback with public opinion. A Web site can act as an interface between government, noise experts, and public. For example, these Web sites can include a mailbox of opinions and suggestions, a mechanism for complaints about specific situations that have not been covered on noise maps, and can be a good medium for distributing public opinions polls and surveys, and in turn, serve as a platform to publicise drafts of action plan projects.

GLOSSARY OF TERMS AND ACRONYMS

Algorithm   Any set of computer instructions to solve complex or repetitive problems. GIS provides a command language interface allowing users to edit and submit programs as a text file containing a series of instructions that can be executed as one command.

Attribute data   A set of nongraphic data (quantitative or qualitative) that describes the characteristics of a feature.

Buffer   An area created around a specific feature in the map. This area is assumed to be spatially related to that feature.

Cadastre   Map information concerning the ownership's rights in an area.

CBA (cost–benefit analysis)   Refers to the estimation of the equivalent monetary value of the benefits and costs to the community of projects to establish whether they are worthwhile.

Centroid   A point used to represent a feature and where its attributes are assigned.

Choropleth map   A map showing coloured areas so that each colour symbolises a class (or value) of the mapped phenomenon.

Clustering   A process that helps to identify spatial patterns of distribution of certain phenomena. The mapping of clusters is particularly useful when action is needed based on the location of these clusters.

Control points   The set of points in the real world, whose positions and elevations have been precisely and accurately determined to be used as reference for the construction of map objects.

CRS (coordinate referencing system)   A reference frame to locate the position of a feature on the earth's surface.

DBMS (database management system)   The set of tools that allows GIS to maintain access and manage attribute data files.

DEM (digital elevation model)   The features that contain information about the geometry of the features, except land.

DSS (decision support system)   In general, DSS is a computer-based information system designed to help decision makers, first to compile and manage useful information, and then to identify and predict problems and make the best decisions.

DTM (digital terrain model)   The topography; the features that contain information about the geometry of land.

Environmental sustainability   Development that meets the present environmental needs without compromising the future ones. Environmental sustainability should be at the core of urban policies, for example, developing strategies for a more sustainable transport system in environmental noise.

Easement area   It is an area of noise protection (sometimes known as a buffer zone) that is situated around an infrastructure so that future land uses have to be compatible with the noise received from the infrastructure. The isolines from noise maps delimit where sensitive uses and residential developments are either compatible or banned.

Feature   A single real-world entity that is part of the model and can be distinguished spatially from other entities of the same characteristics.

Geoprocessing operations   GIS automated analysis of geographic data such as overlay analysis and topology processing.

Georeference   The necessary connection between locations and sizes in a digital map and the real world.

GIS (geographic information system)   A mixture of hardware, software, geographic data, and personnel designed to capture, manage, analyse, and display all forms of spatially referenced data.

Hot spot   An area where an environmental conflict is highlighted by the use of tasks programmed in GIS (for example, areas where the noise score is largest).

Land-use map   Human uses of a certain area of land displayed in a map.

Macro-area   The extended area that benefits from action plans due to the characteristics of noise measures taken. The macro-area could contain one or more hot spots and their surroundings. Extending the measures against noise from hot spots to macro-areas is sometimes inevitable, and sometimes it is easy and inexpensive.

Map   An abstract representation of the spatial distribution of geographical features in terms of a recognizable and agreed symbolism.

Metadata   Literally "data about data." This kind of data describes the content and format of the entire data set. Metadata usually includes the date, methodology used in the collection of data, map projection, scale, resolution, accuracy, and reliability of information.

Mobility urban traffic plan   Part of the transportation planning that encompasses the possibility for the traveller to decide when and where to travel, by being aware and making use of information set for optimizing the journey. Transportation planning includes public transportation, prevention of automobile congestion, designing roadways, and other policies that improve the quality of life, such as the promotion of sustainable transport and manage the demand for car use by changing travellers' attitudes and behaviour.

Noise control zoning or noise zoning   The reclassification of land-use areas in the city, according to the environmental noise quality objectives for each use.

Noise environmental impact assessment   Studies that evaluate through simulation the noise impact of infrastructure projects and sensitive developments. It starts with investigating existing noise conditions as a basis for comparison with expected future conditions.

Noise score   An index that attempts to estimate the number of highly annoyed persons in an area. As the term annoyance takes into account the noise level, the number of people exposed to noise must be weighted so that noise problems in different areas may become comparable.

Noise-sensitive receptors   Some land uses are more sensitive to noise than others due to the types of population groups and activities. Examples of noise-sensitive receptors include schools and hospitals.

Object   A digital representation of a real-world entity.

Overlay   Process of spatial analysis by placing spatial elements from one thematic layer with elements from others. This operation is possible when scale, projection, and extent are the same.

Overlay analysis   The process of combining and reanalysing spatial information from at least two layers to obtain new information.

Query   A spatial or logical question asked to DBMS or GIS that leads you to find specific geographical information.

Raster model   A data structure for digital maps that represents the real world using features built with a composition of points regularly distributed in a network.

RDBMS (relational database management system)   A DBMS capable of connecting different databases thanks to a common key attribute (for example, the same object ID).

Sliver   A gap between two lines. It occurs when the end of the line undershoots the map feature to which it is intended to be connected.

Spatial analysis   The process of modelling and interpreting spatial data with the purpose of assessing, predicting, and understanding spatial phenomena. There are four types of spatial analysis: topological overlay and contiguity analysis, surface analysis, linear analysis, and raster analysis.

Spike   Occurs when the end of the line overshoots the map feature to which it is intended to be connected.

SQL (Structured Query Language)   A standard language interface to RDBMS. A language employed by database users to retrieve, modify, add, or delete data.

Thematic layer   A digital map designed to show a specific theme.

TIN (triangulated irregular network)   A vector topological data structure for digital maps that represents the DTM using a sheet of continuous connected triangular facets, usually based on a Delaunay triangulation of irregularly spaced nodes.

Topology   In a vector model, topology is the description of the spatial relationship between geographic features in the same layer. These relations are used for spatial operations that exploit adjacency, connectivity, or inclusion between these features.

Urban environmental plan   Part of public policy that integrates land-use planning, transportation planning, and environmental assessment to balance the needs of the people who live in the area with the environmental needs.

UTM (Universal Transverse Mercator)   A standardised rectangle coordinate system. A transverse Mercator projection uses the same system of projection that Mercator uses, but with the map projection centred along a meridian to provide low distortion within a zone around the central meridian.

Vector model   A data structure for digital maps that represents the real world using features built with a composition of points, polylines, and polygons.

REFERENCES

1. European Commission. Directive 2002/49/EC of the European parliament and of the council of 25 June 2002 relating to the assessment and management of environmental noise, http://ec.europa.eu/environment/noise/directive.htm.
2. Paul A. Longley et al. (2003). *Geographic Information System and Science.* Wiley & Sons, Chichester, England.

3. H. Polinder et al. (2008). QCITY Quiet City Transport DELIVERABLE 6.3, Decision support tool, http://www.qcity.org/.
4. H. Heich. (2008). SILENCE (Sustainable Development Global Change and Ecosystems) WP I.4, Upgrade of decision support system, http://www.silence-ip.org/.
5. European Commission Working Group Assessment of Exposure to Noise. (2006). Good practice guide for strategic noise mapping and the production of associated data on noise exposure, version 2, Technical report, http://ec.europa.eu/environment/noise/.
6. IMAGINE. (2007). Good practice and guidelines for strategic noise maps, Work Package 1, Deliverable 8. Document IMA01-TR22112006-ARPAT12. http://www.imagine-project.org/.
7. Peter A. Burrough and Rachael A. McDonnell. (2000). *Principles of Geographical Information Systems*. Oxford University Press, Oxford, UK.
8. IMAGINE. (2007). Specifications for GIS-Noise databases, Work Package 1, Deliverable 4. Document IMA10-TR250506-CSTB05. http://www.imagine-project.org/.
9. The OpenGIS®. (2009). Abstract specifications. Topic 5: Features. Document 08-126. Open Geospatial Consortium. http://www.opengeospatial.org/standards/as.
10. The OpenGIS®. (2006). Abstract specifications. Topic 6: Schema for coverage geometry and functions. Document 07-011. Open Geospatial Consortium. http://www.opengeospatial.org/standards/as.
11. European Commission. (2007). Directive 2007/2/EC of the European Parliament and of the Council of 14 March 2007 establishing an Infrastructure for Spatial Information in the European Community (INSPIRE), http://www.inspire-geoportal.eu/.
12. http://noise.eionet.europa.eu/index.html.
13. Julianna Mantay and John Ziegler. (2006). *GIS for the Urban Environment*. ESRI Press, Redlands, California.
14. HARMONOISE. (2004). Description of the reference model report, HAR29TR-041118-TNO10.
15. Scottish Executive Environment Group. (2005). Facilitation of strategic noise mapping for the Environmental Noise Directive 2002/49/EC implementation, Final report, http://www.scotland.gov.uk/Publications/2005/11/15161923/19256.
16. Jose L. Cueto et al. (2008). Implicaciones del uso de Modelos de Tráfico en el desarrollo de Mapas de Ruido. Tecniacustica, Coimbra, Portugal.
17. Vinaykumar Kurakula. (2007). A GIS-based approach for 3D noise modelling using 3D city models, Master's thesis, University of Warsaw, http://www.gem-msc.org/Academic%20Output/Kurakula%20Vinay.pdf.
18. Henk de Kluijver and Jantien Stoterb. (2003). Noise mapping and GIS: Optimising quality and efficiency of noise effect studies, *Computers, Environment and Urban Systems* 27(1), 85–102.
19. J. Borst et al. (2008). Decision support system for action plans in the framework of European noise directive. Euronoise, Paris.
20. Jose L. Cueto et al. (2010). Decision-making tools for action plans based on GIS: A case study of a Spanish agglomeration. Internoise, Lisbon.

21. Department for Environment, Food and Rural Affairs (DEFRA). (2006). Research into quiet areas: Recommendations for identification. London. http://www.defra.gov.uk.
22. Daniel Naish. (2010). A method of developing regional road traffic noise management strategies. *Applied Acoustics* 71(7), 640–652.
23. Sergio Luzzi and Raffaella Bellomini. (2009). Source receivers distance algorithms and soundscapes based methods for hotspot and quiet areas in the strategic action plan of Florence. Euronoise, Edinburgh.
24. G. Licitra, P. Gallo, E. Rossi, and G. Brambilla. (2011). A novel method for priority indices determination in Pisa action plan. *Applied Acoustics* 72(8), 505–510.
25. Elena Ascari, Claudia Chiari, Paolo Gallo, Gaetano Licitra, and Diego Palazzuoli. (2010). Confronto tra metodi per l'individuazione delle aree acusticamente critiche nel Comune di Pisa. Proceedings of the 37th Convegno Nazionale dell'Associazione Italiana di Acustica, Siracusa, Italy.
26. David Palmer et al. (2009). END noise action plans: Prioritisation matrix, CNMAs and GIS. Euronoise, Edinburgh.
27. M. Petz. (2008). Action planning procedures and realized action plans of municipalities and cities—Results from the implementation of END. Euronoise, Paris.
28. G. Licitra, C. Chiari, E. Ascari, and D. Palazzuoli. (2010). Neighborhood quiet area definition in the implementation of European Directive 49/2002. Proceedings of International Symposium on Sustainability in Acoustics, Auckland.
29. Wolfgang Probst. (2007). QCITY—A concept for noise mapping, ranking, hot spot detection and action planning. 19th International Congress on Acoustics, Madrid.
30. A. Stenman et al. (2008). NERS-analysis extended to include the existence of neighbouring quiet areas. Euronoise, Paris.
31. WHO. (2009). Night noise guidelines for Europe. World Health Organization, http://www.euro.who.int/Noise.
32. H.M.E. Miedema and H. Vos. (1998). Exposure-response relationships for transportation noise. *Journal of the Acoustical Society of America*, 104(6), 3432–3445.
33. Wolfgang Probst. (2006). Noise perception and scoring of noise exposure. International Congress of Sound and Vibration. Vienna.
34. P. Gallo P., G. Licitra, C. Chiari, A. Panicucci, and F. Balsini. (2009). Il rumore delle infrastrutture di trasporto in ambito urbano: necessità di una mappa generale dei limiti finalizzata alla definizione delle effettive aree di intervento e all'assegnazione delle competenze per il risanamento. Proceedings of the workshop "Controllo ambientale degli agenti fisici: nuove prospettive e problematiche emergent," Vercelli, Italy.

Chapter 11

Maps and geographic information systems in noise management

D. Manvell

CONTENTS

Environmental noise management can be described as the following tasks
for a particular area and for a range of noise sources[1]:

- Noise mapping—Generating an overview of noise exposure over a
  defined geographical area, typically as a "snapshot" of the existing
  situation.
- Noise planning and management—Ensuring that noise levels are
  within limits or at target values as defined by noise policy or legislation, typically done through generating an overview of noise exposure
  over a smaller or larger defined geographical area as a snapshot of the
  existing and of one or more potential future situations.
- Reporting and information—Reporting differs slightly from information. Reporting is through official channels to management, clients,
  and politicians; information is typically for the public and for other
  stakeholders and is increasingly being based on Web technology.
- Noise complaints—Dealing with complaints about noise from the
  local population.
- Noise policing—Enforcing the noise policy and legislation through
  active policing of compliance by spot checks, monitoring, or calculations.
- Noise abatement—Reducing noise levels to below limits or target values as defined by noise policy or legislation.

255

- Data handling—Management, archiving, and retrieval of data covering all aspects of management including reports, input data, measurement and calculation results, and noise exposure.

Noise mapping and geographic information systems (GIS) are very useful in the management of noise problems through the geographical registration, presentation, and comparison of different activities, thus strengthening and easing environmental noise management. For example:

- Noise mapping—Here, in addition to registration, presentation, and comparison, GIS also provides a natural and often the primary source of input data and a way to share data and results among different professionals such as acousticians, land-use planners, and traffic management.
- Noise planning and management—Whether in holistic large-scale environmental noise management or smaller scale environmental impact assessments, whether for dedicated noise analyses or as part of a more widespread environmental or socioeconomic assessment, or whether looking at source-specific issues or multimodal analyses where different forms of transport are analysed, noise mapping is critical, and GIS indispensible for the sharing of information, over and above the more general benefits. One example is the Silence project in the Netherlands,[2] where the Dutch Ministry of Transport manages the national road network (Figure 11.1). Silence aims to improve the way in which government officials throughout the country produce and use noise mapping for policy-related, strategic, and trend studies. Silence provides a means of standardisation by enhancing input data management, software environment, calculation standard, and noise planning/management tools. It introduces a new set of standard ways to estimate noise levels, which is suitable both on a national and on a regional scale.
- Reporting and information—Noise maps are, when used wisely, invaluable in helping to explain the noise situation in an area, whether it be within an organisation and its partners or as information to the general public. In Europe, different approaches have been taken in different countries. The European Environment Agency's EIONET graphically presents strategic noise mapping results with the reported population noise exposure statistics.[5] This refers to national examples of public noise map information including that in the United Kingdom[3] and in Denmark.[4] One of the most advanced and impressive approaches to managing is the one taken by the state of Nordrhein-Westfalen in Germany, which uses an integrated database–GIS–calculation–web solution.[6–8] One of the earliest projects to visually present the noise levels in a very realistic 3D model was done

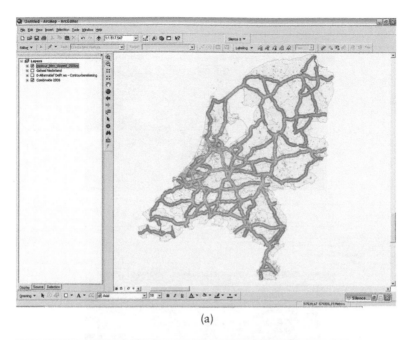

(a)

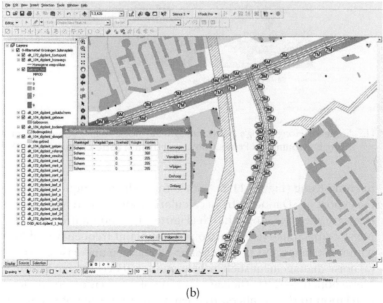

(b)

Figure 11.1 The Silence project in the Netherlands. Silence is a large-scale noise management system for the standardisation of noise mapping, noise assessment,
and development of noise policy and action planning around highways.

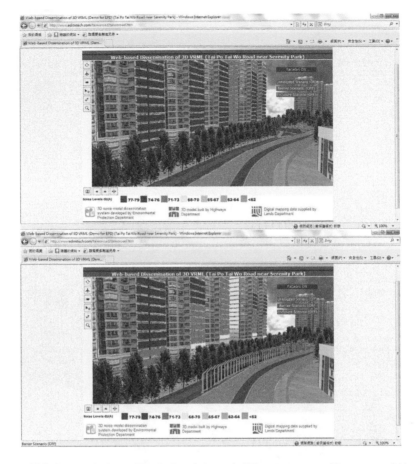

Figure 11.2 Some of Hong Kong's breakthrough 3D noise maps. (Courtesy of Hong Kong Environmental Protection Department.) **(See colour insert.)**

in Hong Kong (Figure 11.2).[9] Since then, the technique has spread to the export of 3D models from calculation software to widely available freeware such as Google Earth (Figure 11.3 and Figure 11.4).[10]

- Noise complaints—Comparison of complaint frequency and location with noise maps and noise levels to strengthen and ease environmental noise management such as helping identify corrective action plans.
- Noise policing—The efficient planning of noise enforcement activities in relation to the numbers, frequency, or usual time of occurrence of complaints or the predicted or actually measured noise levels.
- Noise abatement—Comparison of noise abatement activities with noise maps, action plans, and projected target noise levels that enables action plans to be tracked and their success and efficiency evaluated.

Figure 11.3 Examples of 3D maps for new developments in Amsterdam. **(See colour insert.)**

• Data handling—With large amounts of data required to produce noise maps and with the need to update data, archive historical results, and share data and information among different stakeholders, data handling is a critical issue that, although alleviated through modern and professional geodatabase systems, requires significant management of processes and professional data management such as data grooming, security, access, and backup. Examples of this from the Netherlands include Silence[2] and I2.[11]

Figure 11.4 3D noise maps are now possible in Google Earth. (From Google Earth. With permission.) **(See colour insert.)**

I2 is a system for storing all industrial sources and other acoustic relevant information (profiles, permits, zones) in the Rotterdam Rijnmond area. I2 registers all the developments within its industrial zones. I2 is used by different organisations, including the port authority, DCMR Environmental Protection Agency, and local consultants. The European Union has ratified a directive known as the Inspire directive[12] that defines common formats for the interchange of geospatial data to enable noise, air pollution, population figures, and so forth to be compared not only between countries but between different professions within a city. The work to implement the directive is still under development but the hope is that, in the near future, the Environmental Noise Directive 2002/49/EC[13] will benefit from this.

GOOD PRACTICE WHEN USING GEOGRAPHIC INFORMATION SYSTEMS (GIS) IN NOISE MAPS

GIS data has typically been collected without any consideration of the demands placed by acoustic calculations. Therefore, in many cases, efficient postprocessing of geometry and attribute information are essential before calculating in order to groom the data, align topography and objects with each other (for example, a more accurate definition of motorway junctions including the indication of noise barriers), and to merge data with differing geometrical resolutions.[14] In addition, the data may refer to the situation at different times, causing temporal data differences. Thus, careful evaluation of the model prior to calculation is also required. A classic example used to illustrate this is shown in Figure 11.5. Here, the correctly imported GIS data places roads and buildings in a lake. The lake is generated from a layer called "wetlands" and the dates of the layers are different. Thus, a probable cause of this is that the land was drained and the buildings and roads built after the wetlands layer was defined.

STRATEGIC NOISE MAPS

As defined in the EU Environmental Noise Directive 2002/49/EC,[13] action plans by the member states are "based upon noise-mapping results." Although the strategic noise map need only identify overall population noise exposure figures and overall source-specific noise contour maps, the intermediary data available, such as the façade noise levels and the physical distribution of the population, enable different areas of an agglomeration or a country to be prioritized. In addition, noise contour maps present data for easier understanding (for example, in the United Kingdom) (Figure 11.6).[3]

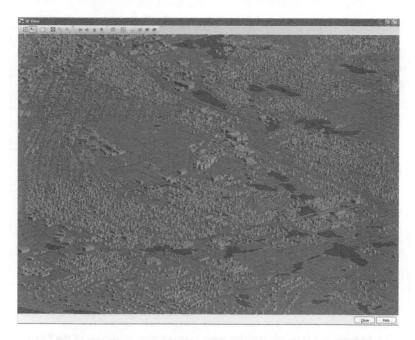

Figure 11.5 Sample data showing that correctly imported GIS data needs evaluation prior to use. **(See colour insert.)**

Strategic noise maps can pinpoint the focus areas of noise policy and action plans (noise "hot spots" and important "relatively quiet areas") to be dealt with. However, as the name suggests, the maps are strategic and thus often not sufficiently detailed to form the basis of detailed action plans. Thus, noise maps are often performed in greater detail as a baseline reference at the start of the action plans. Here, in order to improve the accuracy, additional work is done to improve the quality and accuracy of input data. Particularly where input data is of dubious quality for the purpose in question (for example, a missing overview of road surface type and quality, factors that greatly affect the emission from roads), acoustic measurements are often used. These can be:

- Measurements of specific source issues such as close proximity (CPX) measurements[15] of the road surface enabling individual stretches of road to be more correctly defined (Figure 11.7).
- Of a more general quality assurance nature to identify the areas where data refinement is required. This technique was used in the original Birmingham noise map project[16] that upon its publication in 1999 became an ideal modern reference of the process to be followed and the issues to be met for the entire European Union.

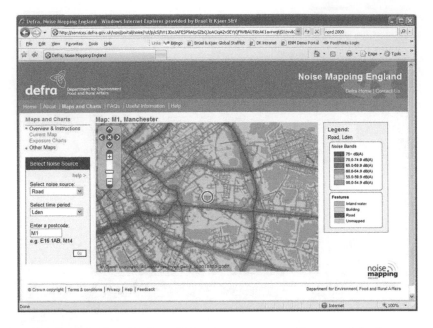

Figure 11.6 Example of a publically available noise contour map. (Courtesy of the Department for Environment, Food and Rural Affairs [DEFRA], United Kingdom.)

- At specifically chosen reference locations that together a general knowledge of both the emission and the topography can provide measurement results that enable source levels to be calculated through the process known as reverse engineering,[17,18] so-called as the source emission is calculated from the level at a location rather than the normal opposite. The measurement locations are determined based primarily on a categorization of the factors needing clarification or

Figure 11.7 An example of a close proximity measurement system. (Courtesy of DGMR.)

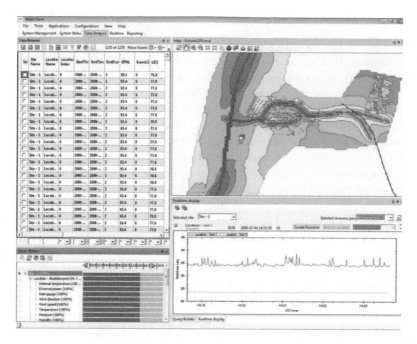

Figure 11.8 An example of a commercial system that manages noise measurements and combines results with calculated noise maps.

improvement. Reverse engineering was very successfully used to make the first strategic noise map of Bucharest from a very meager and inaccurate set of data in a very impressive short duration.[19]

The combined use of measurements and calculations to the benefit of environmental noise management was first proposed by Douglas Manvell et al. from Brüel & Kjær at Internoise 1999 in Fort Lauderdale, Florida.[1] Systems that manage noise measurements and combine the results with calculated noise maps are now commercially available (Figure 11.8).[20] However, a fully blown system that covers all aspects of environmental noise management is, at the time of writing, yet to be developed.

NOISE ACTION PLANS AND NOISE/ ENVIRONMENTAL POLICY PLANS

Setting up some local or regional noise policy or an environmental policy plan is very useful in creating a much better environment. This will improve the quality of living. In this policy plan, goals and objectives can be determined, for example, the number of people seriously annoyed by noise, the number of inhabitants that are disturbed during sleep, and some principles for protecting silent areas of nature.

To improve the acoustical environment it is necessary to investigate the main sources, traffic, trains, airplanes, and industrial activities. It may be necessary also to investigate, for example, the annoyance of the noise from neighbours or from other noise sources/vibrations.

A well-defined communication plan is of vital importance. Inhabitants and politicians have to be informed correctly. Lessons can be learned from successful examples. Evaluation of these examples shows the importance of combining the process of noise mapping directly to the process of action planning and communication to the public. Authorities must realize that they are relatively successful in the development of the noise action plans if they are combining their perspectives with the perspectives of their inhabitants in the development of the noise action plans. It is thereby important that at the start up of the process a complete communication strategy is outthought and applied.

For the future, transform the obligation into a chance to improve the quality for living, working, and relaxing in our main cities. In our perspective we are able to address the issue by not only focusing on solving noise problems but by combining with the other topics and on sustainability. If we succeed in greening our mobility and ways and means of transport, we can successfully address the challenge of ensuring acoustic healthy-living conditions.

FUTURE OF NOISE MAPS AND GEOGRAPHIC INFORMATION SYSTEMS (GIS) IN NOISE MANAGEMENT

In the future, in addition to the spread of data to a more holistic approach where the different aspects can be presented and compared in a variety of ways, one could expect to see the following areas develop as the information technology (IT) infrastructure develops and the knowledge develops:

- Realistic auralisation and virtual reality presentation of the acoustic environment including frequency-dependent propagation and temporal smearing of the signal to give realistic presentations for planning purposes.
- Real-time (dynamic) noise maps enabling single measurement positions to be used to optimise acoustic emissions so that the desired levels are achieved and limits complied to in a cost-efficient manner to the benefit of the producer and community alike. In addition, it can also be real-time in terms of applying measures and instantly seeing the changes in your noise map.
- Annual noise maps to track the progress of action plans and provide better noise management. This is an idea supported by Eurocities.

- With better IT infrastructure (multicore processors, increased power in each computer, and use of networks) and new methods such as Harmonoise,[21] Nord2000,[22] and CNOSSOS,[23] a greater level of detail is possible, with more accurate source descriptions, the ability to calculate more reflections where needed, and more detailed topographical descriptions, thus allowing better results with acceptable calculation times.
- The spread of Web-based hosted solutions from today's information presentation tools to the actual mapping of noise over the Internet allowing calculations to be controlled from any computer anywhere in the world, that is, exploiting the benefits of cloud-based computing.
- The inclusion of building acoustics factors in maps to help determine indoor levels, which may better relate to the noise experienced by the community and thus to better dose–response relationships.
- Further integration of environmental indicators and presentation by using DALY's or Global Footprints.

Although, as indicated, some of the tasks in environmental noise management already widely involve noise maps and GIS to help their resolution, there remain exciting possibilities to further exploit noise mapping and GIS.

The presentation of a noise map is extremely important. All inhabitants directly want to see the noise level at their house so in most cases it is not important to show noise levels on the street but a categorization of houses/dwellings with a certain noise level at the façade (Figure 11.9).

Figure 11.9 An example of a noise map with a categorization of noise levels at the façade. (Courtesy of DGMR.) **(See colour insert.)**

REFERENCES

1. D. Manvell, L. Winberg, P. J. Henning, P. Larsen, Managing urban noise in cities—An integrated approach to mapping, monitoring, evaluation and improvement, Proceedings of Internoise 1999.
2. R. Schmidt, How to manage silence? A large-scale noise management project, Proceedings of Internoise 2010; Silence Web site http://www.silence.nl/introduction.html.
3. Noise Mapping England, Department for Environment, Food and Rural Affairs, http://services.defra.gov.uk/wps/portal/noise.
4. Danish Environmental Protection Agency Strategic Noise Map, http://noise.mst.dk/.
5. European Environment Agency EIONET (official EU strategic noise portal), http://noise.eionet.europa.eu/.
6. A. Czerwinski, S. Sandmann, E. Stöcker-Meier, L. Plümer, Sustainable SDI for EU noise mapping in NRW—Best practice for INSPIRE, International Journal for Spatial Data Infrastructure Research, 2 (2007), 90–111.
7. H. Stapelfeldt, A. Czerwinski, Sustainable environmental data infrastructure in support of acoustic modelling, Proceedings of Internoise 2008.
8. Ministry for the Environment for Nordrhein-Westfalen, http://www.umgebungslaerm-kartierung.nrw.de/laerm/viewer.htm.
9. Hong Kong Environmental Protection Department, Noise page, http://www.epd.gov.hk/epd/misc/ehk09/en/noise/index.html#s1_1.
10. Product Data Predictor Version 7—The Intuitive Solution (Types 7810-A/B/C/D/E/F/G), BP-1602-25, 2009; Brüel & Kjær: http://www.bksv.com/doc/bp1602.pdf.
11. I2, http://www.si2.nl/.
12. Directive 2007/2/EC of the European Parliament and of the Council of 14 March 2007 establishing an Infrastructure for Spatial Information in the European Community (INSPIRE), http://inspire.jrc.ec.europa.eu/ and http://eur-lex.europa.eu/JOHtml.do?uri=OJ:L:2007:108:SOM:EN:HTML.
13. Directive 2002/49/EC of the European Parliament and of the Council of 25 June 2002 relating to the assessment and management of environmental noise—Declaration by the Commission in the Conciliation Committee on the Directive relating to the assessment and management of environmental noise, http://ec.europa.eu/environment/noise/directive.htm and http://eur-lex.europa.eu/LexUriServ/LexUriServ.do?uri=CELEX:32002L0049:EN:NOT.
14. D. Manvell, H. Stapelfeldt, Geometric post-processing of GIS data, Proceedings of Internoise 2003.
15. Method for measuring the influence of road surfaces on traffic noise—Part 2: "The Close Proximity method," ISO/CD-11819-2.
16. A report on the production of noise maps of the City of Birmingham, Department for Environment, Food and Rural Affairs, February 2000.
17. D. Manvell, H. Stapelfeldt, S. Shilton, Matching noise maps with reality—Reducing error through validation and calibration, Proceedings of IOA Seminar "Noise Mapping—Which Way Now?" 2002.

18. D. Manvell, E. Aflalo, H. Stapelfeldt, Reverse engineering: Guidelines and practical issues of combining noise measurements and calculations, Proceedings of Internoise 2007.
19. D. Comeaga, B. Lazarovici, G. Tache, Reverse engineering for noise maps with application for Bucharest city, Proceedings of Internoise 2007.
20. Product Data Environmental Noise Management System Software Type 7843, BP2100-14, Brüel & Kjær, 2009, http://www.bksv.com/doc/bp2100. pdf?document=%2fdoc%2fbp2100.pdf&r=http%3a%2f%2fwww.bksv. com%2fsearch.aspx%3fcategory%3d&searchText=7843.
21. Harmonoise Technical Report HAR32TR-040922-DGMR10 Harmonoise WP 3 Engineering method for road traffic and railway noise after validation and fine-tuning, 2005, http://www.imagine-project.org/.
22. Nordic Environmental Noise Prediction Methods. Nord2000 Summary Report (2001), AV 1719/01, 2009, http://www.madebydelta.com/imported/images/ DELTA_Web/documents/TC/acoustics/Nord2000/av171901_Nord2000_ Summary_Report.pdf.
23. Draft JRC Reference Report 1 on Common NOise ASSessment MethOdS in EU (CNOSSOS-EU), May 2010, http://circa.europa.eu/Public/irc/env/noisedir/ library?l=/material_mapping/cnossos-eu/cnossos-eu_2010pdf/_EN_2.0_&a=d.

Chapter 12

The evaluation of population exposure to noise

G. Brambilla

CONTENTS

As stated in its Article 1, the European Directive 2002/49/EC on the assessment and management of environmental noise, commonly known as the Environmental Noise Directive (hereinafter referred to as END), is oriented "to define a common approach intended to avoid, prevent or reduce on a prioritised basis the harmful effects, including annoyance, due to exposure to environmental noise."

In achieving this objective the END requires estimation of the amount of people living in dwellings exposed to environmental noise, namely, to bands of values of noise descriptors, that is L_{den} and L_{night}, both for agglomerations and major roads, railways, and airports. To comply with this requirement of the END, it is fundamental to have procedures suitable to estimate such data that are the input for developing action plans aimed to reduce the noise impact on the population. The data on population exposure are also useful for the European Commission in monitoring the implementation of the directive in the member states and, at least every 5 years, to draw proposals for amendments and review of the directive itself taking account of scientific and technical progress.

The European Commission's Working Group "Assessment of Exposure to Noise (WG-AEN)" has issued a position paper to assist member states and their competent authorities in undertaking noise mapping and producing the associated data as required by the END.[1] In preparing the position paper, WG-AEN has attempted to find an appropriate balance between the need for a consistent approach across Europe and the flexibility required to meet the demand by individual member states.

Several general issues and specific technical challenges are addressed and 21 toolkits supplementing the recommendations are provided, considering complexity, accuracy, and cost for each item and solution. The Good Practice Guide, known as GPG2, is a good reference and has assisted member states in fulfilling the requirements of the first round of noise mapping. However, national methods laid down in member state legislation may be used, providing proof that they give equivalent results to those obtained with the methods described in the directive.

Any procedure aimed at estimating the exposure of people to environmental noise has to deal with the following:

- Assignment of the receiver points to the façades of buildings
- Assignment of population data to buildings
- Assignment of population data to the receiver points at the façades of buildings

Even though these data seem to be easily obtained, the information on distribution of inhabitants within a building or a block is seldom available also because of privacy regulations that restrict the use of this detailed information. Thus, procedures based on less detailed information have been developed. This has led to noticeable variations of the data transmitted to the Commission by the member states as result of the first round of noise mapping, an example of which is given in Figure 12.1.[2] Details of the population exposure in each country for different type of noise sources inside and outside the agglomerations are available at the Noise Observation and Information Service for Europe (NOISE) Web address (http://noise.eionet.europa.eu/).[3]

This chapter describes the aforementioned problems and the proposed solutions, as well as other procedures available in the literature to estimate the exposure of inhabitants to environmental noise.

THE END METHOD

According to Annex VI of the END, among the data to be sent to the Commission there is the estimated number of people (in hundreds) living in dwellings that are exposed to specific bands of values of L_{den} and L_{night}, with bin width of 5 dB(A), determined 4 m above the ground and at the most exposed façade, separately for road, rail, air traffic noise, and industrial sources. The numbers of inhabitants must be rounded to the nearest hundred (e.g., 5200 = between 5150 and 5249; 0 = less than 50).

This procedure can actually be accurate for detached or semidetached urban configurations, but it could introduce errors in urban areas with tall and large

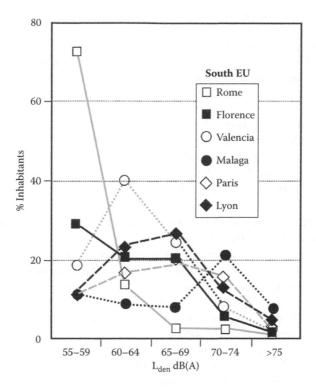

Figure 12.1 Distribution of inhabitants over noise levels L_{den} in some cities in South Europe. (Adapted from M. van den Berg, G. Licitra, EU Noise Maps: Analysis of Submitted Data and Comments, Proceedings of Euronoise 2009, Edinburgh, 2009.)

buildings. In fact, even though detailed data on the distribution of inhabitants among dwellings within the building or the block would be available, the sound level determined 4 m above the ground and at the most exposed façade is unlikely to be representative of the actual exposure of the population.

The locations (assessment points) where noise levels have to be determined for the purpose of producing data to comply with the requirements of the END can be distinguished into two types:

- Points at 0.1 m in front of the façade, spaced 3 m from one another around the building or block façades.
- Grid-based points, such as those used for noise contour mapping, generally spaced no more than 10 m in agglomerations and 30 m in open areas outside agglomerations; especially in urban areas grid spacing less than 10 m (i.e., where buildings face each other across narrow roads) down to perhaps 2 m or variable grid spacing can be more suitable.

In both cases, the noise level refers to the incident sound ("free field") and, therefore, reflections from the façade behind the assessment point shall be excluded, whereas at least first-order reflections from other façades or objects should be considered. The incident sound can be approximately estimated by applying a –3 dB correction to the overall noise levels.

The assignment of noise levels to dwellings differs depending if the block:

1. Consists of a single dwelling or the location of each individual dwelling within the building(s) forming the block is known; in such condition each dwelling is treated as if it is a separate building within the block.
2. Contains multiple dwellings within the building(s) forming the block, but their distribution is not known.

The knowledge of the use of the building is crucial to obtain a satisfactory accuracy in the latter circumstances.

In the first set of cases (known location of dwellings within a building), the highest noise level determined either at intervals around the building façades or at grid points is assigned to the dwelling as the value at the "most exposed façade." For the latter, each grid point surrounding the building is linked to the façade when the area around the grid point (i.e., a square with sides equal to the grid point spacing centred on the grid point) intersects with the façade or is within 2 m of the building (see example in Figure 12.2). This procedure can be automatized in geographic information systems (GIS) and a default value, for instance 10, can be assigned to the grid cell when a grid position is inside the building.

In the second set of cases (unknown locations of multiple dwellings within a building), the lack of data on the location of individual dwellings within the building(s) in the block will inevitably lead to inaccuracy in determining the exposure of each dwelling (and hence, the population exposure). One of the approaches described for the first set of cases, depending on whether façade levels or only grid point calculations are available, can be used to calculate the highest noise level at any point around the whole block. This highest value is attributed to all dwellings in the building(s) as the value at the "most exposed façade." In some circumstances this procedure will lead to an overestimation of the noise level affecting some of the dwellings within the building, particularly during the nighttime due to the fact that bedrooms are highly unlikely to be located always at the most exposed building façade in even the majority of dwellings under analysis.

Regarding the assignment of population to dwellings, detailed data on population distribution over the area are necessary but they may not be available or provided from several sources (e.g., population census, postal code area, etc.), at different levels of detail and for different years, and may not cover all demographic groups. The official census of population is usually the primary data source, but usually the available format is in terms of inhabitants inside

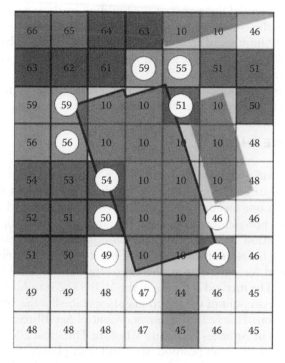

Figure 12.2 Assignment of grid levels to the building, highlighted by white circles. The highest noise level assigned as a value at the "most exposed façade" is 59. (Adapted from European Commission WG-AEN, Good Practice Guide for Strategic Noise Mapping and the Production of Associated Data on Noise Exposure, Position Paper Final Draft, Version 2, 2006.)

the census unit area that likely includes more than one building, namely, a group of buildings (block). Thus, procedures to determine the distribution of inhabitants in the census unit area are required. The simplest one may be dividing the number of inhabitants by the surface of relevant census unit area to get the population density for such area. This procedure can be refined by dividing the number of inhabitants by the sum of surfaces on the map covered by the buildings or, even better, by the residential ones only. However, this approach is rather approximate and further improvements are possible, such as considering the number of floors and the height of the buildings.

If detailed data on the population distribution over the area are not available or cannot be used to satisfactorily estimate the number of people living in dwellings in buildings, toolkits 19 and 20 proposed by WG-AEN[2] may be used in combination and provide a number of options for producing such estimates. For instance, the building area obtained by a GIS multiplied by the number of storeys provides the residential floor area of the building and, hence, the number of residents in the building is given by dividing the residential floor area of the building by floor area per resident, the latter being the entire residential floor area of the mapping area, or sub-area, divided by number of residents, if known, or obtained by national or regional population statistics (e.g., for the Italian legislation[4] the minimum gross floor area per resident is 25 m²).

Toolkit 20 is also suggested to be used to estimate the number of dwelling units and, if necessary, the population per dwelling unit.

THE VBEB METHOD

The END method assigns to each building a single value of noise level and this is assumed to be the exposure for all the inhabitants in the building, concentrated on one façade of the building (the most exposed one).

A different approach, widely used so far, has been proposed in the German calculation method VBEB (*Vorläufige Berechnungsmethode zur Ermittlung der Belastetenzahlen durch Umgebungslärm*, preliminary calculation method for determining the exposure figures caused by environmental noise).[5] The method is based on the VDI 3722 standard but has been adjusted to take account of the requirements specified in the 34th Ordinance on implementation of the Federal Immission Control Act (BImSchV) of 6 March 2006, as well as in Annexes I, IV, and VI of the Directive 2002/49/EC.

Since in general the exact position, size, and floor plan of dwellings are not known, the total number of people living inside specific dwellings is equally distributed over the assessment points located on the building façades. Thus, the value "inhabitants per assessment point" (rounded to the nearest integer) is determined and is assigned to the immission level at that point. Subsequently, the number of people attributed to each façade level has been summed within specific noise level classes. Rules are established to distribute the assessment points along the building façades, an example of which is given in Figure 12.3 where 15 assessment points are identified on 8 façades (plotted by circles in Figure 12.3) and for each point the L_{den} value is given.[5]

For instance, as reported in VBEB,[5] for the building shown in Figure 12.3 having three floors and covering an area of 140 m² on the map, considering a net area of 80% and a floor area per resident equal to 35 m², hence the total number of inhabitants in the building is

$$TI = \frac{140 \times 3 \times 0.8}{35} = 9.6 \tag{12.1}$$

As 15 assessment points have been identified, then the inhabitants for each point are

$$IP = \frac{9.6}{15} = 0.64 \tag{12.2}$$

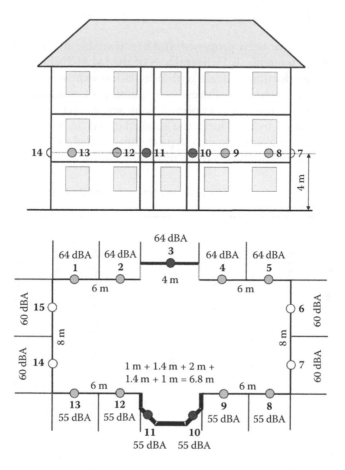

Figure 12.3 Example of distribution of the assessment points along the building façades, according to the VBEB method. (Adapted from *Vorläufige Berechnungsmethode zur Ermittlung der Belastetenzahlen durch Umgebungslärm—VBEB, Federal German Gazette*, p. 4, 137, 20 April 2007.)

Thus, the following distribution of inhabitants in bands of values of L_{den} to obtained:

L_{den} Band dB(A)	Number of Assessment Points (#)	Inhabitants
55–59	8 (7, 8, 9, 10, 11, 12, 13, 14)	5.12 (0.64 × 8)
60–64	6 (1, 2, 4, 5, 6, 15)	3.84 (0.64 × 6)
65–69	1 (3)	0.64

Whereas, according to the END method all the 9.6 inhabitants are exposed to the 65–69 band of L_{den}.

OTHER METHODS

Other methods have been proposed and are available in the literature. A wide and useful reference is Deliverable 8 of the IMAGINE-WP1 Project.[6] Section 5 of the document deals with population exposure and describes in full detail proposed procedures as a function of available input data and gives an indication of the expected accuracy, as well as indicates pre-processing required to transform the input data set in a suitable format. Table 12.1 summarizes the proposed procedures coded in the document as in the last column, namely, from the most to the least accurate:

Table 12.1 Overview of the Procedures Proposed by the IMAGINE Project for Allocating Inhabitants

Location and outline of dwellings	Floor area of dwelling	Number of dwellings in each building	Storeys of buildings	Height of building	Topological outline of building	Use of buildings	Number of inhabitants for each dwelling	Number of inhabitants for each building	Number of inhabitants for each TA	Suggested procedure
Yes							Yes			A
Yes							No	Yes		B.a
Yes	Yes						No	Yes		B.b
No		Yes	Yes		Yes		No	Yes		C.a
No		Yes	No	Yes	Yes		No	Yes		C.b
No		No	Yes/No	Yes/No	Yes		No	Yes		C.c
Yes							No	No		D.a
Yes	Yes						No	No	Yes	D.b
No		Yes	Yes/No	Yes/No	Yes		No	No	Yes	E
No		No	Yes		Yes	Yes	No	No	Yes	F.a
No		No	No	Yes	Yes	Yes	No	No	Yes	F.b
No		No	No	Yes	Yes	No	No	No	Yes	F.c

Source: C. Chiari, A. Iacoponi, D. Van Maercke, Guidelines and Good Practice on Strategic Noise Mapping, IMAGINE–WP1 Final Report, Deliverable 8, Report IMA01-TR221I2006-ARPAT12_D8, 2007.

Note: TA = topological area = street/block/statistical sector/neighbourhood; bland = not relevant; Yes/No = known/unknown.

A   *Most detailed estimation*, as location and number of dwellings and number of inhabitants in each dwelling are known

B.a *Number of inhabitants per dwelling taken constant*, being known location and number of dwellings and number of inhabitants per building

B.b *Inhabitants distributed proportionally to floor area of dwellings*, being known location, floor area, and number of dwellings and number of inhabitants per building

C.a *Position of dwellings within buildings*, being known the number of dwellings for each building, the number of storeys, and the number of inhabitants for each building

C.b *Position of dwellings but number of storeys unknown*, being known the number of dwellings for each building, the number of inhabitants, and the height for each building

C.c *Position of dwellings but number of dwellings unknown*, being known the number of inhabitants per building

D.a *Number of inhabitants per dwelling taken constant*, as in B.a but in this case the number of inhabitants is known in a topological area (street/block/district)

D.b *Number of inhabitants per dwelling proportional to floor area*, being known location, surface, and number of dwellings, and the number of inhabitants in a topological area (street/block/district)

E   *Population per building proportional to the number of dwelling units*, being known the number of dwellings in each building, the height or number of storey of the building, and the number of inhabitants in a topological area (street/block/district)

F.a *Population per building proportional to liveable area*, being known the use of each building (or its part) and the number of storeys and the number of inhabitants in a topological area (street/block/district)

F.b *Population per building proportional to the volume*, being known the use of each building (or its part) and its height and the number of inhabitants in a topological area (street/block/district)

F.c *Use of buildings unknown*, but height of each building and the number of inhabitants in a topological area (street/block/district) are known

In addition to the aforementioned procedures, two other methods are described in the following:

- "Average level exceedance" method, hereinafter referred to as the ALE
- "Nearest grid point" method, hereinafter referred to as the NEAR

In the ALE method,[7] the incident sound level is determined for each of the N assessment points equally spaced (usually 10 m apart) and located

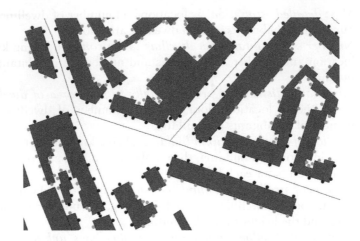

Figure 12.4 Example of the ALE method. The black lines represent road segments.

along the building façades (like the grey and black dots in Figure 12.4). Then the arithmetic average $\overline{L}$ of the N levels is calculated, that is,

$$\overline{L} = \frac{\sum_{i=1}^{N} L_i}{N}$$

(12.3)

and only the assessment points having sound level $L_i > \overline{L}$ (black dots in Figure 12.4) are considered in the calculation of population exposure. All the inhabitants in the building block are equally distributed among the assessment points with $L_i > \overline{L}$ to obtain a cautionary estimate of the population exposure as that proposed in the END method.

The method has the feature to distribute people depending on noise levels (that is their average) that may be different for time periods. It is particularly suitable in dense urban areas where the distribution of dwellings in the block is unknown and the block has at least one long façade parallel to the sound source line (road or railway).

In the NEAR method,[8] the assessment points are distributed along the building façades at equal distances of 3 m (white circles in Figure 12.5) and the sound level at the grid point nearest to each of these points is assigned to it (black circles in Figure 12.5) and reduced by 3 dB to obtain the incident sound level. Then the population in the building is equally distributed among the assessment points, if necessary at every floor and where such data are known.

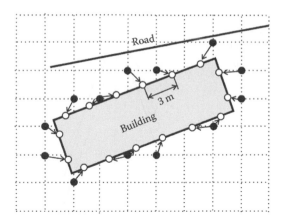

Figure 12.5 Example of the NEAR method. (Adapted from M. Arana et al., Using Noise Mapping to Evaluate the Percentage of People Affected by Noise, *Acta Acustica United with Acustica* 95, 550–554, 2009.)

COMPARISON OF METHODS

To compare some of the aforementioned procedures, they have been applied to the strategic noise map of the entire territory of Pisa Municipality, a town in the Tuscany region in Italy with 89,694 inhabitants in the national census of 2001. The map was calculated by a 5 m square grid at 4 m above the ground. The distribution of population in the buildings was computed according to the procedure F.b (population per building proportional to the volume) mentioned in the previous section and described in the IMAGINE-WP1 Project.[6]

The results obtained by the END, VBEB, and ALE methods applied to the 42,297 buildings in an area of 185 km² are reported in Figure 12.6, showing the distributive and cumulative functions of the percentage of population exposed versus ambient noise L_{den} in 1 dB(A) intervals.[9] All the distributions are similar to the Gaussian one and those corresponding to the END and ALE methods are narrower (standard deviation 5.1 and 5.2, respectively) than that of VBEB (standard deviation 5.9). As expected, the estimate by the END method is the most cautionary and the ALE method provides results among those by the VBEB and END procedures. As shown in Figure 12.6a, the largest percentage of population exposed is observed at $L_{den} = 60.5$ dB(A) for the END method, whereas VBEB and ALE show the maximum at 56 and 58.5 dB(A), respectively. The cumulative distributions plotted in Figure 12.6b show that the observed differences are significant,

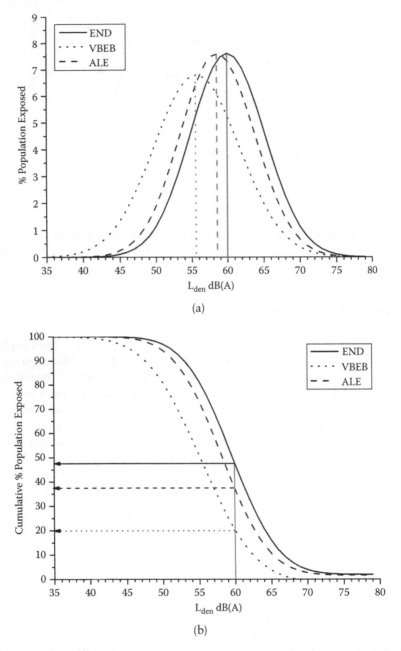

Figure 12.6 Distributive (a) and cumulative (b) functions of the percentage of population estimated to be exposed versus ambient noise L_{den} in 1 dB(A) intervals in Pisa.

for instance, at the exposure of 60 dB(A) L_{den} the END gives an estimate of 47% of population, whereas the VBEB and ALE methods provide 20% and 37%, respectively. The implications of such differences on economical and health protection issues are rather evident. Thus, the need to harmonise the procedures in the estimate of population exposure to ambient noise is important not only to compare data from different areas but also for a more reliable and realistic approach to this important aspect of health and environment protection.

CONCLUSIONS

Most of the procedures proposed to estimate the percentage of people exposed to noise described in this chapter are nowadays applicable straightforward as they are implemented in commercial software packages for calculation of outdoor noise propagation. In addition, they can be calculated automatically by tools available in geographic information systems. However, as it has been shown, they provide different results due to differences in the façade noise level determination and in the assignment of population exposed to this level. These differences are also a consequence of the quality of the input data available, which largely influences the choice of the method, too.

It has to be noted that usually more attention is put to the accuracy in the determination of noise levels than that regarding the distribution of population in blocks, buildings, or dwellings. If this behaviour can be understood from the point of view of acousticians, it cannot be excusable in terms of action plan aims. It is desirable that the accuracy of noise level determination would be similar to that of the distribution of population[10] and both as accurate as possible considering the objectives of the calculation. The END method recommended by the European Commission is, likely, easy to be applied and understood by decision makers and the general population. However, the obtained population exposure is significantly overestimated in dense urban areas where tall buildings are usually present. In these circumstances the German VBEB method seems more adequate, but it requires more detailed data on population distribution. Further developments on the procedures to estimate the noise exposure of population are expected from the activity of Working Group 10 established in the CNOSSOS-EU Technical Committee.[11]

It should not be forgotten that the population exposure would be the input data for designing action plans aimed to reduce such exposure on a prioritised basis, as stated in the END. Thus, it is clear the importance of these data and how they can influence the effectiveness of action plans and the corresponding economical resources.

REFERENCES

1. European Commission WG-AEN, Good Practice Guide for Strategic Noise Mapping and the Production of Associated Data on Noise Exposure, Position Paper Final Draft, version 2, 2006.
2. M. van den Berg, G. Licitra, EU-noise maps: Analysis of submitted data and comments, Proceedings of Euronoise 2009, Edinburgh, 2009.
3. Noise Observation and Information Service for Europe (NOISE), http://noise. eionet.europa.eu/ (accessed April 2012).
4. D.M. Lavori Pubblici 2 aprile 1968 n. 1444, Limiti inderogabili di densità edilizia, di altezza, di distanza fra i fabbricati e rapporti massimi tra spazi destinati agli insediamenti residenziali e produttivi e spazi pubblici o riservati alle attività collettive, al verde pubblico o a parcheggi da osservare ai fini della formazione dei nuovi strumenti urbanistici o della revisione di quelli esistenti, ai sensi dell'art. 17 della legge 6 agosto 1967, n. 765, Gazzetta Ufficiale Repubblica Italiana n. 97, 16 April 1968.
5. Vorläufige Berechnungsmethode zur Ermittlung der Belastetenzahlen durch Umgebungslärm—VBEB, Federal German Gazette, p. 4, 137, 20 April 2007.
6. C. Chiari, A. Iacoponi, D. Van Maercke, Guidelines and good practice on strategic noise mapping, IMAGINE–WP1 Final Report, Deliverable 8, Report IMA01-TR22112006-ARPAT12_D8, 2007.
7. G. Brambilla, D. Casini, A. Poggi, T. Verdolini, Metodologie per la determinazione della popolazione esposta al rumore delle infrastrutture di trasporto, Proceedings 32nd National Congress of the Acoustical Society of Italy, pp. 165–170, 2005 (in Italian).
8. M. Arana, R. San Martin, I. Nagore, D. Pérez, Using noise mapping to evaluate the percentage of people affected by noise, *Acta Acustica United with Acustica* 95, 550–554, 2009.
9. G. Licitra, E. Ascari, G. Brambilla, Comparative analysis of methods to evaluate population urban noise exposure, Proceedings of Internoise 2010, Lisbon, 2010.
10. G. Brambilla et al., Indicazioni operative per la costruzione dell'indicatore "Popolazione esposta al rumore" in riferimento alla Direttiva Europea 2002/49/CE, Report A.
11. Common NOise aSSessment methOdS in EU Working Group 10 Assigning noise levels and population to buildings.

Part 3

Noise mapping in Europe

Chapter 13

Noise maps in the European Union: An overview

P. de Vos and G. Licitra

CONTENTS

The Environmental Noise Directive 2002/49/EC (further to be addressed as END) asked the European Union (EU) member states to prepare noise maps and action plans for a limited number of agglomerations (those with more than 250,000 inhabitants), major roads (over 6 million vehicles per year), major rail (more than 60,000 trains per year), and major airports (more than 50,000 aircraft movements).[1] This is required for the first round of maps, which was to be submitted to the Commission before 30 July 2007, and only a few member states (MS for short) have done so.

The purpose of the END is "to provide a basis for developing and completing the existing set of Community measures concerning noise emitted by the major sources ... and for developing additional measures, in the short, medium and long term." Moreover, according to Article 1 the aim is extended to "define a common approach intended to avoid, prevent or reduce on a prioritized basis the harmful effects, including annoyance, due to exposure of environmental noise." To achieve these aims, it defined interim methods to be used by the countries without national ones; therefore, not all countries used the same way to estimate exposure, so they may vary a lot.

In the next paragraphs the method of data collection is explained, and analysis of the data will cope with the following questions:

- What is the overall quality of the data in terms of coverage, reliability, and comparability?
- How critical are the exposure data?
- What recommendations can be derived for the second round of noise mapping in 2012 and which is the direction taken by experts?

DATA COLLECTION AND PUBLICATION (31 JULY 2010)

Publication on the Web site

All the files submitted by members are available from the CIRCA reporting Web site. Based upon these files, the European Commission created a data repository, EIONET,[2] with a Web platform NOISE viewer where all the exposure data and some contours maps are published in a geographic and chart format. The Web platform belongs to NOISE, the Noise Observation and Information Service for Europe maintained by the European Environment Agency (EEA) and the European Topic Centre on Land Use and Spatial Information (ETC LUSI) on behalf of the European Commission. The agglomerations are identified and a lot of data are available to be downloaded or visualised as bar plots. Most updated data included ones submitted before 31 July 2010. This data has been downloaded to perform elaborations in a unique spreadsheet and will be analysed in the following sections.

Data of the original spreadsheet contained a lot of exposure data (also obtainable on the interactive Web platform), in particular L_{den} and L_{night} for:

- Road traffic noise in agglomerations
- Railway noise in agglomerations
- Industry noise in agglomerations
- Aircraft noise in agglomerations
- Major roads
- Major railways
- Major airports

Most of these data are also available on national or local Web sites (mother tongue), but the availability on a single platform with the same format is very useful in terms of comparability. However, different methods are to be taken into account to evaluate differences between countries.

Data coverage

In the first round of noise mapping, intended to be concluded in 2007, the following parties were involved:

- 164 agglomerations with more than 250,000 inhabitants
- 82,575 km of major roads having more than 6 million vehicle passages a year
- 12,315 km of major railways with more than 60,000 train passages a year
- 76 major civil airports with more than 50,000 movements a year

As said, only few countries reported data in time, but almost all countries have now completed the first round of mapping.

It must be said that about 75% of EU citizens live in an urban environment, so on a purely statistical basis, one can say that only a small percentage of citizens is likely to live near a major road, railway, or airport. Therefore, it is obvious that the problem of exposure to environmental noise is concentrated in urban environments inside agglomerations.

Thus, results of the agglomeration mapping process are more relevant and here deeper detailed. The most mapped source is road traffic noise, which is also the one with the highest impact on the population exposure. Table 13.1 reports the coverage for each source. It must be noted that Norway, which is not an EU member, reported voluntary data that will be analysed, too. On the other hand, Cyprus and Luxemburg were not involved in the first round of agglomeration mapping.

Other considerations about coverage should be given:

- Belgium, Greece, Malta, and Portugal have not yet reported any data.
- France and Italy did not complete all agglomerations of any source maps.

Table 13.1 Global Data Coverage inside Agglomerations

To Be Reported	Population	Agglomerations
	122,370,218	164
Reported road	85%	131
Reported railways	81%	121
Reported air	70%	97
Reported industry	75%	120

- Romania and Spain have yet to complete railway and aircraft noise maps.
- Germany, Lithuania, and Poland have yet to complete aircraft maps.
- Germany and Spain have yet to complete industry maps.
- Some countries had to map a lot of agglomerations respective to others, but the number was not a factor influencing the percentage of maps done.

Moreover, inside agglomerations, in urban areas, some interpretation differences can be observed. In some countries, there is a tradition to treat noise from streetcars as road traffic noise. In other cases, streetcars and metros have been completely ignored.

For urban rail links, the stations represent serious sources of environmental noise, as they generate noise from public address systems, shunting, cleaning, and many other supporting activities, not to speak of the road traffic involved in the follow-up transport chain.

Also, the industrial sites to be mapped have been interpreted differently. Clearly, it is not the intention of the directive to include noise from minor industrial activity, such as the cooling compressor of the nearest supermarket or the roof ventilation units of an assembly hall, into the noise mapping. A large electric power plant could be situated in the middle of a town and hardly be observed by residents. Heavy industrial activity, such as shipyards, refineries, and steel construction plants, are usually located far from urban living areas. For port areas, specifically mentioned in the directive, it is even more complicated to decide what to include. Vessels and barges being loaded and unloaded, or even just waiting their turn, under inboard power tend to generate a lot of noise, which is then to be considered part of the port activity. The NoMEPorts project has produced a Good Practice Guide on Port Area noise mapping and management, presenting recommendations for mapping port areas.

For the areas outside agglomerations, the definition is quite straightforward and few interpretation problems have occurred.

However, despite interpretation differences, the noise mapping process is quite developed and contributed to the development of a noise protection culture in those countries that were still without legislation. However, this first round demonstrated the difficulties to succeed in performing noise maps within the requested time.

ANALYSIS OF AVAILABLE DATA

Data quality and comparability

In the first round, reports were expected on more than 120 million EU inhabitants in the agglomerations of 27 countries. The reporting discipline differed widely, a major nuisance when extracting the reported data is

Section 1.5 of Annex VI of the END (Data to be sent to the Commission), which starts with "The estimated number of people (in hundreds) ..." Oblivious to the fact that the second phrase in the same section explains that this means that the figures should be rounded to the nearest 100, and even gives the example that 5200 means that the real figure could be between 5150 and 5249, quite a few authorities report the number in units of hundreds.[3]

Another source of error is incompleteness of the data. Many member states report exactly what they were supposed to, but others are so incomplete that the remaining data is almost useless.

Moreover, another obstacle in comparing data collected is the difference in method used to evaluate noise exposure level. In fact, not only national methods differ in the standard calculation procedure so that source database and propagation are not the same, but also they may differ in how they assign levels to population. The END suggested evaluating exposure by assigning the noise on the most exposed façade to all the inhabitants, but the German method VBEB[4] distributes population over a ring of receivers around each building. It is obvious that great differences may arise from the method of distribution[5]; furthermore, the method used in calculation and in particular the availability of frequencies-dependent propagation (British method CRTN[6] propagates only 500 Hz frequency) may lead to large differences on exposure.

Thus, it is clearly visible from reported data that it is not possible to evaluate "which is the noisiest country." Figure 13.1 and Table 13.2 show differences in road noise exposure inside agglomerations according to each method.

Even considering variance between agglomerations, German and British cities reported borderline values respective to other distributions. This should be taken into account when looking at the values reported in the next paragraphs.

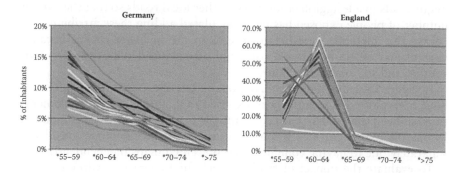

Figure 13.1 Percentage of inhabitants exposed to L$_{den}$ bands in each agglomeration in Germany and England (VBUS [7]+VBEB and CRTN calculation methods).

Table 13.2 National and Zones Average Values with Accuracy (95% Confidence Level)

Country/Region	55–59 dB(A)	60–64 dB(A)	65–69 dB(A)	70–74 dB(A)	>75 dB(A)
England	28 ± 4%	53 ± 6%	6 ± 1%	2.4 ± 0.8%	0.3 ± 0.2%
Germany	11 ± 1%	7.2 ± 0.8%	5.2 ± 0.4%	2.4 ± 0.4%	0.5 ± 0.2%
North Europe	17 ± 3%	15 ± 4%	10 ± 3%	5 ± 2%	0.8 ± 0.6%
South Europe	21 ± 6%	23 ± 5%	17 ± 4%	8 ± 3%	2 ± 1%
East Europe	21 ± 3%	21 ± 3%	13 ± 3%	6 ± 2%	1.1 ± 0.6%

This small variance within German and British agglomerations is not only due to having a national standard method but also to a common approach in the mapping process. In fact, in England all maps have been processed by the national Department for Environment, Food and Rural Affairs (DEFRA) following the advice given in the Good Practice Guide,[8] the position paper produced by the European Commission.

The following section will illustrate results in terms of exposure of each source mapped.

Exposure due to different noise sources

Road traffic noise

Road traffic noise has been the most studied and mapped source, obviously due to its social impact on public health. The most critical points in road mapping inside agglomerations are the use of different methods, as already mentioned, and the extension of the network considered. In fact, many of the first submitted results have been focused only on principal roads neglecting secondary roads whose traffic flow is difficult to estimate. Moreover, some local authorities separately studied the impact of major roads inside agglomerations and other local roads so that the overall number of people exposed has been calculated adding these attributes. Of course this method overestimates the exposure in lower classes because citizens may be counted twice instead of being considered exposed to two sources and so, to a higher value.

In Figure 13.2 and Figure 13.3, national levels of exposure (L_{den} and L_{night}) have been plotted. The number of exposed inhabitants in each agglomeration has been summed for each band and the percentage has been computed. In the following figures each country of the United Kingdom is considered separately due to having computed and presented maps independently.

To evaluate the impact of roads in each agglomeration, without being influenced too much by the mapping approach, the percentage of inhabitants

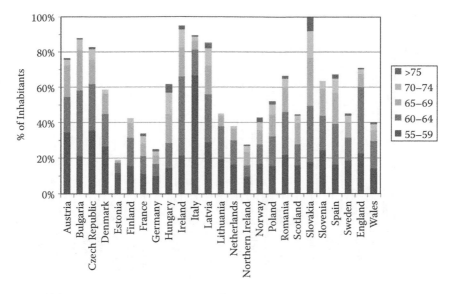

Figure 13.2 L$_{den}$ road noise exposure in different countries.

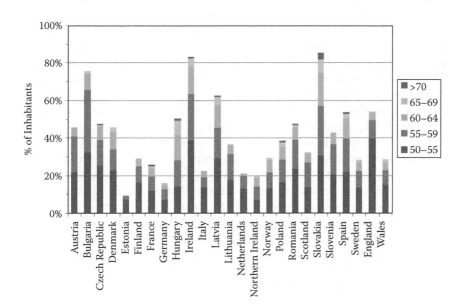

Figure 13.3 L$_{night}$ road noise exposure in different countries.

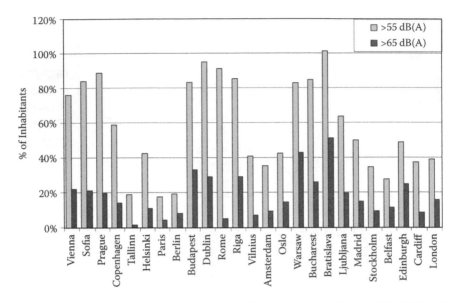

Figure 13.4 Capitals exposure: percent of inhabitants exposed to L_{den} >55 dB(A) and >65 dB(A).

over 65 dB(A) agree with the total percentage reported (inhabitants over 55 dB(A)) has been evaluated. Figure 13.4 reports both percentages for agglomerations of capitals. Looking at Figure 13.4, it is possible to observe that some capitals, especially in northern countries, have a low percentage of exposure; in southern and eastern areas the situation is more critical. This may be due to the development of public transportation networks and to social behaviour and to modern car fleets and the availability of big roads around cities far from buildings.

Then, considering all cities, the amount of people in higher bands (over 65 dB(A) for L_{den} and over 60 dB(A) for L_{night}) have been compared to the total reported (people over 55 dB(A) or over 50 dB(A)) so that a histogram has been produced over all 131 agglomerations. This graph, shown in Figure 13.5, will be hereafter referred as high over total histogram (HTH). The cities that reported a portion in higher bands greater than half (the last band in the previous graph) are Malaga (both L_{night} and L_{den}) and, only for L_{den} exposure, Alicante, Barcelona I, Bilbao, Bratislava, Budapest, Edinburgh, Grenoble, Timisoara, and Warsaw. These agglomerations can be somehow considered as noisy because, regardless of the absolute percentages of people exposed, there is a bad distribution of levels.

Another way to identify differences in the approach and to evaluate exposure is to calculate the ratio between the percentage of exposure in the first two bands: the ratio between 60–64 and 55–59 has been computed and

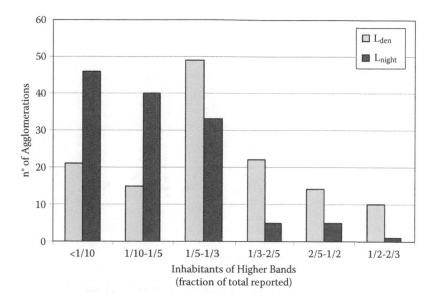

Figure 13.5 Road noise exposure, HTH graph.

plotted against total percentage reported. Figure 13.6 shows how for the lower percentage of people reported (i.e., less noisy cities[*]) the ratio is low. It is lower than one until about 30%. For percentages over 80%, reported by almost all UK cities, the ratio increases rapidly (i.e., inhabitants exposed to 60–64 are much more than the ones exposed to 55–59). From this graph it is clear that, excluding German and British agglomerations, the ratio varies within a small range of values.

Railway noise

Railway noise has been reported from 121 agglomerations. However, some agglomerations reported no people exposed so it is not clear if there are not people exposed over the threshold of reporting or there are none exposed at all. The agglomerations reporting no exposed people are three for L_{den} (Bialystok, Córdoba, and Gijón) values and six for L_{night} (Bialystok, Zaragoza, Murcia, Palma de Mallorca, Córdoba, and Gijón).

Exposure to railway noise is overall quite low. In fact, except for Lyon, Katowice, and Bratislava, all agglomerations reported a percentage lower than 40% over 55 dB(A) for L_{den} and lower than 25% over 50 dB(A)

---

[*] The less noisy result may be due to a mapping process that considered only principal roads, or distributed people around façades, or to good building disposition so that the majority of buildings are screened by other ones.

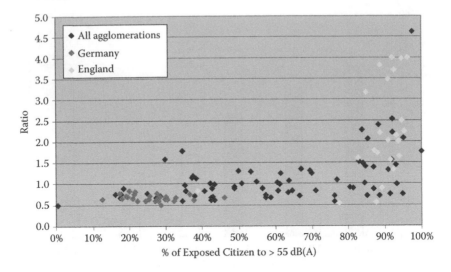

Figure 13.6 Ratio between the number of exposed citizens to 60–64 and to 55–59 dB(A).

for L_{night}. Moreover, the exposure decreases linearly along bands, which is due to railways being far from buildings so that only few people are exposed to high values. In Figure 13.7, percentages of lower bands are plotted against total percentage reported. Here also there is a clear linear tendency.

As already done for roads, a HTH graph has been set up to identify the portion of citizens exposed to the higher levels with respect to total reported (see Figure 13.8). Agglomerations with a portion greater than

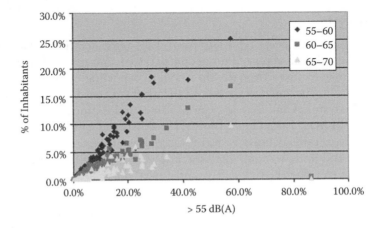

Figure 13.7 Linear relation between exposure in each band and total reported.

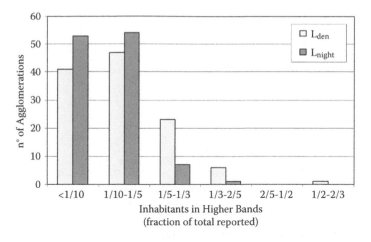

Figure 13.8 Railway noise exposure, HTH graph.

one-third in terms of L_{den} are Brno, Edinburgh, Grenoble, Oslo, Prague, Rome, and Vienna. Regarding L_{night}, agglomerations with a portion greater than one-fifth are Bratislava, Edinburgh, Glasgow, Lodz, Marseille, Metz, Rome, and Vienna.

Finally, total percentages of exposure are reported for each country in Figure 13.9. It must be remembered that some agglomerations considered also urban trams and light railway in noise mapping so that exposure is considerably increased by local transport.

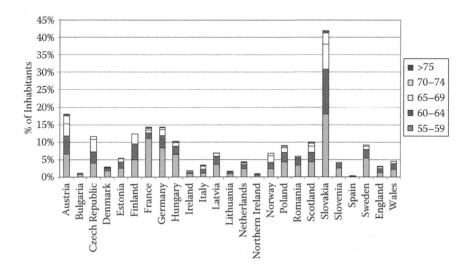

Figure 13.9 Country exposure to railway noise.

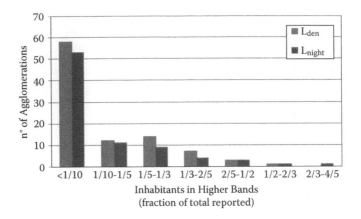

Figure 13.10 Industrial noise exposure, HTH graph.

Industrial noise

Industrial noise has been reported by 120 agglomerations. However, 25 reported no one exposed within the reporting threshold level. Figure 13.10 shows HTH graph for industrial noise. There are generally good distributions. In particular, only four cities have a proportion higher than two-fifths for L_{den} values (Brasov, Brighton, Frankfurt, and Iasi) and for night values Southampton as well had a high percentage in higher bands.

The low impact of industrial noise is probably due to the fact that industries are known for other pollutions convincing local administrations to create industrial areas far from residential ones. Moreover, industrial machineries that make noise are often inside industrial buildings or are screened by other buildings in the area.

In Figure 13.11 are the percentages of exposure for those agglomerations that reported at least 3% inhabitants over 55 dB(A). If we look directly at countries (Figure 13.12) we notice that very low percentages of inhabitants are reported (less than 3%) and that there are some countries that seem not to have considered industrial noise at all. Scotland is not included in the figure. In fact, Glasgow and Edinburgh reported high percentages so that the national exposure is higher than anyone else reaching 25% of inhabitants over 55dB(A).

Aircraft noise

Aircraft noise refers to data compiled by agglomerations about air noise. Data from minor airports are reported; 97 agglomerations reported data on aircraft noise but 38 declared no exposed inhabitants within threshold values in terms of both noise indicators. Within agglomerations that are

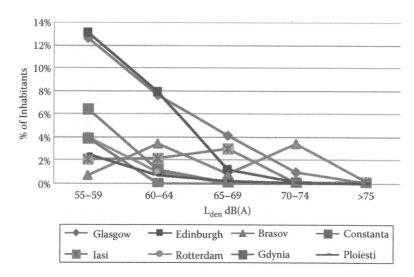

Figure 13.11 Industrial noise exposure for noisy agglomerations.

exposed, only 12 of them have a percentage of inhabitants exposed to L_{den} values greater than 55 dB(A) which surpass 3% (see Figure 13.13). Data of Grand Lyon should be revised considering that it declared that 86% of the population is exposed to 55–59 dB(A), so it is not reported in the figure.

Moreover, only Bucharest, London, and Malaga have anyone over the 75 dB(A) band; but there are none exposed to higher bands. Finally,

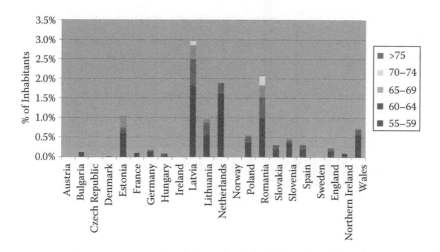

Figure 13.12 Industrial noise exposure in different countries.

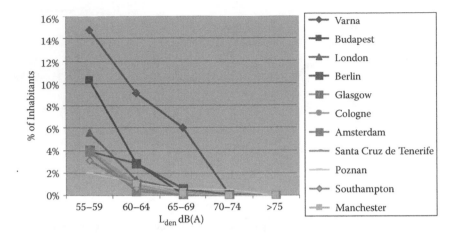

Figure 13.13 Aircraft noise: most exposed agglomerations (L_den indicator).

people exposed at L_{night} levels within the threshold have been reported by only 37 agglomerations.

In Figure 13.14, exposure of each country is presented. It must be remembered that this figure represents only the aircraft noise exposure inside agglomerations and not the overall impact of aircrafts. Noise from major airports will be presented in the following section.

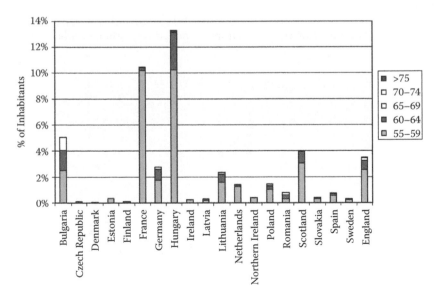

Figure 13.14 Aircraft noise exposure inside agglomeration (L_den indicator) in different countries.

Major infrastructures

Major infrastructures have been mapped according to the END requirements. In particular, all members have to report major roads, 20 members have to report major railways, and 19 have to report major airports.

Major roads data have been provided by 24 countries and railways by 19 countries (only 2 inside agglomerations). Regarding major airports, all countries reported at least some data but sometimes only an indication of the extension of the area exposed to different bands has been provided.

The exposure to major infrastructures in the EU database is given both for inhabitants outside agglomerations and the whole exposed population. However, the whole exposure is expressed more synthetically by number of people exposed to L_{den} values >55, >65, and >75 dB(A).

The major difficulty in analysing these data is the extension of infrastructures. In fact, infrastructure length or annual movements are not always available so that there are not sufficient parameters to compare exposure in different countries or airports. An attempt could be made to verify the portion of exposed people in higher bands as done in agglomerations.

Figure 13.15 shows the distribution in each country for roads and Figure 13.16 shows the distribution for railway infrastructures. Countries that have a portion of high roads levels exposed greater than two-fifths are Belgium, Finland, France, Germany, Hungary, Italy, and Slovakia. Regarding railway exposure countries with a portion of high levels greater than one-third are Belgium, Italy, Luxemburg, Norway, and Portugal.

Moreover, we create the HTH graph for major airports, which is reported in Figure 13.17. Airports with a fraction in high bands higher than one-tenth are Gran Canaria and Turin international airports (all series), Malaga airport (outside agglomerations, all indicators), Copenhagen

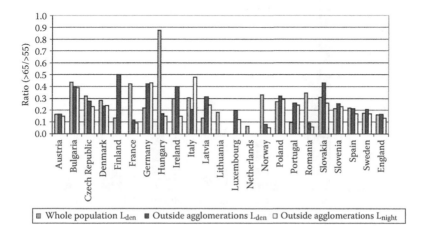

Figure 13.15 Major road noise: fraction of highly exposed for each country.

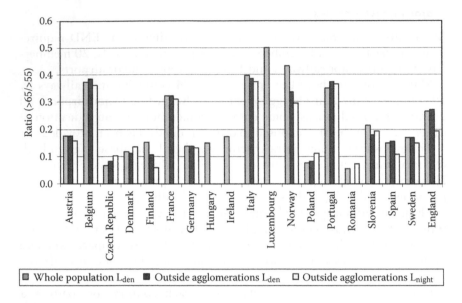

Figure 13.16 Major railway noise: fraction of highly exposed for each country.

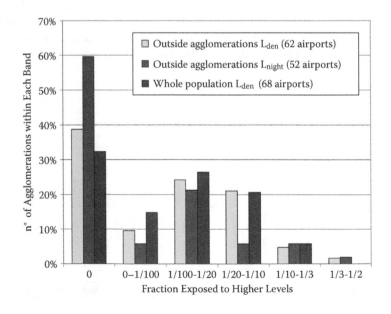

Figure 13.17 Major airport noise exposure, HTH graph.

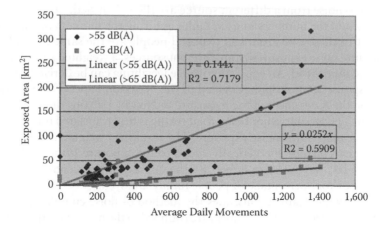

Figure 13.18 Relation between exposed area extension and daily movements.

airport (L_{den} values), Lisbon airport (only L_{night} values), and Paris Orly airport (whole population).

Other data reported for major airports is the area in squared kilometres exposed L_{den} >55 and >65 dB(A). This is directly related to the intensity of the flight traffic of the airport. Figure 13.18 shows the linear relation between area exposed and daily traffic (averaged from annual movements).

Assessing impact of major infrastructures, one of the challenging issues is to manage the presence of many sources, which are managed by different companies. In specific cases, there would be a risk of accumulation of noise from different sources, for instance, where major roads run near airports, or where infrastructure corridors include both a major road and a major railway. The segmented approach of different competent authorities for different sources implicates a serious risk of suboptimisation. One infrastructure manager could be tempted to take action, deciding that a particular site was to be considered a hot spot requiring noise mitigation. If the other source was the dominant one, the mitigation would be very inefficient. For such cases, the application of a common indicator for cumulative noise exposure or multiexposure should be enhanced (see later).

In urban areas, the aforementioned occasion is likely to occur much more frequently. Not only will there be common impact of noise from major roads and railways in combination with urban roads, but there will also be noise from urban sources like industrial sites that almost inevitably coincide with noise from urban road traffic. The risk of suboptimisation will be much larger.

One should be aware of this collaborative effect of different noise sources when interpreting numbers of exposed citizens. Often, one finds tables of numbers of citizens exposed to noise from one of the four sources, for example, road traffic noise. In a different table, the numbers of citizens exposed to noise from a different source are then identified. One is tempted to simply add the numbers in each table, ignoring the more-than-likely fact that part of these two groups of exposed people are the same individuals.

In adding noise exposures from different sources, a weighted summation is highly recommended. An equal annoyance weighting can be derived from the dose–response relationships provided in the "Position Paper of the Working Group on the Assessment of the Exposure to Noise." Usually, the noise exposure from a given source not being road traffic is first translated into an equal annoyance road traffic noise exposure. The resulting "weighted" noise exposure level can then be energetically added to the exposure from road traffic. Here, one has the choice to either use the curves for "percentage of people annoyed" or for "percentage of people highly annoyed."

The following example illustrates the method of noise cumulation from a road and a railway line. All numbers are rounded to the nearest natural integer.

Road noise exposure level L_{den} = 62 dB
Rail noise exposure level L_{den} = 67 dB

Energetic summation of these two levels would result in 68 dB ($62 \oplus 67 = 68$ dB, where the $\oplus$ indicates energetic or logarithmic summation). However, the noise exposure level of rail noise will cause 10.6% of highly annoyed citizens. The same annoyance score would have been achieved by road traffic noise with a level of 60 dB L_{den}. The "equal annoyance weighted' rail noise level is 60 dB L_{den}. The cumulative noise level then is 64 dB L_{DEN} ($62 \oplus 60 = 64$ dB).

The relevance of this method emerges when one starts defining mitigation measures for this situation. One would tend to start with the railway line, as it causes the higher exposure. A reduction of 5 dB of railway noise would cause an overall reduction from 68 dB to 65 dB ($62 \oplus 62 = 65$ dB), so an effective reduction of 3 dB. However, in terms of equal annoyance, the 64 dB would be reduced to 63 dB ($62 \oplus 55 = 63$ dB), so an effective reduction of only 1 dB. In conclusion, it would probably not be the best decision, in this case, to start reducing at the railway only.

OTHER ISSUES CONCERNING THE NOISE MAPPING PROCESS

Noise mapping reporting format

The implementation of the strategic noise mapping activity starts with the collection of relevant data. For road traffic noise, this data includes per stretch of road:

- Number of vehicles per category and per unit of time, during the average day, evening, and night

- Speed of these vehicles per category, during the average day, evening, and night
- Possibly: distribution of these vehicle flows over different lanes of the same direction
- Road surface type

This data can be derived from a traffic model, which is available in most of the larger cities. Such traffic models, however, do not supply all of the needed data in the proper detail. Default correction factors are usually applied, for instance, to derive the average week hour intensity (which is needed for the noise calculation) from the peak rush hour intensity (which is the usual outcome of the traffic model). Such correction factors are available in many countries, and the Good Practice Guide on Noise Mapping provides some recommendations in case they are not. However, the justified use of such standard correction factors may sometimes be questioned. In such cases, additional data will have to be collected, possibly in the field, to support the corrections.

For railways, it may be slightly easier to collect the data, as there is a time schedule available, which defines the exact traffic flows over a year's time.

Industrial source data may be difficult to collect, as there are competition restrictions involved. In the Netherlands, industrial sites are assumed to emit noise precisely in accordance with limits specified in the license to operate (noise permit). Although this is not always the case, it is an efficient first approach to map industrial sites.

When producing a noise map, the noise exposure of a building block can be characterised by allocating a single receiver point to the façade of that building block. According to the directive, the receiver point should be selected "at the most exposed façade" (Annex I). Note that the most exposed façade for industrial noise may differ from the most exposed façade for road traffic noise. This definition therefore requires a detailed investigation of every building block, which is usually beyond the scope of the noise map. Again, the Good Practice Guide gives some recommendations.

More complex is the obligation stated in Annex VI of the directive:

In addition, it should be stated, where appropriate and where such information is available, how many persons in the above categories live in dwellings that have:

- special insulation against the noise in question, meaning special insulation of a building against one or more types of environmental noise, combined with such ventilation or air conditioning facilities that high values of insulation against environmental noise can be maintained,

- a quiet façade, meaning the façade of a dwelling at which the value of L_{den} four metres above the ground and two metres in front of the façade, for the noise emitted from a specific source, is more than 20 dB lower than at the façade having the highest value of L_{den}.

The data on special insulation may be hard to acquire. In some situations visual inspection can be sufficient to decide whether a façade has been treated with an acoustic absorbing material, but the effort to collect such information through visual inspection is beyond the scope of noise mapping. In the Netherlands, large-scale noise insulation programs have been carried out between 1980 and 2010, but generally there is no record of where these projects have been carried out. Part of that work was actually achieved in the precomputer era.

An additional complication is that there are hardly any dose–response relationships for exposure to noise inside a dwelling. The large majority of noise annoyance refers to exposure inside a house, but it is related to a dose (i.e., noise exposure) of the outer façade. When relating it to interior noise, one would introduce significant cultural and climate differences, for instance, due to sleeping with either open or closed windows. All these complexities lead to the fact that special insulation is often ignored in noise mapping, in spite of the directive's requirements. However, when a hot spot is defined and noise mitigation is considered in a draft action plan, it is highly recommended to check whether any façade insulation has been installed before considering a new action at the same spot.

Some relaxation of noise annoyance can be achieved by supplying a quiet façade to a dwelling, which is otherwise exposed to high noise levels. This is the typical situation in many city streets, where the front façade is heavily exposed to traffic noise, but the backyard is relatively quiet thanks to screening by other building blocks. The way to identify such quiet façades as suggested in Annex VI includes a complex computation step, assessing the noise exposure levels at both the most exposed façade and basically any other façade of the same building, assessing whether or not there is a 20 dB difference. This is an elaborate computational step with very limited accuracy. After all, noise level assessment in the deep shadow of a building block may introduce large inaccuracies. Therefore, the effect of a quiet façade is most often ignored, in spite of the directive's requirements. Irrespective of this, it is highly recommended to create quiet façades in new urban planning, as it is a useful and cost-efficient way to achieve a better sound quality and reduce the health effects of excessive sound exposure.

Once a map has been produced, it should be used to produce the tables required in the data flow, that is, numbers of exposed citizens, in classes of 5 dB. The assessment of these numbers is not straightforward either. Most cities have demographic figures available and can make an estimate of the

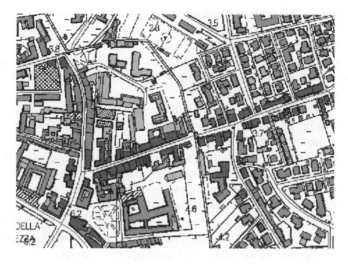

Figure 13.19 Coloured map showing most exposed façades. **(See colour insert.)**

number of citizens per dwelling. The next step would be to assess the number of dwellings in a building block. If there is no better data available, one can use the outer dimensions of the building block to estimate the number of dwellings but the number of dwellings assessed with this method has a large uncertainty. At the top end of the assessment is the computation of the noise exposure level of the most exposed façade of that building block. Often this step is based on a GIS cross-section of the geometrical plan, showing the building blocks in 2D, and the isophones or equal noise exposure contours. Figure 13.19 is an example of such a contour map. There is some flexibility of interpretation in deciding whether a given building block is in one shell between two contour lines or in another. In the strictest interpretation, the whole building block is considered to be exposed to the higher class as soon as one point of its outer dimensions is located within the respective contour shell. Obviously, such an interpretation leads to an overestimation of the number of exposed citizens.

As a way of presenting the current situation, the contour map is a suitable instrument. It shows clearly the loud spots in an urban area, allows people to have an impression of their noise situation, can clearly express the effect of future changes, and reflects the inaccuracy that is inherent in noise exposure levels on a large scale.

On the other hand, when defining hot spots for action plans, the contour map is probably not the preferred presentation. It draws attention to loud spots, but these do not necessarily require mitigation, as there may well be nobody living there. For designing action plans, the "dwelling exposure map" as shown in Figure 13.19 is a more effective presentation format.

It shows coloured building blocks, the colour being representative for the noise exposure (most exposed façade) of that block, and draws attention to the sites where many people are exposed to high noise levels.

More sophisticated representation formats have been developed and proposed, for example, in the EU project Q-City. Reference is made to this project for further detail.

There is no obligation or even recommendation in the directive itself with respect to the presentation format of the maps. The directive merely requires submitting the exposure data in tabular format. However, there is a distinct task for the actors, namely, to inform the public. Various presentation forms can be more or less suitable for this task. For this purpose, many cities have chosen to present the map in an interactive way, comparable to or based on a Google Earth application. This allows citizens to zoom to their own dwelling to find out their own noise exposure levels. For information purposes, this presentation is to be preferred. Care should be taken that this presentation format suggests an accuracy that does not really exist.

At the other end of the spectrum are printed maps (in pdf format) of a whole city, which give very little information to the individual citizen but are better suited for reproduction than interactive maps. If such pdf maps are applied, it is recommended to limit the scale to 1:10,000 or larger.

The noise map is the basic tool to assess the number of exposed citizens in classes of 5 dB. For this assessment, one can ask whether it is sufficiently accurate and sufficiently complete. The accuracy of the assessment depends on the accuracy of the input data and of the prediction method used. In principle, the prediction method is prescribed in the directive; for the first round of mapping, the common interim method should have been used. Here, the actor has no influence on the accuracy of the method, other than collecting a set of input data that has sufficient accuracy. With respect to the completeness of the data, there are some serious issues to be considered. These can best be discussed by looking at the distribution of exposed citizens over the various noise classes.

In an ideal hypothetical case, with a homogeneous distribution of the population density and one single line source (road or railway line), and assuming geometric spread of the acoustic energy as the only propagation attenuation, the distribution has the shape shown in Figure 13.20.

The bands of area between contour lines of 5 dB step size show a width that increases with the distance to the line source, and given the supposedly equal distribution of the population, the number of people within a contour band decrease with a factor of approximately 3 for every step from one contour band to the next higher exposure.

Recognizing this behaviour, we come to the somewhat shocking conclusion that by starting the assessment at 55 dB, we are ignoring the far majority of citizens in the city, that is, those who are living far from any main road.

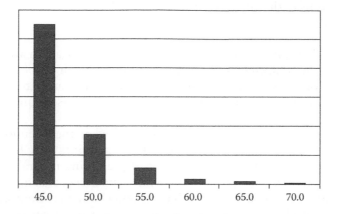

Figure 13.20 Theoretical exposure shape due to a single linear source.

This effect is enhanced by the input data. In most urban traffic models, the purpose of the model is to assess congestion during rush hours and to find alternative routing that produces a better flow. To this effect, most traffic models concentrate on the main routes, where traffic flows are heavy. The finer arteries of the urban transport network are usually ignored. These are the streets with less traffic but also with a narrow cross-section (not seldom street canyons) where even minor traffic flows may well lead to noise exposure levels of 50–55 dB. If such arteries are ignored in the traffic model, they will neither show in the noise map nor in the dataflow tables.

On the other hand, the effect is weakened by excess attenuation aspects, such as air absorption, meteorological effects, and ground absorption. These tend to lead to bands between consecutive noise contours that are narrower than indicated in Figure 13.20. The step between the numbers of residents in two consecutive classes of noise would be smaller than a factor of 3, and the curve would therefore flatten. This effect is moderate for a main road in a rural area; in an urban situation it would be strong, particularly at the first line of buildings.

Quiet areas

The follow-up of the noise mapping is the drafting of an action plan. In an action plan, the tendency is to concentrate on sites where large numbers of people are exposed to high levels of noise. In doing so, there is a risk of ignoring other sites, where noise quality is still good. This issue is addressed in the directive. According to Article 1c, the obligation of the member state is, among others, to preserve environmental noise quality where it is good.

To this effect, the directive defines what should be understood by a quiet area.

> Article 3(l): quiet area in an agglomeration shall mean an area, delimited by the competent authority, for instance, which is not exposed to a value of L_{den} or of another appropriate noise indicator greater than a certain value set by the Member State, from any noise source.

This definition raises some questions as to what the intention of the commission has been. Quiet areas outside urban zones have been established in several countries as an element of preserving natural beauty or protecting wildlife. Such areas are generally experienced as quiet as long as man-made noise is not dominating the soundscape (a soundscape is a representation of an observed sound quality in a given surrounding) in such areas. Natural sounds, such as wind noise in trees, bird song, and others are no threat to the noise quality. Therefore, there is little sense in setting a noise exposure limit—in whatever indicator—in order to certify or designate quiet areas.

Quiet areas inside urban zones represent an important value in city life, as they offer relaxation from the everyday hustle and noise in a city. Such quiet areas can be city parks as well as inner gardens in housing blocks. The areas can be characterised by the fact that the exposure levels are significantly lower than in the surrounding area, but this is merely a relative indicator, not an absolute noise exposure limit.

The obligation to preserve quiet areas is part of the full definition of Article 1c, stating that the member state is to adopt action plans based upon the noise mapping results. This suggests that the directive expects quiet areas to be identified on the basis of the noise mapping result, then designated by the member state as something worthy of preservation and so becoming part of an action plan. This is contradictory to the aforementioned statements.

It is certainly useful to pay attention to "areas where the noise quality is good" and try and preserve these. The standard mapping approach, however, is not the most suitable approach to prepare for such preservation.

Information to the public

Besides the production of strategic noise maps and the drafting of an action plan, informing the public is the third important task for city and agglomeration authorities required by the END. Informing the public is required to raise more awareness for the issue of environmental noise and a reason for the harmonisation of the approach. The citizen should be allowed the capacity to compare his situation with that of fellow citizens in other agglomerations or other countries of the European Union.

The third reason why the citizen should be informed is that he is given an important role in drafting the action plans. After all, the European Commission has refrained from setting a harmonised noise reception limit, even though some member states have specifically requested so. The reason is that the directive is based on the principle that noise is mainly a local problem that should in first instance be solved locally. Only a local judgment can make out whether it is fair and reasonable to spend a certain amount of money for noise mitigation. The directive therefore requires noise maps to be made available to the general public and requires public consultation on noise action plans.

In many practical cases, local authorities are reluctant to involve the public in what they feel is their responsibility. In some cases, the publication of noise maps has been well hidden, announced in the smallest possible announcement in a local newspaper. A complication adding to this is that the general public is often more concerned about the sources of noise that raise complaints (such as bars and discotheques, youngsters with mopeds, dogs barking, and roosters crowing) than the sources of noise that are potentially dangerous to health (road and rail traffic, air traffic, and industrial noise). In fact, when presenting mapping results to the public, inevitably the discussion comes up about the choice of the four sources to be mapped. People tend to consider these sources not the most relevant when it comes to noise annoyance. The general public tends to accept these sources as an inevitable part of urban life. To the general public, it is hard to understand why so much attention is paid to noise sources they would never complain about.

The recommended response to these citizens would be to state that the directive focuses on environmental noise with a potential hazardous effect, that is, impairing public health. Such effects only occur after a long period of exposure to noise, and the mentioned sources attracting complaints are those that occur only incidentally and usually for a short period of time. Although the general public considers them as particularly annoying, their health effect would probably be negligible.

It is a big challenge for future rounds of noise mapping to involve the general public more effectively in the process of noise mapping and action planning, and to profit from the advantages that such an involvement may have for the local authorities and politicians.

CONCLUSION

Data provided within the first round of mapping are now almost complete and only a few countries have not produced any data yet. However, comparability has not been reached due to different methods applied and different local scenarios so that it is hard to understand data.

In fact, differences are expected not only because of the mapping procedure but also due to the specific context. Agglomerations definition, population density, and mostly the infrastructures networks influenced results. It is obvious that those countries with large open countryside may be less noisy than the ones that have small towns all along the railway path. Moreover, an important role is played by the age of cars and wagon fleets. Unfortunately all these factors are not so clearly visible in the results because of the different procedures followed. In following the steps of requirements of the END, a common procedure should be applied that follows existing position papers written by the European Commission as the Good Practice Guide, version 2[8] and the Presenting Noise Mapping Information to the Public,[9] but also the common assessment method,[10] which is still under definition.

Improving comparability will help develop common actions to protect citizens from the harmful effects of noise exposure and it will also allow policy to benefit quieter countries.

REFERENCES

1. European Parliament and Council. Directive 2002/49/EC of 25 June 2002. Official Journal of the European Communities, 17-07-2002.
2. Noise Observation and Information Service for Europe, http://noise.eionet. europa.eu/.
3. M. van den Berg, G. Licitra, EU-Noise Maps: Analysis of submitted data and comments, Proceedings of Euronoise 2009, Edinburgh, Scotland, October 2009.
4. Vorläufige Berechnungsmethode zur Ermittlung der Belastetenzahlen durch Umgebungslärm—VBEB (preliminary calculation method for determining the exposure figures caused by environmental noise), in Federal German Gazette, 20 April 2007.
5. G. Licitra, E. Ascari, G. Brambilla, Comparative analysis of methods to evaluate urban noise exposure, Proceedings of Internoise 2010, Lisbon, Portugal, June 2010.
6. Department of Transport/Welsh Office HMSO (UK), Calculation of Road Traffic Noise (CRTN), 1988.
7. Vorläufige Berechnungsmethode für den Umgebungslärm an Straßen (VBUS) (preliminary calculation method for environmental noise at roads), in Federal German Gazette No. 154, 17 August 2006.
8. European Commission Working Group. Assessment of exposure to noise. Good practice guide for strategic noise mapping and the production of associated data on noise exposure, version 2, August 2007.
9. European Commission–Working Group Assessment of Exposure to Noise, Presenting Noise Mapping Information to the Public, 2008.
10. Draft JRC Reference Report (Contract no. 070307/2008/511090). Common NOise ASSessment MethOdS in EU (CNOSSOS-EU), version 2, May 2010.

Chapter 14

Noise maps from different national assessment methods

Differences, uncertainties, and needs for harmonisation

S. Kephalopoulos and M. Paviotti

CONTENTS

HISTORICAL NOISE ASSESSMENT METHODS
AND THEIR RANGE OF APPLICABILITY

In environmental noise assessment, noise mapping methods are used to assess the levels of noise over a large amount of positions forming a grid of elements, eventually equidistant one from each other. The first noise calculation methods in Europe were developed in the 1980s, and most likely at that time the objectives of the assessments were focussing on the most exposed locations in an as far as possible advanced manner. Although calculations were considered to be made eventually via the use of computers, evidently the way the first methods were written was triggered by the need to allow hand calculation of noise at single locations. Calculations at those times could already benefit from the first extensive measurements aimed at characterising the noise sources, typically near the source itself, which were then used as test cases to verify the correctness of the calculations themselves. Until recently, the testing against source data and simple cross-sections and only next to the source has remained as the most used strategy to prove the validity of the existing noise assessment methods in such limited conditions. For several methods, some combinations of source parameters and propagation parameters, which are difficult to effectively handle by means of physical descriptions, were modelled by the use of correction factors. This is, for example, the case of the correction applied to the noise produced by trains on the basis of the commercial definition of the type of track, which is nowadays instead implemented as a specific model depending on physical parameters such as roughness, pad, stiffness, and sleeper type. The same applies to meteorological corrections that were making use of an average correcting factor, whereas nowadays modelling of the meteorological effects is based on physical parameters like wind speed and temperature gradients.

The reason why a better modelling approach with a wider range of applicability was not searched for several years could be also attributed to the fact that primary interest was for the hot spots, these being the building's closest area to the source (with focus on the building's façade facing the source), and also the fact that the noise reduction was to be achieved with the placement of noise barriers next to the hot spots.

If a method (preferably as simple as possible) was capable of modelling a specific national situation (which mainly was created for) and by good approximation of the effect of barriers close to the noise source, the method had already met its objectives and justified its raison d'être.

The range of validity of a method, although sometimes stated in the methods' documentation, however, is often not considered and the methods have therefore been also applied for situations other than those for which they were originally developed, beyond their range of validity.

PARTS OF A NOISE ASSESSMENT METHOD

Source

The definition of the source part of a noise assessment method is done on the basis of the most influencing parameters for the most common situations to be assessed. Some specific conditions differing from the most common ones, though not frequent and not easy to model, are still tried because they represent cases of serious noise disturbance (e.g., although the squealing of trains in urban situations is not easy to model and it is not a commonly encountered situation, it is given consideration because it causes specific noise problems).

The noise source is mostly defined as a sound power. It can be expressed as sound power of single vehicles or other noise elements (e.g., a single car, a single fan in an industry), as sound power per length of noise source (in this case the source is treated as a line source), or it can be defined in terms of sound pressure at a reference position, near or far from the noise source itself (e.g., sometimes the road traffic noise source and almost always the aircraft noise which is described mixing source and propagation effects in a single tabulated value). It can therefore be concluded that there exists different ways of representing the noise source depending on the assessment method used and the noise source considered. This has created difficulties when assessing the differences among the different definitions of noise sources according to the various noise assessment methods.

In all cases, the source contribution is described by values obtained as a set of simple mathematical formulas and parameters, the latter being either the values of the database associated to the method (coefficients) or the input values necessary to implement the formulas for the specific source condition/location (user input).

A general list of the elements considered to model the source contributions is drawn, usually corresponding to a specific formula or correction coefficient described in the method, to have a quick overview of the components of the source specificities in existing noise assessment methods.

Tables 14.1, 14.2, 14.3, and 14.4 show the specificities of the source definition in noise calculation methods for road traffic, railway traffic, industrial, and aircraft noise, respectively.

Table 14.1 Road Traffic Noise Source Specificities

Classification of vehicles	Aerodynamic noise
Speed dependence	Bridges
Acceleration/deceleration (Traffic flow)	Tunnels
Gradients	Viaducts
Road surface type correction	Crossings
Tyre type correction	Segmentation of the source
Engine noise/Exhaust noise	Source(s) position

Table 14.2 Railway Traffic Noise Source Specificities

Wheel roughness	Track/support structure classification
Rail roughness	Bridges
Classification of vehicles/locomotives	Tunnels
Rolling noise/speed dependence	Viaducts
Engine noise/speed dependence	Crossings
Aerodynamic noise/speed dependence	Segmentation of the source
Squeal noise	Source(s) position
Braking noise	

Table 14.3 Industrial Noise Source Specificities

Point source definition	Area source definition
Line source definition	Sound power and directivity (database)

Table 14.4 Aircraft Noise Source Specificities

Segmentation (function of aircraft performance and track)	Source directivity
Aircraft performance and flight profile as a function of air parameters, aircraft type, engine type, TOW (database)	Dispersion of tracks
Aircraft noise as function of performance (database)	Ground operations

Table 14.5 Noise Propagation Specificities in Noise Assessment Methods

Geometrical divergence	Reflections
Atmospheric absorption	Diffractions/screening obstacles
Terrain profile	Modelling of meteorological influence
Ground effect	

Propagation

Propagation presents a wider set of approaches than the source definition, since its modelling has traditionally been the most challenging part of the entire noise mapping process. The propagation part of a noise assessment method accounts for the following physical phenomena associated with the sound propagation: geometric divergence of sound with distance; air absorption due to mechanical energy dissipation; ground absorption due to ground impedance; ground reflection; screening, reflection, and diffraction of buildings and other built-up objects (like noise barriers); tall vegetation absorption; and influence of meteorological conditions on the sound ray curvature along the propagation path.

In Table 14.5 the noise propagation specificities under which the physical phenomena are grouped together are outlined. In some methods few effects, such as those linked to barriers, are expressed in combination with other effects (i.e., the ground effect).

The segmentation technique, which is a methodology used to divide the source into single parts on which to perform the calculations, and the definition of the source by means of a point source or line source, lead to differences in the overall calculation of the noise propagation. For this reason, the segmentation is considered as also part of the propagation and not only of the source definition.

Table 14.5 presents a list of the main noise propagation specificities as typically described in existing noise assessment methods.

NOISE ASSESSMENT METHODS USED IN EUROPE

Major methods

In Europe, noise assessment methods were developed in the past years mainly by national teams of researchers and only in a few cases by international teams (e.g., ISO 1996, ICAO Doc 9911-ECAC Doc. 29, and HARMONOISE/IMAGINE). The description of these methods can be found either in the form of technical reports or as national standards (see references 1 to 17). These methods were addressing the general purpose of assessing noise levels at specific locations for a given configuration of the noise source and propagation parameters. They are widely implemented in software and typically used for assessing noise levels in association with national databases. The existing methods vary in terms of their complexity and capacity of modelling specific situations. Some may be simple though not capable of describing many specific conditions (e.g., CRTN allows for performing point-to-point calculations relatively easily even by hand but does not handle different ground reflections nor meteorological propagation); others may be more capable of handling various cases of interest (e.g., RVS allows for detail analyses of the possible combinations of multiple reflections on the façades of the buildings), but they are not simple enough to be implemented coherently (i.e., following the same approach) from one situation to another as the final choice for several implementation aspects is left to the end user.

To get a better understanding about the commonalities and the differences among the components of existing noise methods, the specificities of these methods were analysed and compared to a certain level of detail mainly in four studies.[18-21]

The main noise assessment methods widely used in Europe, Japan, and the United States are reported in Table 14.6.

Minor methods

Tendentiously, minor methods are not officially reported nor are reviewed at national or international levels by panels of experts. They are the result of small offices or small-medium enterprises know-how being embedded in software. This might be more typical, for example, for infrastructure managers of major roads, major railways, and major airports. Sometimes they were found to consist in adaptations of preexisting standards (widely used as starting point is the ISO 1996 for road traffic, railway traffic, and industrial noise, and the ECAC Doc 29 for aircraft noise).

ELEMENTS OF THE EXISTING NOISE METHODS LEADING TO DIFFERENCES IN THE NOISE ASSESSMENT

There are components of existing noise assessment methods concerning the description of the several acoustically relevant conditions of the noise source and the physical effects determining the noise propagation to which one can mostly attribute the differences observed in the outcome of a noise assessment performed by these methods. At this point it is important to clarify the following issue: When applying a method, some parameters of the method may have a major impact on the final result (i.e., the noise level calculated at the receiver point) compared to others. This is the so-called sensitivity of a specific method to its parameters. In a past study,[22] it was

Table 14.6 Noise Assessment Methods used in Europe, Japan, and United States for Assessing Environmental Noise

Road traffic noise
ASJ RTN 2009 (Japan)
CRTN (United Kingdom)
HARMONOISE/IMAGINE* (European Union)
NMPB 2008 (France)
Nord 2000 (Denmark, Finland, Iceland, Norway, Sweden)
RLS90/VBUS (Germany)
RMW (The Netherlands)
RVS (Austria)
Sonroad (Switzerland)
TNS (United States)

Industrial noise
HARMONOISE/IMAGINE (European Union)
ISO 9613 (European Union)

Railway traffic noise
CRN (United Kingdom)
HARMONOISE/IMAGINE (European Union)
Nord 2000 (Denmark, Finland, Iceland, Norway, Sweden)
Onorm 305011 (Austria)
RMR (The Netherlands)
Schall 03 / VBUSch (Germany)
Semibel (Switzerland)

Aircraft noise
AzB 2008 (Germany)
ECAC Doc. 29 3rd rev.-ICAO doc. 9911 (European Union)
HARMONOISE/IMAGINE (European Union)

Note: Text in parentheses indicates where the methods are developed and used, in general, mandatory.

*The HARMONOISE/IMAGINE methods have been developed by a European consortium funded by the EU Framework Programs Five and Six, and represent the first tentative method for developing the future noise assessment methods in Europe.

shown that different methods often show the same kind of sensitivity to the same input parameters, thus proving that a shift in the value of the most sensitive parameters would cause in both methods a corresponding shift in the final results. In this case, one may conclude that these methods both reflect the same prioritisation of the most sensitive parameters, however, no general conclusion can be drawn on the effect on the potential difference among the noise levels calculated with these methods.

When comparing a method to another, it is necessary to look at the specific contribution to the different results (i.e., to the point-to-point calculation) by the formulas describing a specific effect affecting the source strength or the propagation calculation. This should be performed, as much as possible, by describing the same physical situation of interest using an identical or equivalent set of parameters.

Before proceeding with a detailed analysis of the possible origins of the differences observed in noise assessment, it should be mentioned that there is still a set of elements that are recognised to be identical among the various methods, whereas others may be completely different. To the former category, for example, belongs the case of geometrical divergence and air absorption, which are identically described in most of the existing methods following the ISO 1996. This might also be the case for some expressions of the source strength, like for aircraft noise where data of the same database (e.g., the public Aircraft Noise and Performance (ANP) database accompanying the ECAC Doc 29 3rd edition, and ICAO Doc 9911 guidance documents on airport noise) is mainly used.

To the latter category belongs the case of the treatment of reflections in the presence of obstacles along the direct path from source to receiver. Some methods (e.g., HARMONOISE) include all reflections on the ground: before, between, and after the obstacles along the propagation plane. Other methods, in the presence of a barrier, neglect all ground reflections (e.g., RLS-90).

Set of input values

Another source for the differences in the results of the assessment consists in the databases used in association with the national methods, which in most cases are expressed following national definitions of the classes used (e.g., classification of vehicles for the road traffic noise, classification of trains depending on the commercial names instead of the acoustical parameters for the railway traffic noise), and that also include different relevant effects (e.g., some methods include the correction for the road gradient and may consider the accelerated traffic as opposed to others that do not). Finally, it is worth mentioning that expression of basic parameters such as the road traffic speed may also differ. Examples are given in the following formulas for the RLS (Equation 14.1), CRTN (Equation 14.2), and RVS (Equation 14.3):

$$Cspeed = 27,7 \cdot 10 \cdot \log[1 + (0,02 - v)^3] \tag{14.1}$$

$$Cspeed = 33 \cdot \log\left[v + 40 + \frac{500}{v}\right] - 68,8 \tag{14.2}$$

$$Cspeed = 26,2 \cdot \log\left[\frac{v}{50}\right] \tag{14.3}$$

Formulas used for the description of the source and propagation

Based on the studies conducted in the past a list of the major known causes of the differences in the noise assessment has been identified and attributed to the source and propagation parts of the methods. For noise propagation these are the definition of the number of reflections used; the formulation for the ground absorption; the formulation of the diffraction around obstacles; and the technique used to derive the propagation paths. This latter, for example, based on the method used it may imply the inclusion or exclusion of the ground reflection as it happens with the ISO 9613 and the NORD 2000 methods. The ISO 9613 approach considers that if the direct sound ray is interrupted by a set of at least two screens, the diffraction is calculated and the ground in between the screens shall be neglected. Instead, NORD 2000 considers all ground reflections between each single obstacle encountered along the propagation path.

Differences in noise assessment that are attributed to noise source concern the values used for the speed dependent source strength definition in case of road traffic and railway traffic noise, and the whole aircraft noise and performance data for the aircraft noise.

Another major difference in the noise source definition consists in the inclusion or exclusion of the first reflection of the ground into or from the single emission value representing the source.

Some other differences are in most cases of minor importance, like the methodology used to describe the ground impedance or "ground factor," the meteorological effect for all those situations either near the source or in urban situations, and the inclusion or exclusion of lateral diffraction or not (e.g., CRTN does not include lateral diffraction, whereas ISO 9613 does).

Differences in the noise source affect all calculated locations and have mostly a shift effect on the values at these locations.

In case of a direct propagation there is not much of a difference, however, if obstacles are present such that far source segments become relevant, the modelling of the segmentation may also become a relevant source contributing to the differences in the assessment observed. For this reason a thorough analysis of the results focussing only on simple cross-sections is not necessarily the best approach to understand the reasons for the differences observed.

Another problem in terms of direct comparability of the methods is the fact that, for example, some methods grouped the barrier effect and the ground effect into a single expression, and this makes it difficult to straightforwardly compare these methods, since the comparison could not be made separately effect by effect (i.e., the formula used for one effect in one method could not directly be compared to another formula used for the same effect in another method).

ANALYSES OF THE DIFFERENCES IN NOISE ASSESSMENT

Available quantifiable parameters

The possible parameters to be quantified in assessing differences in noise assessment when using various methods are many, if it is considered that each single element of a noise calculation method is to be analysed separately (e.g., one would analyse how the influence of the road gradient is accounted for in the various methods, and separately how the percentage of heavy vehicles is modelled by them). However, these specific analyses if done by considering each single element separately would normally lead to only a partial picture of the differences observed in the overall assessment results. Moreover, the different parameters sometimes combine each other compensating for different effects that could not be effectively distinguished.

L_{eq}

An alternative is to opt for a series of full calculations performed over a set of specific conditions (i.e., test cases), so as to verify the combined effects of the different elements of a method and its implementation altogether. This means to focus on the overall result of a calculation, in terms of L_{eq} calculated in specific assessment points (receivers). Differences can then be addressed among the different methods for this set of test cases.

Spectra

Similarly, it is possible to look at the details of the spectral levels calculated and their differences when using various methods. Moving the interest to the spectra is useful if an analysis of differences is required to understand a specific effect of, say, low frequencies, physiological effects of tonal components, or effectiveness of an action planning. In particular, the last one is actually the main reason that pushes modellers and authorities to perform calculations in third octave bands, since different noise reduction measures, like noise absorbing asphalts, rail grinding, and noise barriers, are actions whose effectiveness depends on the spectral components of the source, and therefore spectral results are required to assess the effectiveness of the action plans.

Population exposure

An alternative way to quantify the differences in the overall assessment is to investigate on the effect that the different methods may have on the estimation of the population exposure. Population exposure is referred to as the number of people in a given area, exposed to a certain noise level (usually given per noise bands of 5 dB as defined by the Environmental Noise Directive 2002/49/EC).[23]

In fact, it may be argued that even if two methods may locally (i.e., at a level of a single building) give different results in terms of population exposure (e.g., the noise levels in one building calculated by means of method A are higher than the levels calculated for the same building by method B and therefore the exposure of the population estimated by the two methods for that building will be different), the same would not necessarily apply when estimating the population exposure for a large area. In this latter case, differences in terms of the estimated population exposed in various noise bands may be counterbalanced across the various parts of the large area regardless of the method used.

Source sound power definition

Another indicator that might be relevant when comparing the various assessment methods is the source sound power. It is often useful to compare how the source part is modelled by the different methods, as it may be based on different databases and mathematical formulas. This is of special interest when the methods share equivalent propagation parts, as this would help elucidate on source data conversion issues from one method to another and on the differences in the assessment that can be attributed to the source and propagation parts of the methods. Even if such a comparison might seem to be straightforward to perform, at the moment there is no international standard to define noise sources that in the existing assessment methods have rather different indicators to determine the source strength. For example, sound power may be given per vehicle, or per metre line, or a sound pressure at a reference distance is used to quantify the source strength. Moreover, the first reflection of the noise produced by a source on the ground may be included in the source definition.

Overall, it can be concluded that comparison of the source strength among the various methods is not straightforward. Comparisons performed in the past have been mostly based on the full method (source and propagation modelling) at close by distances to the source so as to analyse the differences in the source strength modelling.

Uncertainties in the determination of differences among noise assessment methods

Interpretation of the standard

The different methods typically have a similar structure, in the sense that they model each "physically separable" contribution by means of a set of equations, parameters, and procedures (e.g., the air absorption, the ground reflection, etc.).

One issue of major concern is how well a method is documented. In the past it happened that subsequent versions of a method and its associated documentation had to be produced to better cope with the requirement of the unique and clear interpretation of the method's description.

Also, some formulas sometimes presented "bugs" for a specific combination of conditions or known limitations in the modelling, which again have not been always well documented. In such cases the option often adopted was to use correcting factors or alternative procedures not based on physical modelling to better match the calculations to the measurements. For example, the NMPB method foresees a correction to be introduced for objects in which one dimension is smaller than 0.5 m if they are more than one, although barriers, as like any screening object, are modelled in the general part of this method.

These corrections have to be interpreted in the software that performs the calculations and decisions have to be taken at the programming level, both left to the experience of the software experts and not always done on a thorough and transparent manner. Again, as an example, methods like NMPB and RLS 90 introduce specific procedures for road in trench. The software has to "interpret" when to consider a cross-section a trench and when to consider it as a standard cross-section, eventually with a screening object of a given height.

The things become even more complicated if we consider that real cases are much more complex to model compared to the ideal situations, which are mostly used for validating the existing methods.

Software implementation

The description of the methods is commonly translated into software, which then becomes the interface for interpreting the method documented on paper and making it easily accessible and manageable to the end user.[24]

There are differences in the outcome of the noise assessment due to the different ways the calculations are performed for a given noise assessment method in various existing software. Part of the differences consist in the programming choices made to perform the computation, including the strategy regarding the segmentation technique, which is not always mandatory (e.g., the segmentation technique is a critical issue in aircraft noise modelling, and it is typically described in terms of minimum and maximum segment length in methods for road traffic, railway traffic, and industrial noise mapping). This also includes the use of software proprietary techniques to get noise values without strictly following the method on each of the assessment points with the ultimate objective of speeding the calculation time.

Another part of the differences accounts for errors potentially introduced by the different sound rays traced by the so-called geometrical engine of the software. The geometrical engine interprets the digital terrain model (DTM), the objects, and finds possible rays between the source and the receiver. Different ray positions in the space means that different cross-sections are identified and calculated by the software (since all methods establish formulas for performing point-to-point calculations across a given cross-section). Because software codes are not publicly available and the choices made for the calculations and the segmentation strategy are not usually reported in the documentation accompanying the software, the only feasible way to estimate the differences arisen by the implementation of the same method in various software is again by performing calculations on appropriately chosen test cases and then comparing the results.

Such an exercise was performed in the past by the Joint Research Centre of the European Commission and included four different commercial software. Test cases were developed and implemented into these software for which a strict protocol related to the parameters and input used, including software calculation options, was followed. In this section, an example is given concerning the different software implementations for a few simple situations, mainly representing road traffic noise on a flat terrain with focus put on the receiver positions showing larger differences.

In Figure 14.1, points in light grey denote a standard deviation between 1 and 2 dB due to different software implementation, and points in dark grey a standard deviation between 2 and 3 dB. All uncoloured points resulted in standard deviations of less than 1 dB.

As it can be seen, the standard deviations were higher than 2 dB at locations where noise barriers were present and also in the case of building reflections. The same result was also encountered in other situations, and it seems reasonable to conclude that barriers are among the most critical parameters to consider when looking at software differences. Still relevant

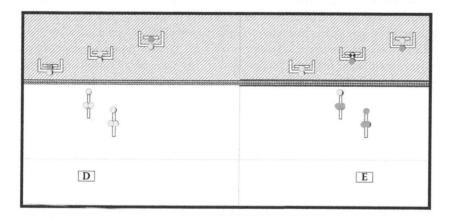

Figure 14.1 Plots of an example of road cross-sections for flat terrain. In light grey the points at which standard deviation was found between 1 and 2 dB, and in dark grey those between 2 and 3 dB.

but less critical are the reflections and screening of buildings. Propagation in a free field is absolutely consistent as implemented in the software tested.

The outcome of the calculations performed by the four software tested in other situations not presented here showed that for road traffic, railway traffic, and industrial noise[25] the standard deviations between these software calculations is lower than 2–3 dB for all assessment locations.

For aircraft traffic noise, the situation is different and the outcome of the software tested (on the basis of the ECAC Doc 29 2nd edition method and the same database) show good agreement. This can be in part attributed to the fact that the approach for modelling aircraft noise propagation in existing noise mapping methods is oversimplified (i.e., no screening and ground impedance are considered and the segmentation technique is fixed). It can therefore be expected that existing implementation of aircraft noise in various software will not result in differences in the noise assessment.[26–28]

Different definition of the input parameters

Another issue of concern, which might complicate the analysis and the interpretation of the differences in the noise assessment when different methods are employed, is in the nonequivalent definition of the input parameters in existing noise assessment methods. For example, some methods account for meteorological differences and include classifications based on three or four classes of road vehicles as opposed to others that do not or they include a vehicle classification based only on two classes of vehicles. For aircraft noise, an example is given on the differences in the overall takeoff noise depending on the database used for aircrafts. In Figure 14.2, a graphical comparison is made between calculations using two different databases. The first is given specifically for each combination of aircraft type and engine type, and the second is made only with groups of aircraft types where the specific aircraft type and engine type is mixed with others considered to be equivalent.

Table 12.7 presents the changes in areas between the two databases for each aircraft used for the comparison, in the case of a straight departure (ST), a turned departure (TU), and an approach (AP).

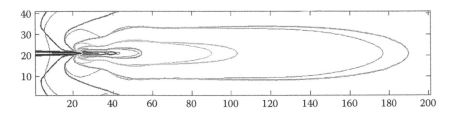

Figure 14.2 An example on the effect of choosing two different input databases for the determination of an aircraft takeoff noise footprint.

Table 14.7 Percentage Change of the Calculated Areas above 55 dB L_{den} for Each of the Single 19 Aircrafts and for Straight Departure (ST), Turned Departure (TU), and Approach (AP) by Means of the Interim Method with the ANP Database versus a National Calculation Method

	DHC 830	BAE 300	EMB 14L	F100 65	737 300	737 400	737 500	737 700	737 800	757 RR	A319	A320	A321 23	MD 82	MD 83	767 300	777 300	A340	747400
Straight departure	533	15	9	10	0	11	3	0	−29	−28	0	−8	−10	8	0	13	6	13	7
Turned departure	657	35	8	8	7	9	8	2	−31	−23	11	7	−4	15	18	11	11	16	30
Approach	0	16	6	14	0	6	0	28	12	0	0	6	−5	21	27	6	4	7	5

Table 14.8 User Settings Influence Final Results

	Motorway	City
Standard deviation	1.8 dB	1.7 dB

Source: Acts of the 37th AIA Conference, Siracusa, (Italy), May 2010.

Software settings

The effect of different ways in handling the settings of the noise software and sometimes the different interpretation on how to use a specific input parameter in connection to the field measurements may also cause differences in the calculated results. This happens since end users are allowed for handling some of the software options, for example, for reducing the overall calculation time regardless of the quality of the final result. Moreover, traditionally each software has developed its own semantics regarding the calculation parameters triggering the software calculation engine. The Italian environmental agency has tested the end user effect by providing some tens of end users with strict calculation test cases to perform. The set of test cases were previously developed by the Joint Research Centre of the European Commission in the context of the exercise on equivalence among the existing national noise assessment methods against the interim ones. The result was that on average the standard deviation of the results depending on decisions of the end users was about 1.3 dB to 2.8 dB (Table 14.8).

Objectives of existing noise assessment methods according to the Environmental Noise Directive (2002/49/EC)

The methods in the annex to the European Directive 2002/49/EC are intended to be used to assess the number of people exposed to different bands of noise levels and to prepare maps of noise levels at the façades of buildings in the European countries. The "results" referred to in the Environmental Noise Directive (END) are the L_{den} and the L_{night} levels, and not in terms of the number of people exposed, notwithstanding that the latter is part of the results to be submitted through noise mapping and might be the most relevant indicator for a correct exposure assessment.

In general, other factors—in addition to those presented in this chapter—may contribute to the differences observed for a given noise mapping study, which still affect the overall number of people exposed as required by the END. For example, the quality of the input values (e.g., DTM, buildings, positions of sources, source traffic data, screen heights, local meteorological data, etc.) and the strategy used to attribute noise levels to the façades as well as people to the buildings may significantly impact on the differences observed.

Unfortunately, Directive 2002/49/EC is not so specific for addressing a minimum quality of the noise maps to be produced in the foreseen rounds of noise mapping in Europe nor does it address the quality of input data and the methodology used to assign population and noise levels to the buildings.

The need to extend the methodological concept of existing noise mapping methods to include data quality requirements, appropriate guidance in operating the methods (including the handling of software settings), and a common strategy for attributing population and noise levels to buildings has already started to be discussed among the European noise experts in the context of the development of the common noise assessment methods in Europe (CNOSSOS-EU). These are important issues to take on board as noise mapping, besides being a costly process, it also triggers the estimation of number of people exposed to noise and the associated health risk and has strong implications on the huge resources made available if action plans are based on wrong estimations.

It is interesting to look at the overall picture of the differences in noise assessment in the European Union, as estimated from a comparison among major European cities of the same dimension and with respect to the same noise source (i.e., road traffic) using the data reported by member states during the first round of noise mapping in Europe.

It is worth noting that the population exposure to more than 55 dB is between 15% and 95%. This result appears quite astonishing and motivates noise experts to start elaborating ways to effectively improve the reliability of the noise mapping process in Europe.

Overall evaluation of the differences in noise assessment

Before proceeding with an overall evaluation on the differences in noise assessment, it would be instructive to have a look at graphs showing results of road traffic noise modelling. In Figure 14.3, the differences of the interim method for road traffic noise (NMPB) and one national method (SRM II) are depicted. These differences have been corrected to account for possible errors in software use. The differences between the two methods are to be interpreted as the difference in noise levels predicted between SRM II and NMPB, thus a negative value of, say, −4 means that the modelling with the SRM II results in 4 dB less than the NMPB. To evaluate the influence of the software, one should look at the constant lines, which represent ±1 standard deviation of potential errors due to the software.

The two graphs in Figure 14.4 represent the differences among seven noise calculation methods, for which the same DTM, buildings, ground properties, and source properties were used. The first graph shows the overall differences including the modelling of both, the source strength, and propagation, whereas the second graph shows the expected difference among the seven

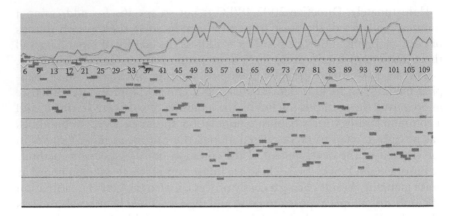

Figure 14.3 A ±1 standard deviation of the implementation in software of the interim method NMPB (continuous lines) and results obtained using a national method SRM II (dashes) for each assessment position for road traffic noise test cases. Each horizontal line corresponds to 2 dB.

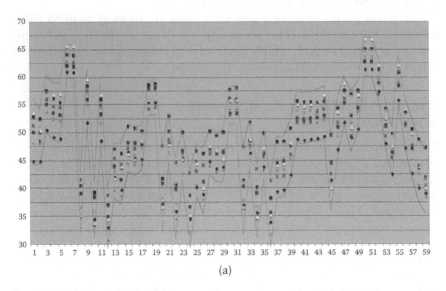

(a)

Figure 14.4 (a) The calculated values in 59 receiver positions with 7 well-known noise calculation methods for a specific situation and using the same software and calculation settings by one end user. Only the calculation method is changed.

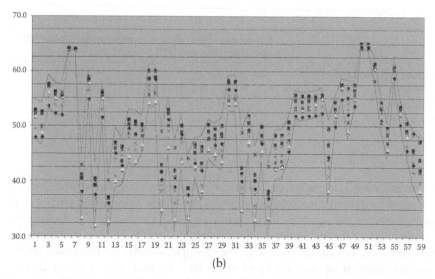

(b)

Figure 14.4 (*Continued*) (b) The same values but artificially shifted so that all values match for close-to-source receiver positions, somehow cancelling the differences only due to the source definition. The bottom and the top continuous lines represent a ±1 standard deviation of the methods' population of values.

methods only due to the propagation. For performing the calculations for this latter case, the source strength for each of the seven methods correspondingly was set at 7.5 m from the source in all seven different methods tested.

It is acknowledged that generalising the effect (the difference) if choosing a specific method and database with respect to others is not possible. Nonetheless, thanks to the studies performed by different noise experts some overall numbers on the potential differences can be drawn based on the literature previously presented. Moreover, the analyses have been so far wide enough thanks to the international teams of experts and the number of end users involved in the development and the implementation of the test cases. Based on the reasonable assumption that source definition, propagation, software implementation, and the end user's options/interventions are independent sources of potential errors to the final result, an overall standard deviation associated to the final result can be derived when different methods, software, and level of technical preparedness are employed. A simple approach is to take the standard deviations discussed previously as partial uncertainties and then obtain the overall uncertainties.

The generalised conclusions that can be derived in terms of estimated uncertainty are presented in Table 14.9 for two situations representing the best and the worst case scenarios based on the available know-how today.

When the source and propagation parts cannot be considered totally independent (for example, if reflection is included in the source but the first reflection

Table 14.9 Estimation of Uncertainties Encountered in Noise Mapping Calculations

Element	Standard Deviation (best Case/Simple Situation) [dB]	Standard Deviation (Worst Case/Complex Situation) [dB]
Source definition and source database	3	5
Propagation	3	6
Software	1	2
End user	1.2	2.8
Overall	4.1	7.9

of propagation is excluded, or when corrective values are introduced into the source definition to counterbalance effects due to the propagation part) it shall in principle be necessary to derive sensitivity coefficients and use these in the combination of the uncertainties. Because it is difficult to go deep in the correlation analyses, one can alternatively, for example, assume for combined source and propagation an uncertainty of just a little bit more than the one derived from past analyses, like 4 dB to 6 dB, so that the overall standard deviation now becomes 4.3 and 6.9, slightly less than what is considered in the table.

To estimate the effect of such uncertainties on the final results, the indicators discussed earlier can be used.

The equivalence exercise performed by the Joint Research Centre used three indicators on the basis of which the following three differences have been evaluated:

1. *Weighted percentage of* $L_{national\,method}$ at single assessment points that fall outside the error boundaries estimated by using different software.
2. *Overall averaged difference* L_Δ (considering the error boundaries estimated by using different software and allowing possible compensations between points where the difference $L_{interim} - L_{national\,method}$ is positive and points where the same difference is negative within the same noise band).
3. *Percentage differences of people exposed* to several bands: 25–30 p_{25-30}, 30–35 p_{30-35}, 35–40 p_{35-40}, 40–45 p_{40-45}, 45–50 p_{45-50}, 50–55 p_{50-55}, 55–60 p_{55-60}, 60–65 p_{60-65}, 65–70 p_{65-70}, 70–75 p_{70-75}, >75 $p_{>75}$.

The assumptions under which these indicators can be used are the following:

- The noise maps performed under the END requirements are run on a very large number of cross-sections, therefore the number of buildings affected by different noise levels can be assumed to be a continuous function of the levels (also, this corresponds to the fact that a shift in the noise contours will continuously increase/decrease the number of buildings affected).

- The people are attributed to the same buildings regardless of the noise mapping method used, hence, if a method—with respect to another one taken as reference—attributes higher values to that building, then all population therein present will be exposed to these new values.
- The number of people affected by noise levels and living in the buildings is very large (and therefore it is possible to assume that it is continuously distributed among buildings).

Indeed, if the levels obtained using national methods are different from those calculated by the interim methods, the obvious (and banal) consideration that "the people will nevertheless remain assigned to the same dwellings" (since people are assigned to dwellings regardless of the calculation method used) would lead to the following solid conclusion: the differences between the results of the two methods would be equivalent when the noise levels are considered or the number of people exposed to these noise levels is considered.

An example can be useful to understand the peculiarity of the aforementioned issue. Let us suppose that 10 people are living in a dwelling that, according to method A, is affected by a noise of $L_{den} = 52$ dB on the façade, whereas method B estimates that the same façade (same cross-section) is affected by $L_{den} = 57$ dB. Obviously, the difference in the results between method A and method B could either be regarded as a +5 dB L_{den} difference between method A and B or, alternatively, as 10 more people exposed to the next 5 dB L_{den} band (10 people less in the 50–54 band and 10 people more in the 55–59 band).

For the two methods presented, this corresponds to the following differences among the population exposure assessed: for the population exposed to more than 65 dB, given the number of people calculated with method A is x, that with method B will result $0.3x$, and given the number of people with more than 55 dB according to A is y, according to B will reduce to $0.65y$.

PAST COMPARISON STUDIES AMONG EXISTING NOISE ASSESSMENT METHODS

There are three major analyses known in literature that identify the differences among implementations of existing noise assessment methods: two studies conducted in the Netherlands[20,21,29] and one study performed on behalf of the UK government.[22] The main conclusions from these studies are summarised in the following.

The larger study[20,21] was conducted on a set of the most commonly used national noise assessment methods, detailed analyses were performed on the different parts of the methods, and efforts were made to evaluate separately the effects of the source part (e.g., speed influence, type of asphalt, acceleration correction, etc.) from those of the propagation part (e.g., ground absorption, barrier, meteorological correction, etc.).

Concerning road traffic noise source emission, it was concluded that considerable differences occur for the emission models of cars, where up to 6.2 dB(A) differences were found for light vehicles and 4.4 dB(A) for heavy vehicles. It was also concluded that the speed dependence is the most relevant parameter affecting the final result along with modelling the same conditions by means of one or more line sources, the asphalt type, and the road gradient.

Concerning the propagation part, it was concluded that meteorological conditions as well as the ground effect are modelled differently, having substantial effect especially in combination with the presence of noise barriers. An overall effect cannot easily be assessed, but for each single component (i.e., diffraction, ground absorption, meteorological correction) of the propagation, commonly up to 5–10 dB(A) differences among methods for road and railway noise assessment can be observed.

In the study between the Dutch method and the NMPB '96 method, it was concluded that the differences due to the database itself, that is, the differences near to the source, were about 1 dB. The differences due to the propagation could be in the order of ±3.7 dB around an average mean difference of 1 dB in city centres, and of the order of ±2 dB but with a larger averaged mean difference between methods of 8.3 dB in open areas and in the case of highways, and finally of about ±1 dB but with a large averaged mean difference of 6.4 dB in suburban areas with small and large buildings.

NEEDS FOR HARMONISATION

Large discrepancies among results obtained by different calculation strategies

As it was outlined in the previous chapters, the discrepancies among the use of different methods, input databases, software, and nontrained software end users on the overall calculations can be considerable. These discrepancies are high in the sense that they still do not allow the safe comparison of different locations, given that the number of people exposed might be double or half the estimated one depending on the differences outlined. Moreover, from a perspective of the action plans, the differences, which become uncertainties if these are not quantified in full by tests against widely approved and representative reference cases, do not allow one to safely assume that action plans are accurately prepared. As a consequence, expenditures may easily double, or, on the other side, being too weak with respect to the real citizens' protection required.

Moreover, the exercise on the equivalence of the methods made clear that there undoubtedly do exist differences between the national and the interim methods that are not necessarily overshadowed by different software implementations. Different calculations can easily over- or underestimate the population exposure to different noise levels bands by 50% to 150%, making it impossible to compare data. Noise mapping is therefore still far from ensuring a sufficient quality. Common EU noise assessment methods and validated and updated EU databases of input values are needed to solve all these issues and achieve sufficient quality in results.

Conceptualisation and development of the Common NOise ASSessment MethOdS in Europe (CNOSSOS-EU)

The purpose of the END is to perform an overall assessment of the relevance of noise as a potentially serious health problem for large areas, to identify hot spots, and to quantify overall numbers of people exposed with good accuracy. For specific delimited areas where deep understanding of the problem is required to identify and fine-tune action plans and to specifically evaluate the effectiveness of these action plans when implemented, precise determination of the noise levels to which people are exposed is an asset. In this case, for using the existing noise assessment methods with reasonable effort, requirements on calculation power and availability of input values should be kept reasonable as well.

A "fit for purpose" approach for appropriately using noise assessment methods in a multitargeted assessment framework in the member states of the European Union is under elaboration by the European Commission's Directorate General for the Environment (DG ENV) and the Joint Research Centre (JRC) in the context of the development of the Common NOise ASSessment MethOdS in Europe (CNOSSOS-EU).

For the development of this common noise calculation methodological framework and its associated databases the following criteria were adopted:

- Ability to consider differences in noise sources among EU regions (i.e., specific features that vary due to environmental factors like road and railway surface maintenance, specific regulations in force, techniques used to prevent noise, road, and railway networks and aircrafts)
- Ability to consider meteorological effects
- Easiness of implementation
- Availability free of intellectual property rights
- Integration of up-to-date scientific evidence
- Availability and quality of input data
- Fulfilment of the requirements of the END

The roadmap for the preparation of CNOSSOS-EU includes:

- The exercise on equivalence of existing noise assessment methods in the European Union
- The definition of the target quality and input values requirements for European noise mapping
- The establishment of requirements and criteria for the screening, rating, and preselection among existing assessment methods in the EU, United States, and Japan that best cover the needs and requirements of END
- The conceptualisation of a "fit for purpose" framework, which allows a two-step application of CNOSSOS-EU according to the objectives of the assessment (i.e., strategic noise mapping or action planning)
- The selection of the components of the common noise assessment methods through a series of workshops, benchmarking/testing, and other ad-hoc meetings with European noise experts
- The drafting of the CNOSSOS-EU methodology along with guidelines for its competent use in connection with data requirements and in line with the two-step approach
- The in-depth consultation, review, and finalization of CNOSSOS-EU together with the EU member states
- The preparation of the operational part of CNOSSOS-EU
- The long-term planning for assisting the EU member states in reliably implementing CNOSSOS-EU starting from the third round of noise mapping in Europe

The first draft of the CNOSSOS-EU methodological framework that can be used for strategic noise mapping purposes or for action planning came up in May 2010. Almost 200 scientists contributed with suggestions, discussions, proposals, and amendments to the draft text of CNOSSOS-EU. Since June 2010, the advancement of the CNOSSOS-EU framework is being performed through a formal process with the EU member states under the umbrella of the Noise Regulatory Committee managed by DG ENV and the technical coordination of the JRC.

The common noise assessment framework (CNOSSOS-EU) will allow for a coherent, transparent, optimised, and reliable use for strategic noise mapping (first step of application, mandatory) and action planning (second step of application, voluntary) in relation to the data requirements, their quality, and availability, and last but not least, in terms of flexibility to adapt the national databases of input values, thus ensuring a smooth transition from existing national methods to the common methods.

For strategic noise mapping, a simplified version of CNOSSOS-EU can be used with less input data requirements along with default data where appropriate; a more sophisticated version of the CNOSSOS-EU with associated increased requirements of input data can be used on a voluntary basis for noise planning purposes. The CNOSSOS-EU framework would help increase the consistency among the action plans adopted by the EU MS on the basis of the results of the noise mapping as stated in Article 1 of the END and also would allow a better evaluation of the effectiveness of the action plans and the development of a basis for Community measures by the commission to reduce noise emitted by the major noise sources (Article 1.2 of the END). This would also allow the EU member states to concentrate more on the reliable implementation of a common tool (i.e., CNOSSOS-EU) and its further development and maintenance, thus optimising their efforts instead of coping with different assessment noise methods used for different purposes and with different capabilities and range of applicability, which is a highly demanding task in terms of both resources and time.

The European Commission, the European Environment Agency, and the EU member states aligned to the requirements of the END (Article 1.1) are intensifying their efforts for facing at best the big challenge and opportunity to:

- Make available to the European citizens reliable information on the noise levels they are exposed to and the associated health implications
- Draw appropriate action plans for preventing and reducing exposure to harmful levels of noise

REFERENCES

1. Calculation of road traffic noise, Department of the Environment and Welsh office, 1988.
2. Guide du bruit des transports terrestres—prévision de niveaux sonore, Centre d'Estudes des Transports Urbains, 1980.
3. Doc 9911—Recommended method for computing noise contours around airports ICAO, First edition, International Civil Aviation Organisation, 2008.
4. Calculation of railway noise, Department of Transport, United Kingdom, 1995.
5. Semibel, Version 1, modele Suisse des emissions et des immissions pour le calcul du bruit des chemins de fer, Manuel d'utilisation du logiciel, 1990.
6. SCHALL 03 2006, Richtlinie zur Berechnung der Schallimmissionen von Eisenbahnen und Straßenbahnen Entwurf, Stand: 21.12.2006, Germany, December 2006.
7. Verkehrslärmschutzverordnung, 16. BImSchV, Bundesministeriums der Justiz, Germany, 1990.
8. Reken- en meetvoorschrift wegverkeerslawaai 2002, The Netherlands, 2002.
9. H.G. Jonasson, S. Storeheier, Nord 2000. New Nordic prediction method for road traffic noise, SP Rapport 2001:10, Borås 2001.
10. H.G. Jonasson, S. Storeheier, Nord 2000. New Nordic prediction method for rail traffic noise, SP Rapport 2001:11, Borås 2001.
11. Reken- en meetvoorschrift geluidhinder 2006, Staatscourant 21 December 2006, nr. 249, The Netherlands, 2006.

addition to making provisions for the assessment and management of noise pollution, the legislation includes instructions that the information generated during these processes be disseminated to the general public. Both strategic noise maps and resulting noise action plans are covered under these instructions; indeed, public consultation and participation is noted as being a key element in the production of action plans. Furthermore, the directive aims to progressively raise awareness of the potentially harmful consequences of exposure to excessive noise by ensuring that information on environmental noise and its effects is freely available to the public. To quote directly from the legislation, all material must be made available to the public using "the most appropriate information channels" and should be "clear, comprehensible and accessible."[1]

Although no deadlines or preferred methods are detailed for these actions, the intent of these sections of legislation is clear, the public is the key stakeholder in the noise management process and all information should be communicated in an effective manner. However, this key step in the noise mapping and action planning cycle is not as straightforward as it may seem. Disseminating "comprehensible and accessible" noise prediction data to the public at large presents considerable difficulties, largely due to the technical nature of noise descriptors. This task is essentially a two-step process, the first step being to educate the intended audience and equip recipients with enough knowledge to interpret and understand the material being presented. Only then can the relevant information be effectively communicated.

THE ROLE OF THE PUBLIC

The study of noise, as distinct from acoustics, is not an exact science. Although no less technical in nature, the treatment of noise must allow for the unquantifiable and widely varying effects of human perception. While the generation and propagation of noise may be repeatable throughout a range of conditions, the perception of all acoustic phenomena is inherently subjective, and none more so than noise. The level of annoyance caused by environmental noise from any particular source can vary from person to person and can depend largely on the listener's environment and attitudes to noise. Consequently, the task of quantifying noise annoyance and relating this information in an impartial manner becomes a complex problem. But why tackle this challenging issue in the first place? What are the benefits of relating this highly technical information to the public at large? An examination of the roles of major stakeholders in the noise management process may illustrate the potential gains of making this complex data more widely accessible.

A stakeholder can be any group that makes a significant contribution to the process or is likely to be affected by any resulting actions. Some of the key stakeholders include the noise mapping and action planning authorities, environmental groups and agencies, politicians, transport authorities, local government councils, land use planners, and, of course, the general public. An effective implementation of the Environmental Noise Directive, that is, generating strategic noise maps, proper analysis, subsequent action planning, and eventual noise management involves participation from a number of these groups and organisations. Initially, the responsibility lies with the designated mapping authority, typically a government body such as an environmental department, city council, or transport authority. These groups produce noise maps for their own functional areas, which can then be used for the generation of noise mitigation strategies. It can safely be assumed that within the noise mapping authority there is an appropriate level of expertise needed to fully understand and interpret noise mapping results. The contribution made by the noise mapping authority, that is, the generation and interpretation of strategic noise maps, is a prerequisite of the directive. Effective participation in the noise assessment and management process is assumed and little incentive is needed to promote engagement. Perhaps at odds with all other stakeholders, the noise mapping authority does not require noise predictions to be communicated to them, but rather are responsible for disseminating the information generated during the mapping phase.

Closely linked to the mapping authority is the noise action planning authority, the body responsible for processing noise mapping results and generating short- to medium-term policies to manage environmental noise levels. These authorities may be designated from any of a number of relevant organisations, for example, city councils, local authorities, transport authorities, or environmental departments. In some cases the noise mapping authority and the action planning authority are one and the same. The policies and actions that are recommended are based primarily on strategic noise maps. Generating an action plan requires the interrogation and analysis of noise mapping data on a detailed and localised scale, and so a minimum level of technical expertise must be present within the action planning authority. Accordingly, noise maps can be delivered to the relevant body in an "undiluted" form and little or no extra information is required to ensure effective communication. In order to fulfil the brief of the action planning authority, the designated organisation must have a high level of familiarity with the concepts and metrics involved in the noise mapping process, making the transfer of knowledge relatively straightforward. Similarly, with land use planners who may have input into an action plan, a certain level of expertise in noise pollution and other environmental issues can be reasonably assumed. The relevant information for these groups consists of strategic noise maps, as produced by the relevant mapping authority, and any accompanying technical analysis. This is the purpose and audience for which noise maps are intended.

The aforementioned stakeholder groups can be expected to have moderate familiarity with the concepts involved in producing and interpreting noise mapping results. At least a general acquaintance with noise measurements and statistics should be present among staff in these bodies. Processing community noise issues is a standard remit of many of these groups, meaning noise measurements and associated descriptors are not novel concepts. Each of these groups makes a significant contribution to the implementation of the END; however, the effectiveness of these contributions is limited without participation from the largest stakeholder group, the general population. As with many environmental initiatives, public engagement is a vital part of any effective noise management strategy. Participation by the general public in the noise management process has the potential to yield highly beneficial results. Aside from an increased level of acceptance and confidence in noise control legislation, the input from the public, which can be gained as a result of effective communication, can help to produce far more efficient and practical noise policies. The most effective noise management plans will be informed by feedback from the local population. Local knowledge, experience, and attitudes toward environmental noise will lead to optimised action plans that have the potential to produce noticeable benefits for those affected.

Public participation in the directive's noise management cycle happens primarily during the action planning phase. Strategic noise maps are generated by the designated mapping body and then made available to all interested parties who may have some input into the contents and goals of a noise action plan. In the case of local citizens this input can go further than just influencing the contents of an action plan. Local citizens also have the power to directly influence the execution and as such the success of noise mitigation measures. This stems from the fact that the public at large is the greatest contributor to environmental noise pollution. If relevant authorities can succeed in communicating this fact along with information about the effects of excessive noise pollution, then responsibility for improving the noise climate begins to move back toward the individual. Involving and engaging the public through the communication of relevant and accessible data increases awareness of these issues, leading the way to a shift in attitudes toward environmental noise and the behaviour patterns that generate noise pollution. These bonuses of education and awareness can potentially lead to the ideal scenario of reducing environmental noise pollution through changes in behaviour such as a modal shift toward more environmentally friendly means of transport. These goals are more easily achieved by tailoring information to the target audience, which in this case could span a range of social groupings.

Participation and engagement by the public regarding local government issues is regularly confined to a small subgroup of the population. Quite often interest from members of the general public is highest among an older, more highly educated demographic. Dissemination of environmental data such as strategic noise maps will never be an all-encompassing activity concerning every member of the public. A more realistic goal is to attempt to raise awareness of the effects of noise pollution on a wide scale and effectively communicate noise data to a smaller, more interested group. Nonetheless, despite any disproportionate representation among interested individuals with regard to age, means, or education, noise information should be presented in such a manner as to be accessible and comprehensible to the widest audience possible.

MAKING A USEFUL CONTRIBUTION: WHAT DO PEOPLE NEED TO KNOW?

In order to make a valid contribution to the noise management process, members of the public must be aware of all relevant information. This does not necessarily equate to the widespread publication of noise maps. For members of the public to fully understand noise mapping information, they must be made aware of what a noise map displays, how it was produced, and its implications and limitations. This means understanding what quantities a noise map is showing, what the descriptor (L_{DEN}, L_{NIGHT}) is measuring, and the source of the noise. This "hard data" should be accompanied by a caveat that strategic noise maps show predictions only, sometimes based on limited information. For example, mapping procedures might consider only major roads or rail lines, possibly neglecting nearby noise sources that are below the threshold for consideration. Noise maps are tools intended for a specific purpose. They provide an indicative insight into how environmental noise may be generated from a particular source. Whether maps take the form of graphical contour plots or wide-ranging geographic information system (GIS) data sets, prediction methods can provide large amounts of very "attractive" data, the limitations of which are not always apparent. It is important to note that the results are not intended to be a precise calculation of the total noise exposure at any given receiver point, and the temptation to rely too heavily on noise maps should be avoided. This is a point that should be made clear to the public, but also kept in mind by land use planners and regulators alike.

Supplying this contextual information can serve to frame mapping results and any subsequent analysis, helping to convey the function of the noise map more clearly. In addition to presenting information in context, any publications must take account of the large variance in audience implied by the umbrella term "general public." This means addressing the various issues associated with communicated technical noise data to a nontechnical audience.

PROBLEMS WITH COMMUNICATING TECHNICAL DATA

Consider a resident of any town in any EU member state. This citizen is an "average" member of the community with a reasonable level of education, but little mathematical or technical expertise. In line with progressive environmental policies, strategic noise maps are produced for this town and surrounding major roads and rail lines. All maps are published online and in the offices of the relevant authority, in compliance with the Environmental Noise Directive. Submissions regarding environmental noise pollution, based on these maps, are sought from the general public to aid in the production of a noise action plan for the area.

Now consider the difficulty this ordinary member of the public might have in making a valid contribution when presented with concepts such as decibels, frequency dependent A-weighting, and varying logarithmic units. The technical nature of acoustic measurements (e.g., logarithmic scales) is, for many, a fundamental barrier to understanding the topic. Add to this the fact that the decibel is not a fixed unit, but a ratio that may refer to any of a number of noise descriptors. References to dB(SPL), dB(A), dB(A) L_{DEN}, dB(A) L_{NIGHT}, or dB(A) L_{EQ} are all valid descriptions of noise levels. But are they valid for communicating information to such a wide and varied audience? Such a range of descriptors and statistics makes for an intimidating and perhaps impenetrable read. Merely providing accompanying documentation containing formulae for calculating these statistics, often the default approach of the technically minded, will do little to aid those of limited mathematical training. The decibel, though ubiquitous in several technical fields, may prevent many from fully understanding noise information. There are, however, no specific obligations for member states to provide any material beyond strategic noise maps, in any form, and the associated action plans. Member states meet their regulatory requirements by publishing noise maps in a manner that makes them widely available (i.e., online). This raises the question of whether simply making this data available follows the spirit or the letter of the law. Is a strategic noise map, even one accompanied by prediction methodology, "clear, comprehensible and accessible?"

Presenting noise maps in graphical form such as colour-coded contour plots represents information in quite an intuitive manner. The concept of colour coding from high to low is easily understood by people of all ages and all levels of education. Overlaying this contour plot on a satellite image or a map is an ideal way of displaying variations in magnitude over a geographical area. Plots are independent of language or literacy, and can range from large-scale overviews to detailed local information. Graphical noise maps, as a medium for information, are well suited to reaching a wide audience. The difficulty then lies with the quantities that strategic noise maps are plotting. The L_{DEN} indicator in particular can be misleading, and any tendency to take L_{DEN} as a direct measure of loudness, rather than an indicator of annoyance, must be avoided. A full understanding of what a noise map is showing is key to accessing the information and subsequently making a meaningful contribution. This essentially means that metadata, or information about the quantities being plotted, must also be supplied. This additional information should educate the intended audience and help present the noise maps in the proper context. The L_{NIGHT} indicator poses less of a problem, as it is closer to an average noise level for the specified time period. Nonetheless, it should be made clear that the purpose of the L_{NIGHT} descriptor is as an indication of sleep disturbance, not a direct measure of sound levels. Calculations of L_{DEN} and L_{NIGHT} both compute long-term averages, the effects of which should be made clear. Energy averaging, particularly during the nighttime, may mask very annoying but infrequent sound sources. Highlighting the scope of a strategic noise map, its intended purpose, and its limitations promotes more realistic understanding of what is being shown. Perhaps more important than communicating what a noise map *is* will be informing the public what a noise map *is not*. When viewed as a guidance tool for generating more effective noise policies, and not a prediction of total noise levels at every dwelling, the benefits of strategic noise maps can be properly realised.

MAKING INFORMATION MORE ACCESSIBLE

So what approach can offer a solution to these problems? How can this information be made available to the widest audience possible? First consider the desired outcome of properly communicating noise data to the public. Beyond the obvious legislative requirement to make information available, there are numerous benefits to going further than the basic regulatory obligations and fully engaging the public at large. Perhaps the most obvious of these is increased awareness of the issues involved in environmental noise management. The availability of information will raise awareness of noise as a pollutant and also its potentially undesirable effects in terms of personal health and possible economic impacts, for example, increased stress or property pricing.[2-4] Coupled with this awareness of noise pollution is an understanding of the associated legislation, particularly the scope and aims of environmental noise regulations. This helps to define the distinction between dealing with noise nuisances and long-term protection of the acoustic environment. By properly clarifying the nature of environmental noise and the goals of the noise mapping and action planning procedure, the public can gain an understanding of its own role in the process. This role involves meaningful input during action planning as well as participation in any resulting initiatives. Noise then becomes a local issue where public participation is a two-way process with interested individuals contributing to the generation of new policies and, ideally, playing a part in their implementation.

If this community-led approach to environmental protection is somewhat idealistic, a wider awareness of noise pollution issues at least opens the way for a more achievable goal: a realistic and practical noise policy. In an economic and political climate where overambitious environmental policies can be relegated to low-priority issues, particularly for such an intangible pollutant as noise, generating feasible and affordable noise management actions is vital. A better understanding of environmental noise pollution, its sources, effects, and possible treatment measures opens the way to developing realistic and practical action plans that can be properly implemented.

SOME RECOMMENDED REMEDIES

With these desired results in mind it is possible to determine what an interested member of the public needs to know in order to access the information presented in a strategic noise map and formulate a useful and relevant response. The first step, as discussed earlier, is to educate, thereby framing any hard data in its proper context. Some of the key topics to be conveyed include:

- The scope of noise legislation and action plans, highlighting the difference between environmental noise and community or nuisance noise. Environmental noise regulations derived from the END are focused on noise from transport systems and industry, and aim to address noise pollution with medium- to long-term policies. The associated process should provide a framework where the value of quiet spaces, especially in urban areas, should be made clear. This value is, perhaps, difficult to quantify, but when located in areas of public amenity such as a park, these quiet spots represent a valuable community resource. Raising awareness about excessive noise levels should also promote an appreciation of quieter places.

- Information on the sources of noise pollution is also a key issue, highlighting the likely impact major transport and industrial systems have on surrounding communities. Chief among these sources is traffic noise. This must be followed up with information on ways to minimise noise exposure, presenting the problem and possible solutions in tandem. It is important, however, to present realistic and achievable goals using methods that are economically feasible. A practical noise policy must represent a balance between necessary infrastructure and environmental responsibility; this means specifying noise management goals that can actually be achieved, not ones that are chosen as the best acoustic solution by software predictions.

This is the background information that sets out the scope of this legislation and particularly a noise action plan. Once this is clearly established it is possible to present strategic noise maps as a tool for guiding the formulation of noise management policies. But some extra metadata is still necessary to make a noise map an effective tool for communicating data to a nontechnical audience. Some of the most important topics are outlined next.

- Clearly outlining what a noise map shows is a crucial point. The L_{DEN} unit should be seen as an indicator of noise annoyance, which takes into account the time of the day, and the L_{NIGHT} as an indication of possible sleep disturbance.
- At least as important as defining a noise map is clarifying what a noise map does not show. It is not necessarily a direct plot of the noise level at a given location. The maps are plots of annual averages and, as such, may mask certain intermittent but intrusive noise events.
- It is important to point out that noise maps consider acoustic emissions only from a single source, which exceed certain qualifying thresholds, and so do not represent a complete sound level estimate.
- Also of great importance is an easy-to-understand explanation of the decibel scale. Qualitative examples are often more effective than mathematical ones and conveying knowledge of what constitutes a small change and what represents a major or noticeable change is the key point.

This information, when presented in appropriate language, should allow an interested individual with no prior knowledge of these concepts to gain some understanding of strategic noise mapping plots. The designated authorities for noise mapping and action planning are required to provide an accompanying summary along with all noise maps produced that covers the most important points of all data generated. It is arguable that this background information can be considered as being of sufficient importance to be included, since it goes some way toward making the data more clear and comprehensible.

There then remains the issue of presenting the strategic noise maps themselves. The selection of the correct level of detail and the appropriate publication channels is an important consideration in ensuring accessibility. With proper accompanying explanations and instructions, strategic noise maps can and should be published in full detail, ideally in electronic format. This allows interested parties to access and interrogate the information generated during noise mapping. A good example of this can be seen on the UK's Department for Environment, Food, and Rural Affairs (DEFRA) Web site,[5] where useful and clear information is supplied along with easily accessible noise maps. Employing appropriate Web-based technologies to present data in an engaging manner can also help generate public interest. This includes GIS tools, 3D maps, and searchable noise plots. The Paris noise map initiative provides a good example of an engaging user interface.[6]

In many cases the publication of fully detailed noise maps may not be necessary and could represent an overly complicated medium. In such cases the use of less detailed, alternative maps are a straightforward way of communicating relevant information. For example, in instances where a legal or guideline noise limit exists, an exceedance map immediately identifies hot spots. Such a map shows whether levels in a given area lie above or below a certain threshold. Providing information at varying levels of complexity and detail allows communication of relevant data to a more varied audience, spanning various ages, education levels, and degrees of technical competence. The Dublin noise mapping Web site provides a good example of publishing noise maps at varying levels of detail to more clearly communicate to a wide public audience.[7]

SUMMARY

The average city dweller is exposed to an astounding variety of noises from all manner of sources. Road traffic, railways, industry, and the din of so many people going about their everyday lives are just some of the factors that contribute to the complex soundscapes in which city inhabitants are immersed. Exposure to excessively high noise levels can have detrimental effects on the well-being of the resident population. Methods of predicting and measuring noise levels are ever improving. However, though the quality of the noise data may be improving, one fundamental problem still remains: how to communicate this technical information to the general public.

Accessible publications regarding noise data and the necessary contextual information help to promote more meaningful public engagement in the noise management process. Informing the public about the effects of noise exposure and the value of noise mitigation is the first step in promoting participation. The correct choice of content, medium, language, and channels for publications on noise data is vital for effective communication to such a wide and varied audience. Getting this balance right can allow interested members of the general public to make a valid and useful contribution to more effective and progressive noise policies.

NOTE

I have attempted to keep this article as free from technical content as possible. At the heart of what this chapter is about is the ability to discuss the more "human" side of noise pollution without relying on statistics and technical references. As such, the contribution is less of a high-level analysis and more a discussion of what information is necessary and what data causes headaches. The obvious challenge when communicating with a wide and varied audience is honing the ability to talk about noise in a qualitative and accessible way, omitting the more onerous concepts while still imparting an understanding of the end effect. The key to doing this is to stop approaching the issue from the point of view of the acoustics expert, the town planner or the policy maker. Instead, consider the position of the citizen on the side of that busy road, or living beside that airport, aware of the problem but lacking in useful information. The statistics, tools, and even the language we use to convey that information must be carefully chosen to ensure that the public can be engaged and noise pollution can be properly treated as the wider community issue that it is.

REFERENCES

1. European Commission (2002). Directive 2002/49/EC of the European Parliament and of the Council of 25 June 2002 relating to the assessment and management of environmental noise, OJ L 189, 18/07/2002, Luxemburg, 2002.
2. A.S. Haralabidis et al., Acute effects of night-time noise exposure on blood pressure in populations living near airports, *European Heart Journal* (2008) 29(5):658–664.
3. J.P. Nelson, Highway noise and property values, *Journal of Transport Economics and Policy*, May 1982, pp. 117–138.
4. P. McDonald, D. Geraghty, I. Humphreys, S. Farrell, Assessing the environmental impact of transport noise using wireless sensor networks, *Transportation Research Record: Journal of the Transportation Research Board*, 2008, pp. 133–139.
5. DEFRA Noise Mapping Portal, http://services.defra.gov.uk/wps/portal/noise.
6. Paris Noise Map Display Portal, http://www.v1.paris.fr/commun/v2asp/fr/environnement/bruit/carto_jour_nuit/.
7. Dublin Noise Mapping Portal, http://www.dublincity.ie/waterwasteenvironment/noisemapsandactionplans/pages/default.aspx.

Chapter 16

Which information for the European and local policy makers?

L. Maffei

CONTENTS

The awareness that noise is one of the main environmental problems in Europe was probably well known by the local policy makers and local communities before the presentation of the Green Paper on Future Noise Policy of the European Commission (1996), but, without any doubt, this document has represented a milestone as far as it stated that the "local nature of noise problems does not mean that all action is best taken at local level" and that a "framework based on shared responsibility involving target setting, monitoring of progress and measures to improve the accuracy and standardisation of data to help improve the coherency" is needed to have a successful noise abatement policy.[1]

The Directive 2002/49/EC[2] of the European Parliament and of the Council relating to the assessment and management of environmental noise (END) was the natural consequence of the Green Paper as far as its "aim to define a common approach intended to avoid, prevent or reduce on a prioritized basis the harmful effects, including annoyance, due to environmental noise" is considered.

A key feature of the END is stated in Article 9.1 where is strictly states that

> Member states shall ensure that the strategic noise maps they have made, and where appropriate adopted, and the action plans they have drawn up and made available and disseminated to the public in accordance with relevant Community legislation, in particular Council Directive

90/313/EEC of 7 June 1990 on the freedom of access to information on the environment, and in conformity with Annexes IV and V to this directive, including by means of available information technologies.

Making available all data with the information action is meant to enlarge the consciousness of the problem and to share responsibilities. For this it is however necessary to go over a simplified definition of "public" as "one or more natural or legal persons ... or their associations, organizations or groups" and to consider in that category local policy makers (local authorities, politicians, technicians, designers, urban planners).

The Working Group on the Assessment of Exposure to Noise (WG-AEN) in its position paper on "Presenting Noise Mapping Information to the Public"[3] underlines that the public is the main stakeholder and the main subject to involve in the action planning. However, without adequate information on real possibilities and on the application limits of all data coming from noise mapping, there is the risk of a limited interest among the stakeholders and, on the other side, overly high expectations that the action planning activities can realistically determine community noise control. As far as this can be true, the full involvement of policy makers is as much important to make propositions and to guarantee the achievement of common real interests and not only virtual projects.

In the European Union END and in the following position paper of the WG-AEN, much emphasis is given to the modalities and to the level of detail that should be used to present the acoustic data but less attention is given to how this information should be handled at different levels and by whom and who should have the responsibility to establish environmental policies. It is not explicit how the output of the noise mapping could determine the policy makers' decision and influence the action plans, how much significance should be given to numerical data, and how much significance should be given, instead, to their interpretation.

The END leaves to member states several degrees of freedom. No noise level limits are imposed and the classification of quiet areas, together with the activities to achieve their protection and preservation, is given totally to the member states.

The END has had and will have in the future a strong economic impact on a community. This economic impact is not only related to the short term in which it is requested, although with a large effort, to create noise maps and action plans and to make several data available, but it will be determined mainly on a long term when action plans will be transformed in executive projects and then they will be realized.

As far as reducing the ratio cost–benefits for environmental issues, a key question is: *How relevant is the information imposed by the END for European policy makers, local policy makers, and the population to reach the general aim of reducing noise annoyance in Europe?*

INFORMATION FOR EUROPEAN POLICY MAKERS

Annex VI of the END is extremely precise on the information requested by the commission. For agglomerations, besides graphical representation of strategic maps, the key information is the estimated number of people living in dwellings that are exposed to five bands of value of L_{den} and L_{night} separately for noise from road, rail, and air traffic, and from industrial sources. In addition, where appropriate and where such information is available, it is asked how many people of the above live in dwellings that have special insulation against noise combined with ventilation or air conditioning facilities and/ or live in dwellings that have a quiet façade, meaning a L_{den} or L_{night} 20 dB lower than at the façade having the highest level. This information should be accompanied by noise-control programmes that have been carried out in the past and noise measures in place, and by a summary of the action plan. Same information is requested for major roads, major railways, and major airports.

The data received from the member states will permit the commission to perform several actions in the short, medium, and long term. In the short term it will be possible to evaluate the extension of the community noise annoyance at the European scale in terms of number of people exposed, the grade of extension of noise insulation inside dwellings, and the grade of influence on the overall problem of the specific sound sources (road, rail, air traffic, and industries). Also significant will be the information on the level of attention that member states have given in the past to control the community noise.

All these data, although treated with a statistical approach, will represent the basis for medium-term actions of the commission and the European Parliament and Council. These actions, implemented in terms of new directives and recommendations, will have as main criteria the reduction of the harmful effects and the cost-effectiveness ratio. They could mainly consist of:

1. The requirement to label all outdoor equipment with guaranteed noise levels.
2. The introduction of additional measures for a reduction of the environmental noise emitted by means and infrastructures of transport (e.g., promoting the use of low noise surfacing; road-worthiness tests to include specific noise testing of in-use vehicles; negotiated agreements between the railway, car, and air industries and communites on target for noise reduction and measures to ensure maintenance of in-use equipment).
3. The introduction of taxes and charges (e.g., landing fees for noisy aircraft; road use that differentiates according to noise emissions of vehicles; a variable track charge that would enable the infrastructure fee for the use of track to be differentiated according to the noise levels of the wagons and train freight charges).

4. The introduction of economic incentives to encourage noise reductions (e.g., grants to purchase low-noise goods and vehicles).

5. Encouraging member states to apply several community financial instruments through which cooperative ventures between member states and particularly local authorities are supported and where noise abatement could be given a higher priority (e.g., Life, InterReg, Eurocities).

6. Support of the research in noise abatement through specific priorities inside the Framework Community programme.

7. The promotion of exchanges of experience and dissemination of good practice.

Since data received by member states should be "reviewed and revised if necessary when a major development occurs affecting the existing noise situation, and at least every five years," in the long term, the community will be able to monitor progress and verify the effectiveness of all the actions implemented at European and local level.

With END, a virtuous mechanism is then started. The harmonized information continuously sent to European policy makers permits the implementation of new actions that can be monitored, verified, and adapted to the changing situation.

INFORMATION FOR LOCAL POLICY MAKERS

A deeper discussion is needed on the impact that the information requested by END can have on local policy makers' activities. It is up to them to safeguard public health, answer to the population needs of comfortable habits (urban and extraurban), determine investment planning, implement projects, and evaluate cost effectiveness.

It is at the local level, despite the multiplication of European networks (e.g., Eurocities), whose aim is the exchange of experiences and best practices, that the main problems determined by the noise pollution are faced and resolutions found.

Local policy makers can find practical difficulties in handling the data requested by END. First, the numerical data referred to the noise exposure of each building or of a single receptor or of a large area, despite how accurate it can be, has not a great significance if it is analysed alone. It should be transformed, with the use of other tools, in the identification of critical situations. These critical situations should be then inserted inside a priority environmental plan, which contemplates several aspects but that puts at the centre of the actions the overall satisfaction of the population.

In fact, if local policy makers should afford any critical situation (e.g., noisy street) with projects based on a single environmental parameter (e.g., noise levels), there is the high risk that the implementation of a solution (e.g., noise barriers or traffic limitations) could be in contrast with other environmental parameters (e.g., natural light contribution to dwellings or citizen mobility).

The "acoustic" action planning should be tackled according to a priority scale in which the data referred to noise pollution is connected and correlated to data that contemplates other environmental parameters and problems, such as air pollution, mobility, energy, and also takes into account the real expectations of the population.

This multidimensional way with which the local noise pollution control policies should be implemented is strongly suggested by the technical and scientific community involved with soundscape studies.

The soundscape approach, concentrating on the way that people consciously perceive their environment (namely, the interactions between people and sounds), might open novel perspectives and provide further insights toward fighting noise. In this approach, complementary to the noise control engineering techniques, the participation of people is fundamental along with their involvement that complies with the requirements issued by the European directive on the assessment and management of environmental noise. A strong synergy is then required between local policy makers and local population.

Which are the most appropriate communication channels to make this synergy successful? On 25 June 1998 the community signed the UN/ECE Convention on Access to Information, Public Participation in Decision-Making and Access to Justice in Environmental Matters (the Århus Convention). Among the objectives of the Århus Convention is the desire to guarantee rights of public participation in decision making in environmental matters in order to contribute to the protection of the right to live in an environment that is adequate for personal health and well-being. Directive 2003/4/EC of the European Parliament and of the Council of 28 January 2003 on "Public access to environmental information"[4] and the following Directive 2003/35/EC of 26 May 2003 "Providing for public participation in respect of the drawing up of certain plans and programmes relating to the environment"[5] are important milestones towards the application of the principles of the Århus Convention.

However, for the application of the aforementioned principles the tendency during past years was oriented to the pure information, as for "information" is intended the transfer of a message to one or more subjects who are different from the one that keeps the knowledge.

The population's participation process needs something more extensive. We can talk then of "communication" as that comparison moment between the subject that is informing and that is waiting for a response from an interlocutor to verify if the contents of the message have been effectively transferred and attended results have been reached.

This population participation process can be realized only if some fundamental requirements are present, such as:

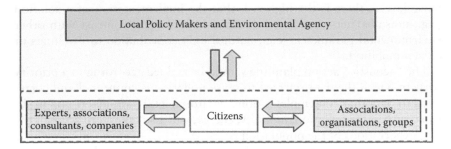

Figure 16.1 Direction of communication.

- Bidirectional communication
- Willingness of the subject that is informing to confront the receiver (public)
- Availability and capacity of the public to value the information in order to accept it or refuse it
- The system's transparency

Two integrated preferential direction for the communication should be identified (Figure 16.1):

- Vertical communication—From local policy makers to the citizens and vice versa
- Horizontal communication—Between single citizens, associations and/or groups, experts

Although the communication channels could be of different types (press, TV, workshops and conferences, printed documents such as pamphlets, guidelines) it is evident that modern Web channels can better fulfil the requirements and typologies.

All over the world several Web communication initiatives on the environmental noise issue have been built, but several deficiencies, that could make these initiatives not successful, can be registered:

1. Differences in the approach typology (bureaucratic, informative, educational, or participative)
2. Difficulties in the navigation architecture
3. Lack of information oriented to specific targets of citizens
4. Use of technical terminology not accompanied by a thematic glossary
5. Predominance of textual information instead of graphic information, which is more intuitive

Recently a study by the Second University of Napoli analysed 41 official Web sites of local administrations (cities) (Table 16.1), published in several

Table 16.1 Local Administrations

Number	Nation	Cities	Internet Address	Population
1	Australia	Sydney	http://www.cityofsydney.nsw.gov.Au	4,198,543*
2	Australia	Melbourne	http://www.melbourne.vic.gov.au	3,700,000*
3	Australia	Perth	http://www.cityofperth.wa.gov.au	1,507,900*
4	Australia	Hobart	http://www.hobartcity.com.au	202,000*
5	New Zealand	Wellington	http://www.wellington.govt.nz/	179,466
6	United Kingdom	London	http://www.london.gov.uk	7,512,400
7	United Kingdom	Birmingham	http://www.birmingham.gov.uk	1,001,200
8	United Kingdom	Liverpool	http://www.liverpool.gov.uk	447,500
9	United Kingdom	Leeds	http://www.leeds.gov.uk	715,235
10	United Kingdom	Manchester	http://www.manchester.gov.uk	441,200
11	United Kingdom	Bristol	http://www.bristol.gov.uk	398,300
12	United Kingdom	Edinburgh	http://www.edinburgh.gov.uk	463,500
13	United Kingdom	Glasgow	http://www.glasgow.gov.uk	580,700
14	United Kingdom	Belfast	http://www.belfastcity.gov.uk	277,391
15	United Kingdom	Sheffield	http://www.sheffield.gov.uk	516,100
16	United Kingdom	Oxford	http://www.oxford.gov.uk	134,248
17	Ireland	Dublin	http://www.dublin.ie	505,739
18	France	Paris	http://www.paris.fr	2,153,600
19	France	Lion	http://www.lyon.fr	466,400
20	France	Bordeaux	http://www.bordeaux.fr	230,600
21	France	Nice	http://www.nice.fr	347,100
22	France	Nantes	http://www.nantes.fr	270,251
23	France	Lille	http://www.mairie-lille.fr	212,597
24	France	Strasbourg	http://www.strasbourg.fr	273,100
25	France	Toulouse	http://www.toulouse.fr	390,350
26	Espana	Madrid	http://www.munimadrid.es	3,099,834
27	Espana	Zaragoza	http://www.zaragoza.es	649,181
28	Espana	Bilbao	http://www.bilbao.net	349,972
29	Espana	Barcelona	http://www.bcn.es	1,605,602
30	Espana	Valencia	http://www.valencia.es	807,396
31	Espana	Malaga	http://www.ayto-malaga.es	560,000
32	Espana	Las Palmas Gran Canaria	http://www.laspalmasgc.es	354,863
33	Portugal	Lisbon	http://www.cm-lisboa.pt	564,657
34	Argentina	Buenos Aires	http://www.buenosaires.gov.ar	3,034,161
35	Republik Singapura	Singapura	http://www.gov.sg/	4,608,595

(Continued)

Table 16.1 Local Administrations (Continued)

Number	Nation	Cities	Internet Address	Population
36	Canada	Vancouver	http://www.city.vancouver.bc.ca	603,000
37	Canada	Québec	http://www2.ville.quebec.qc.ca	532,329
38	Canada	Montréal	http://ville.montreal.qc.ca	1,812,723
39	USA	New York	http://www.nyc.gov	8,168,388
40	USA	San Diego	http://www.sandiego.gov	1,223,400
41	USA	Boston	http://www.cityofboston.gov	589,141

*Population of the agglomeration.

languages: English, French, Spanish, and Portuguese.[6] All Web sites had a specific section dedicated to "environmental noise." The analysis and the evaluation were done using six codes. To each of them is attributed a specific score by a jury.

> ARCH (Architecture of the Web site)—Examine the facility and intuitiveness for accessing the pages dedicated to environmental noise starting from the homepage, the articulation of the different sections, the number of pages, and the use of links.
> COM (Communication)—Take into account the presence of different forms of communication (e.g., audio, video, interactive) that can be useful for the citizens' understanding of the information, the presence of sections dedicated to specific targets (e.g., noise around schools or noise in the district), the type of communication language.
> CONT (References)—Examine the presence of communication sections from the citizens to local policy makers (e.g., complaint section, citizen opinion section on noise control projects), and the presence of specific references (offices, agencies, consultants, associations).
> EDU (Education)—Consider the presence of technical glossaries for a better understanding of the information, the presence of educational initiatives for conscious and responsible behaviour toward noise reduction, the presence of awareness campaigns against noise.
> JUS (Justice)—Consider the presence of a section dedicated to legislative documents (laws, codes) on noise, the possibility to download them, the presence of information on citizen behaviour in case of noise pollution problems (e.g., how and to whom to address the complaint, direct access to the conciliation judge).
> TEC (Technic)—Take into account the presence of technical documentation elaborated by the local policy makers and consultants (e.g., strategic maps, action plans, environmental studies) but also friendly use information on noise control strategies at a small scale (e.g., how can I build my house with protection against external noise).

The results of the study confirmed that, up to now, there is not a homogeneous approach to the communication among the examined cities with the exception of the UK cities—London, Leeds, Birmingham, and Bristol. These cities undergo similar approaches in the education and communication towards citizens and this is probably due to the fact that they identify as a reference for the documentation and the approach of three well-known agencies: Environmental Protection UK (NSCA), Department for Environment, Food and Rural Affairs (DEFRA), and Health and Safety Executive (HSE). Less homogeneous is the situation in France and in Spain, although there are examples of interesting approaches (e.g., Paris).

More information can be obtained from the analysis of the single codes. The code ARCH underlines that 71% of the Web sites present a simple access to the noise pollution section but 76% of them dedicate to this problem only less than five pages. The code COM reveals that 86% of the Web sites only use text communication and for 73% of them the communication only has an informative approach. Most of the Web sites contain online forums with which citizens can pose questions to the public administration.

Regarding the EDU code, only one city presents a technical glossary and 12 cities present information on best practices. For the IUS code, 83% of the Web sites publish several legislative documents but only 54% permit their download. Finally, the TEC code shows that there is a generalised lack of technical documentation on the Web sites; so far only 24% make noise maps and noise control plans available. None of the examined Web sites interface the noise data with other environmental parameters (e.g., air quality, mobility) or present multisensorial maps (e.g., soundscape maps).

SUMMARY

In summary, the END has triggered a virtuous mechanism that will lead to more uniform and successful results in the fight against environmental noise in the medium and long term.

Although for European policy makers, the importance is evident that the huge acoustic data required by the END will have on their actions; for the local policy makers it will be extremely difficult to handle and to make use of the action plans with so much information.

A synergy between local policy makers, experts, and citizens is strongly recommended, and it should be built up. This synergy should go over the simple data information required by the END. The use of appropriate communication channels offered by the Web can be the right answer only if appropriate and modern communication criteria, that take into account the interests, expectations, and feelings of the population, will be used for their implementation.

REFERENCES

1. Future Noise Policy. European Commission Green Paper. Commission of the European Communities. Brussels, 1996.
2. Directive 2002/49/EC of the European Parliament and of the Council of 25 June 2002 relating to the assessment and management of environmental noise. Official Journal of the European Communities of 18 July 2002.
3. Presenting Noise Mapping Information to the Public. A Position Paper from the European Environment Agency Working Group on the Assessment of Exposure to Noise (WG-AEN). March 2008.
4. Directive 2003/4/EC of the European Parliament and of the Council of 28 January 2003 on public access to environmental information and repealing Council Directive 90/313/EEC. Official Journal of the European Communities of 14 February 2003.
5. Directive 2003/35/EC of the European Parliament and of the Council of 26 May 2003 providing for public participation in respect of the drawing up of certain plans and programmes relating to the environment and amending with regard to public participation and access to justice Council Directives 85/337/EEC and 96/61/EC. Official Journal of the European Communities of 25 June 2003.
6. Research Study 06-005-APAT: New methodologies for the evaluation and control of environmental noise in urban sites with the involvement of the population. Final report. Second University of Napoli–APAT, May 2009.

Chapter 17

From noise maps to critical hot spots

Priorities in action plans

W. Probst

CONTENTS

THE STEPWISE DEVELOPMENT OF AN ACTION PLAN

According to Directive 2002/49/EC, noise maps are only a first neverthe-less important step to develop effective and sustainable noise reduction measures and to decrease the exposure in the vicinity of traffic sources in agglomerations. It is further necessary to show the areas where noise reduc-tion measures should be planned and executed with first priority based on the strategic noise maps. It is obvious that these "hot spots" are not neces-sarily the areas with highest noise levels, because the real urgency to take action depends on the number of people exposed. Even areas with high noise levels may not be a severe problem if nobody lives there.

This problem can be solved by using a noise rating system, where num-bers of residents and values of noise indices characterising their exposure are merged to an area-related single number "noise score."

Figure 17.1 shows the stepwise procedure to develop an action plan as it was investigated and proposed in the frame of the project Quiet City.[1]

    Step 1—Basis of each noise evaluation of larger areas and agglomera-tions is the noise map, including the relevant noise indicators L_{den} and L_{night} for each building or dwelling.

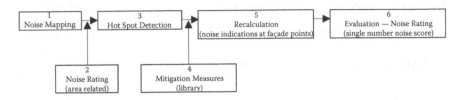

Figure 17.1 The basic methodology to develop an action plan based on strategic noise maps.

Step 2—A ranking methodology and an additive noise score, NS, is needed to characterise an area with any number of dwellings by a single number depending on the noise exposure and the number of people exposed.

Step 3—The hot spots are areas where the area-related NS is larger than 90% of all values occurring. These hot spots are a strong and valuable indication for areas where mitigation measures should be planned.

Step 4—A catalogue of mitigation measures has been developed containing data sheets with the description, a valuable estimation of the achievable noise reduction and the cost, the limitations of the method and, if possible, reference applications.

Step 5—After implementing the measures according to an action plan in the 3D model, the noise map is recalculated. This is done separately for all scenarios alternatively discussed.

Step 6—By calculation of the NS value and summing it up in the relevant area for all regarded scenarios each can be characterised by the total NS. This allows one to rank the different solutions and to decide about the best alternative.

The described procedure is based on the existence of an accepted method of evaluating and scoring noise exposures and a method to derive the hot spot areas according to step 3.

NOISE SCORING METHODOLOGIES

The noise score is a single number that represents the "social weighting" of an unwanted situation due to high noise exposures. Different concepts for the calculation of such an NS value have been proposed and published. In the following, only some "one-equation" concepts are considered because they can directly be applied on existing noise mapping projects.

Concept 1: A linear relation

$$NS = \sum_i n_i \cdot (L_i - L_R)$$

(17.1)

where n_i is the number of persons in group i, L_i is the exposure level of group i, and L_R is the limiting or reference value.

This pragmatic and simple approach with linear weighting allows one to compensate an increase of the exposure of inhabitants by x dB(A) by a similar noise reduction for another group by the same amount of x dB(A), independent of the absolute height of the noise levels to which these two groups are exposed. With respect to equity it is questionable if a level increase from 70 dB(A) to 75 dB(A) of a person should be ranked equal to the level decrease from 55 dB(A) to 50 dB(A) of another person.

Concept 2: Percentage of highly annoyed people (%HA)

$$\%HA = 9.87 \cdot 10^{-4}(L_i - 42)^3 - 1.44 \cdot 10^{-2}(L_i - 42)^2 + 0.51(L_i - 42)$$

$$NS = \sum_i n_i \cdot \frac{\%HA}{100}$$

(17.2)

This equation proposed by Miedema et al.[2,3] for road noise approximates the relation between the mean noise exposure and the annoyance caused by it. It is based on many studies and questionnaires. The NS value calculated is the expected value of the number of highly annoyed people derived statistically. Expression (17.2) produces a relatively low weighting of high noise levels; the NS value is doubled by a level increase of about 8 dB. Therefore, a level increase of 1 dB for people living with an absolute exposure of 70 dB(A) would be accepted based on this assessment method if the exposure of two other persons living with an exposure of 62 dB(A) would be reduced by 1 dB.

Equations similar to Equation (17.2) have been published for the noise sources railway and aircraft, and even a combination of all these noises in a single number quantifying the total annoyance has been used in a rating system applicable with existing noise maps.[4]

Concept 3: Exponential increase

$$NS = \sum_i n_i \cdot 10^{k \cdot (L_i - L_R)}$$

(17.3)

The parameter k defines the slope of the evaluation curve. With respect to road traffic noise it can be shown (e.g., see Probst[5]) that NS is minimised if the car flows are concentrated as much as technically possible if k is smaller than 0.038. This is the case with the highly annoyed concept explained earlier, because Equation (17.2) can be approximated by Equation (17.3) with a k value of 0.03.

These three concepts are only examples. For the detection of hot spots an equation according to concept 3 has been used in the mentioned European Commission project Quiet City.[6] This equation is

$$NS = \left| \sum_{i} \begin{array}{l} n_i \cdot 10^{0.15 \cdot (L^*_{den,i} - 50 - dI + dL_{source})} \quad with \ L^*_{den,i} \leq 65 \ dB(A) \\ n_i \cdot 10^{0.30 \cdot (L^*_{den,i} - 57.5 - dI + dL_{source})} \quad with \ L^*_{den,i} > 65 \ dB(A) \end{array} \right. \tag{17.4}$$

where n_i is the number of persons exposed with level $L_{den,i}$; $L^*_{den,i}$ is the effective noise indicator at the relevant façade at dwelling i; dI is the deviation of mean sound insulation of dwelling i from the mean insulation of all dwellings; and dL_{source} is the correction that accounts for different reaction versus noise from roads, railways, aircraft, and industry.

In agglomerations road traffic is generally the most important noise source and it is convenient to take only this road noise into account. Further, there is in most cases no detailed knowledge about different insulation of dwellings and therefore Equation (17.4) can be simplified for practical applications to

$$NS = \left| \sum_{i} \begin{array}{l} n_i \cdot 10^{0.15 \cdot (L^*_{den,i} - 50)} \quad with \ L^*_{den,i} \leq 65 \ dB(A) \\ n_i \cdot 10^{0.30 \cdot (L^*_{den,i} - 57.5)} \quad with \ L^*_{den,i} > 65 \ dB(A) \end{array} \right. \tag{17.5}$$

where $L^*_{den,i}$ is the effective noise indicator due to road noise at the relevant façade at dwelling i.

CALCULATION OF THE NOISE SCORE FOR EACH BUILDING

According to 1.5 of Annex VI of the directive, the number of people living in dwellings that are exposed to each of the defined bands of values of L_{den} in decibels 4 m above the ground on the most exposed façade shall be determined. It is therefore appropriate to use the same information if the noise

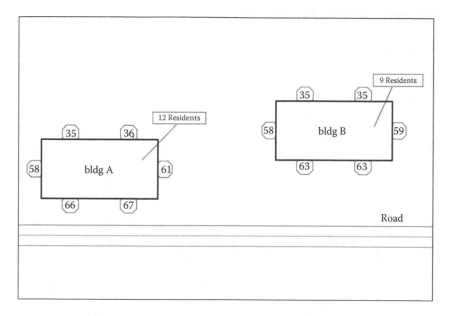

Figure 17.2 Building A and building B with 12 and 9 residents, respectively.

score and hot spots shall be detected. The indicator L_{den} is calculated with receiver points distributed around the façades at all residential buildings at a height of 4 m and the highest value is taken to determine the NS value according to Equation (17.5). Figure 17.2 shows a simple example.

According to Directive 2002/49/EC, the façade levels have been calculated at points around the façade 4 m above ground. The level L_{den} at the most exposed façade is 67 dB(A) at building A and 63 dB(A) at building B. Therefore, the NS values for the two buildings are

$$NS \text{ (building A)} = 12 \times 10^{0.3 \times (67 - 57.5)} = 8495$$

$$NS \text{ (building B)} = 9 \times 10^{0.15 \times (63 - 50)} = 802$$

It shall be mentioned that this is an exponential weighting that takes into account the health effects that may occur if noise exposures exceed 65 dB(A). Therefore, the rounding of the level values may have an influence on the result. It is recommended to use the same rounding with one decimal (e.g., 61.4) for all calculated level values.

After having performed this calculation, the NS value is known for each residential building.

THE DETECTION OF HOT SPOTS

To find the hot spots a map of the area-related noise score is produced. The little crosses in Figure 17.3 are regular-spaced grid points. To get a map of the area-related NS value, a window (e.g., 100 m × 100 m) is located on the map with the first grid point in its center. Then the NS values of all the buildings inside the window are added. Buildings that are intersected by an edge of the window are taken into account proportional to their area inside the window. The obtained sum is divided by the window area and multiplied by a reference area (e.g., 1000 m²) and the resulting area-related NS value is attached to the grid point. Then the window is centred above the next grid point and the procedure is repeated. At the end, the grid shows in this case the distribution of the NS value related to 1000 m².

The resulting map is coloured with red for all NS values exceeding a certain limiting value NS_{limit}. This value NS_{limit} was adjusted in the mentioned pilot projects so that 10% of the agglomeration area was presented as a hot spot.

This procedure transforms a noise map presenting the level distribution, shown as ground cover in Figure 17.4, to a hot spot presentation as shown in 3D in Figure 17.5. This is an understandable basis for all the necessary discussions about noise action plans between noise experts, politicians, and the people concerned.

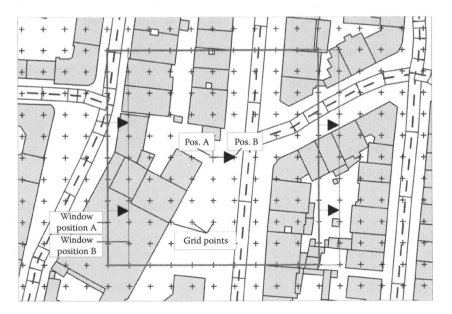

Figure 17.3 Gliding window to calculate the area-related noise score (NS) for one grid point.

Figure 17.4 Noise map projected on the ground in a 3D-view presentation.

The technique presented is only one of many others; the best choice depends on the goals of the exercise and the problems that shall be solved. An automated approach filtering hot spots out of complex built-up scenarios should in all cases be flanked by a thorough inspection of the site before the priorities for an action plan are defined and the recommended measures are evaluated.

Figure 17.5 Map of an area-related noise score in a 3D presentation showing the hot spots in red. **(See colour insert.)**

REFERENCES

1. Quiet City Transport, Project FP6-516420, European Commission, 6th Framework Programme.
2. Miedema, H.M.E., Vos, H. Exposure-response relationships for transportation noise. *Journal of the Acoustical Society of America* 104 (1998), no. 6, 3432–3445.
3. Miedema, H.M.E., Oudshoorn, C.G.M. Annoyance from transportation noise: Relationships with exposure metrics DNL and DENL and their confidence intervals. *Environmental Health Perspectives* 109 (2001), no. 4, 409–416.
4. Miedema, H.M.E., Borst, H.C. Rating environmental noise on the basis of noise maps, Report Deliverable D 1.5, European Commission project Quiet City FP6-516420.
5. Probst, W. Zur Bewertung von Umgebungslärm (Assessment of environmental noise), *Zeitschrift für Lärmbekämpfung* 53 (2006), no. 4–Juli, 105–114.
6. Probst, W. Noise ranking and scoring methodology, Report Deliverable D 1.2, European Commission project Quiet City FP6-516420.

Part 5

Future perspectives

Chapter 18

From noise to annoyance mapping

A soundscape approach

G. Memoli and G. Licitra

CONTENTS

Noise has invariably accompanied people throughout the ages, but its types and the human perception of it have changed over time. Noise in large urban agglomerations is nowadays seen as a factor that greatly impairs quality of life, similarly to air or water pollution. In a recent survey involving 75 European cities,[1] for instance, more than half of respondents agreed that noise was a major problem in their city, with a proportion ranging from 51% in Rotterdam and Strasbourg to 95% in Athens. The same study showed a strong positive correlation between the opinion on air and noise pollution and the perception of a city as a "clean" place, where it would be possible "feeling safe." With the latter being two key indicators of the well-being in a modern city, the perceived impact of noise on the environment where we dwell everyday seems to have a great impact on the quality of life in our cities. In addition to this, unwanted effects like sleep disturbance, loss of concentration and learning difficulties, increased blood pressure and stroke occurrence, annoyance and higher stress have been correlated with high levels of prolonged exposure to noise.[2]

Are European cities as noisy as their inhabitants perceive them? In accordance to the European Commission's Green Paper on Future Noise, more than about 250 million people were exposed in 1996 to A-weighted outdoor levels higher than 55 dB.[3] Conversely, the World Health Organisation (WHO) has recommended a daily level not greater than 65 dB(A) to guarantee the well-being of a population exposed to noise.[4] According to these two studies, the situation in 1996 was quite critical.

As we have seen in the other parts of this book, European Directive 2002/49/EC on the Assessment and Management of Environmental Noise (END) was adopted to define a common approach "to avoid, prevent or reduce on a prioritised basis the harmful effects, including annoyance, due to exposure to environmental noise."[5] To that end, the directive has introduced a common set of tools to cope with the issues related to noise management on the European scale, at least for areas affected by the major sources of environmental concern (i.e., around transport infrastructures and within agglomerates). The European Commission required member states to produce noise maps for the main sources of noise pollution (traffic, railways, airplanes, factories), described the indicators to be used (namely, L_{den} and L_{night}, measured in decibels) and, in later documents, suggested the methodologies to be followed and the algorithms for modelling noise emissions from the different sources. Those noise maps, updated every 5 years, were supposed to constitute the base of local strategies to manage noise pollution, actively pursued through what the END calls action plans. In a nutshell, the END was intended to provide a common strategy across Europe to improve the previously highlighted discrepancy between the real situation[1-3] and the ideal one.[6]

Since 2002, the work of the European Union (EU) Commission has been integrated by the research produced by the Working Groups Assessment of Exposure to Noise (WG-AEN) and Health and Socio-Economic Aspects (WG-HSEA), and in the context of relevant EU-funded projects, which have pointed out the best practices to be shared among the community.[6,7]

In addition to this, the first round of noise mapping (in 2007) and of action planning (in 2008) have highlighted the limits of the noise maps as planning tools. It was found that since the accuracy of predictive algorithms (within commercial software) depends on the quality of input data, noise maps may have a local uncertainty as large as 5 dB.[8] With this limit, noise maps have proven to be very effective in determining exposure in hot spots (i.e., where risks for health are nonnegligible and actions cannot be delayed) but have shown not to work well in quiet areas when the algorithms tend to fail and the perceived noise is the cause of annoyance and stress. Between those two extremes, there is then a "gray area"[9] where average energy levels are not so high to prioritise an immediate action and most of the population resides. Here, as many questionnaire surveys show, the energy impacting on the receiver (as it is measured in decibels) is no longer a sufficient measure of the impact of noise pollution on the exposed population. Here the "quality" of acoustic energy, as it is weighted by the receiver, differentiates "sounds" from "noise" (i.e., "unwanted" sounds). For these areas there is the need of a new acoustical metrology, with the ambitious goal to provide objective measurements of noise perception, as a mere reduction of the noise levels may not be enough.

In this context, questions as to whether it is possible to predict how visitors/dwellers will perceive an acoustic environment, or is it possible to map annoyance, or even is it possible to create perception-oriented action plans are more actual than ever. On the eve of the new deadline imposed by the END for 2012, targeting smaller agglomerates and minor transport infrastructures, and in coincidence with the ongoing review process of the END (started in 2009), it is then worth discussing what appears to be an opportunity for a paradigm shift in noise management.

SOUNDSCAPES: FROM CONCEPT TO ACTION PLANS

The concept of "soundscape"

The concept of "soundscape" was introduced by Murray Schafer, a Canadian composer famous for having started the World Soundscape Project (WSP), an educational and research group initially active during the late 1960s and early 1970s. According to his definition, a soundscape can be both an acoustical environment and an environment artificially created by sound. The word *soundscape* can then be used to describe the ensemble of sounds and noises that we experience in the environment where we live (or, better, in a specific part of that environment like a park, street, supermarket, airport, city centre, or a particular village on the Alps[10]), over which we have little or no control, or a particular type of acoustical composition created by the artist to communicate a specific feeling (and thus artificially controlled).

In his studies,[11] Schafer highlighted the negative approach to the sonic environment implicit in the term "noise pollution" (i.e., we can only defend ourselves by unwanted sounds, preferably through actions on the source, but eventually using barriers) and proposed to search for a more positive approach: the identification in a place of not only the negative sounds but of the desired or expected sounds that are kind to the ear and characterise that particular environment.[12]

Soundscape, then, becomes simultaneously the physical environment (quantified by numerical measurands, like the decibel) and the way of perceiving that environment, through the judgment of the individual (or of group of individuals): a concept to describe the whole acoustic sensation related to a place, either expected or experienced. In this sense, soundscapes are then the new subjects of mapping and, consequently, of action plans.

Soundscape-based action plans

Particularly interesting, in this vision, is the design of solutions for quiet areas, locations where the acoustic quality is good and needs to be preserved[5] and whose presence has proved to be very important for the well-being of people living nearby.[13] In these areas, a wide range of requirements must be met at the same time, as different users could require and expect a different level of "quietness."[14] In addition to this, measures that reduce the overall noise level might be ineffective in places like urban parks, district green spots, and natural areas, either because they are quite invasive to the landscape or because they are not technically possible.

One attempt to solve the problem would be to induce artificial variations in the soundscape, so to create healthier sound environments. The idea would be to contrast the feeling of strain communicated by modern cities to visitors and inhabitants alike (i.e., the individual appears to be unwelcome, as he or she is bombarded by unwanted noise from different sources and directions). This can be obtained using sounds that people regard as generally positive, often taken from the acoustical history of the city, to reduce negative perceptions of noise. More ambitiously, it would be possible to design actions that, maybe at the cost of a slight increase in the noise level, directly affect the soundscape in an area, transforming the perception of its users. A way to do this would be superimposing an "artificial soundscape" to the existing polluted one, whose characteristics depend on the needs of end users.

According to many researchers, this is probably one of the reasons why fountains have traditionally been such popular architectural features in public spaces.[15] They superimpose a sound generally perceived as pleasant (i.e., running water) to an otherwise noise polluted soundscape. Designing a fountain, however, means choosing the number of water jets and their height, as both parameters have an effect on the frequency content of the water-based sound emission in the surrounding areas. How can this be done? Should the architect prefer to mask the traffic noise (e.g., with a lot of water, in the example of the fountain) or add to the existing soundscape some gentle suggestions (i.e., trying to maintain the overall sound pressure level as low as possible)? Answering to these questions requires a detailed knowledge of the effects on humans of artificially induced variations in a soundscape, and how these variations are perceived and used by the residents. This topic has been the subject of many studies in the recent years, so that a platform of knowledge is now well established[16] and ready for pilot action plans. Still, these questions remain unresolved: where pilot studies involving direct or indirect modifications of the soundscape have taken place, there was little or no formal evaluation of their success. In the ideal process of an action plan—preassessment of the soundscape, intervention, postassessment—the last step was usually missing, thus hindering the optimisation of further interventions in different contexts (based on the identification of which design aspects work and which do not).

Nevertheless, the interest in action plans based on the effect of positive soundscapes is growing across Europe.[16] In the UK, for instance, different research projects have been funded by the Research Council (e.g., NoiseFutures, ISRIE, Positive Soundscapes) to promote emerging psychoacoustic research, and local authorities are acting in this direction. An example is the mayor of London's Ambient Noise Strategy "Sounder City" in 2004, which is probably one of the first public policy documents to promote not just noise reduction but positive soundscape management.[17] The concept can also be found in the more recent "London Plan" (i.e., the Spatial Development Strategy for London[18]), which explicitly refers to protecting/enhancing relative tranquility more than merely reducing noise levels.

As many other cities in the UK and across the EU (e.g., the members of Eurocities[19]) followed London's example and included "tranquillity" among their goals, the UK Ministry for Environment (Department for Environment, Food and Rural Affairs, or DEFRA) commissioned in 2008 the study "Research on Practical Applications of the Soundscape Concept in Action Plans,"[20] to identify the important gaps in the soundscape knowledge base, delaying the development of effective actions. This study highlighted the lack of a close connection between soundscape research, design, and planning practice (i.e., the lack of a common language) but, more important, confirmed the absence of more soundscape-specific indicators and tools that could eventually be used for soundscape design.

The challenge then becomes finding suitable indicators to map and design the outdoor environment from a psychoacoustical point of view, describing the different areas of a city by their acoustical fingerprint (e.g., on the scale of perception but using quantitative indicators). New indicators will also open the way to a new generation of noise maps, which will describe soundscapes and have a keen eye to people's perception. Such maps would be crucial to define the goals to be achieved by noise control and action plans.

The need for a new metrology

In 2008, a working group of ISO/TC 43/SC1 was established to begin consideration of a standardised method for assessment of soundscape quality outdoors.[21] The difficulty of the task can be measured by the amount of time initially spent in defining "the entity under study" (i.e., what a soundscape is), as many studies in the literature use the same term in different contexts. Still, most of the studies agree on the centrality of human perception. To use the words of Brown et al., "a soundscape exists through human perception of the acoustic environment of a place."[21] And the relationship between soundscape and individual is complex, with many factors not easily quantified and not necessarily all of acoustic nature.

It is common experience that for the same energy level in decibels a mosquito is more annoying than a passing car. Annoyance studies have also shown that, again for the same energy level in decibels, the annoyance due to aircraft noise is greater than the one due to railway and road traffic.[22] Almost every environmental officer may also confirm that the number of complaints due to mopeds (in city centres[23]) or neighbourhood noise[24] is often greater than that coming from the opening of a new road. It might be an effect related to the implicit "danger" of the event (as the car is confined on the road, while a mosquito or an airplane are not), or related to the frequency content of the noise produced, or even more on the amount of "surprise" (i.e., to the time history of the noise events more than to their average energy), but duration and intermitting character of the noise also have an effect on the annoyance and thus on the definition of "unwanted sounds."

The fact that A-weighting is commonly applied to energy levels is a first step toward taking perception into account (A-weighting tries taking into account the mechanical weighting our ear applies to impinging acoustical energy before the message reaches the brain). Similarly, the fact that when repetitive noise is present, some member states (like Italy) apply a positive correction to measured levels for the purpose of assessing their environmental impact is a step in this direction. According to some studies, it is even possible to predict the percentage of "highly annoyed" in an area by weighting the energy levels (in L_{den}) with appropriate correlations.[22]

However, personal and cultural factors related to the specific situation, to the person's sensitivity, to the relationship with the source national differences (the same level of background noise may be unacceptable in Sweden and essential in Spain) all impact the way we perceive the acoustic environment around us. For this reason, studies on annoyance conducted in countries other than the Netherlands seem to show different correlations than the ones observed by Miedema et al.[25] Finally, factors other than acoustic ones have their own weight in determining the specifics of a site: the presence of greenery in a park[25] or the architectural aspect of a square,[27] for instance, both have an impact on our experience of the time we spend in such places and on experienced tranquillity.[28]

It seems then impossible to find a standardised and finite set of indicators (i.e., measurands) that takes all of these factors into account. Figure 18.1 shows how the complexity of an indicator needs to increase in order to be representative of all the population. When the target is the perception of the individual, the complexity tends to infinity as an infinite number of parameters needs to be taken into account.

Still, having a set of quantifiable indicators is paramount to design actions addressing not only the factors that might cause new pathologies but also the ones that affect well-being.[4,29] In this sense, if it is true that a scale of perception based on average noise levels in A-weighted decibels only represents a small percentage of the population (29%, according to

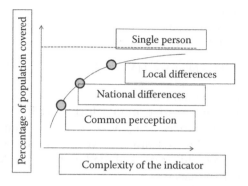

Figure 18.1 A tentative relationship between the complexity of an indicator (or a set of indicators) and the percentage of population covered in terms of predicting their perception.

the discussion at the Applied Soundscapes Symposium[30]), and considering that on the scale of city planning great numbers (i.e., statistics) are what lever decisions, the challenge might be simpler. It might be sufficient to find a minimal set of indicators that covers more population than the mere decibels.

THE QUEST FOR NEW INDICATORS

A key reference in the quest for new indicators is the work of Zwicker and Fastl.[31] These authors collected a significant amount of data, with the intent to describe the processing of sound by the human hearing system, both qualitatively (sound ↔ impression) and quantitatively (acoustical stimulus ↔ hearing sensations like *loudness, sharpness, fluctuation strength, roughness*). They presented a mathematical model for the different hearing sensations, thus producing a first set of indicators and the means to calculate their values. These indicators have been successfully used for more than 20 years indoors, and in particular for the acoustical design of car interiors or the evaluation of noise emissions from commercial products. It is not surprising that the indicators have been looked at as a first candidate for the assessment of outdoor soundscapes.[29,32,33] In this context, the studies of Genuit and Fiebig[32] highlighted a major role of the temporal variations in sound levels, which appeared to be more important than distinct noise levels or other psychoacoustical indicators in predicting perception. Among the psychoacoustical indicators, only roughness and sharpness showed a distinct role, for a fixed amount of loudness, in distinguishing outdoor spaces.[33] For these reasons, Fiebig et al.[34] proposed an indicator (EI, for evaluation index) considering the autocorrelation of the acoustical signal (i.e., the information about its temporal content), its loudness, and a combination of psychoacoustical parameters (roughness and sharpness, containing some insight into its frequency content):

$$EI_{traffic} \sim RA_{50}(FT) + N_5 + S + HMR + HMI \qquad (18.1)$$

where the different components of EI, selected by multiple regression analyses, are respectively relative approach $(RA_{50}(FT))$,[35] loudness (N_5),[36] sharpness (S), hearing model roughness (HMR),[37] and hearing model impulsiveness (HMI). This indicator was very successfully used to predict the perception along a highly trafficked road in the Netherlands but has not been tested on other sources yet.

Since 2003, De Coensel et al.[38] have been constructing a different indicator (slope), also related to the time history of the sound pressure level. In 2005, slope was successfully correlated to the perception assessed by questionnaires in selected urban locations in Italy (featuring traffic and railway noise) and, used simultaneously to a measurement of the sound pressure level, could be used to distinguish soundscapes with a 80% success rate.[39,40] This means, in practice, that by measuring slope at one of the test locations (as prescribed by Licitra and Memoli[39]), 4 times over 5 it was possible to predict the results of a questionnaire survey, later conducted in the same place, designed to assess noise perception.

The details for the calculation of the slope indicator have been described extensively elsewhere,[39,40] but it is worth remembering here that its numerical value represents the exponent S of a power function fitted, using a least squares method, to the power spectrum $G(f_0)$ of the L_{Aeq} time history in the interval [0.02, 0.2] Hz, so that

$$G(f_0) = A \cdot f_0^S, \text{ where } G(f_0) = \hat{L}_{Aeq}(f_0) \cdot \hat{L}^*_{Aeq}(f_0) \qquad (18.2)$$

where $\hat{L}_{Aeq}(f_0)$ is the Fourier transform of $L_{Aeq}(t)$ and $\hat{L}^*_{Aeq}(f_0)$ is its complex conjugate. Under the previous definitions, the obtained frequency of occurrence (f_0) is not related to the signal emitted every second, but to the time history of L_{Aeq} over a fixed amount of time. A peak in the spectrum evidences a repetitive event during the selected acquisition time. The exponent S, then, measures (a) how many peaks (events) are present in the signal and (b) how much those peaks emerge from the background.[40] Borrowing the term from other disciplines, slope hence measures the self-organised criticality of the acoustical energy level. Now, since the power spectrum of a stochastic process is the Fourier transform of the corresponding autocorrelation function (Wiener–Khinchin theorem[41]), the indicators in Equation (18.1) and Equation (18.2) must be correlated. Therefore, these two independent studies agree on the key role of the temporal behaviour in determining perception, but show a different rate of success.

The solution to this apparent conundrum was found testing slope in different countries than Italy: locations in the United Kingdom[42] and in Spain[43] were explored with a methodology similar to the one described by Licitra and Memoli.[39] By putting all the results in the same statistical database, it was found that slope gave a successful prediction only in 60% of the cases. Looking back at the model sketched in Figure 18.1, this reduced percentage of success is probably due to national differences. Therefore, the limit of "common perception" (targeted by an indicator not bound by national boundaries) must at least be 60%. The building of slope also shows that, subject to additional field tests, it is possible to create an acoustic-only set of indicators with a powerful predictive power.

Zwicker and Fastl[31] confirm what we said earlier: in addition to acoustic features of sounds in particular, aesthetic and cognitive effects may play an essential part in the crucial judgment of what is noise. For this reason, other indicators have been proposed, including a visual component.[26,27,29]

Testing the validity of different acoustical indicators

Once an indicator that seems to work is found, there is always the temptation to stick to it, without considering the more classical ones. A comparison with more classical indicators, however, is needed to understand the advantages and the limits of using the new one. In particular, slope has been compared in previous studies[40] with the number of emergences (defined as the number of peaks over background) and with quantities derived from statistical analysis, but a comparison of a more psychoacoustical nature was certainly needed.

The experiment was conducted in 2007 at Giardino Sonoro Limonaia dell'Imperialino, an urban garden in Florence facing a very high traffic road.[44] At the time, it was an open laboratory where a team of garden, sound, and light architects was showing its artistic creations and sound compositions, usually in the shape of Plexiglas and fabric diffusers that incorporated sound and light into a green environment. The comparison between slope and the most famous psychoacoustic indicators from the literature (loudness, fluctuation strength, roughness, sharpness, and unbiased annoyance[31]) was attempted when the soundscape was artificially improved (by adding positive sounds). The relative variation of the different indicators when the sound was switched on is reported in Figure 18.2. In this figure, "+" and "−" report a minimum variation (positive and negative, respectively), while "Ø" and "++" represent a null change and a well detectable increase, respectively. Colour has also been added to Figure 18.2 to reflect the change in perception: a positive change in slope means a change

	Ref	Path	Road	Wall	Pos2	Pos4	Pos6	Pos8	Pos14	Pos15
L	Ø	+	++	++	++	Ø	+	++	++	++
Sh	Ø	Ø	Ø	Ø	Ø	Ø	Ø	Ø	Ø	Ø
FS	Ø	Ø	+	Ø	−	+	Ø	+	Ø	Ø
R	−	Ø	−	−	Ø	Ø	Ø	−	Ø	−
UA	+	+	++	++	++	+	+	++	Ø	Ø
L_{Aeq}	Ø	+2	+6	+8	+6	+1	+4	+7	+7	+7
SL	−12	+70	+50	+80	+20	+80	+50	+10	+30	+30
	Fixed positions				Pathway				Road path	

Figure 18.2 Variation of the different indicators when the soundscape was changed: loudness (L); sharpness (Sh); fluctuation strength (FS); roughness (R); unbiased annoyance (UA), L_{Aeq} (change reported in dB) and slope (SL). "Ref" position corresponds to a free field microphone. The increasing number (Pos2, Pos4, Pos6, etc.) describes the increasing distance from the start of a soundwalk in a park. (Adapted from Memoli G., Licitra G., Cerchiai M., Nolli M., Palazzuoli D., Measuring Soundscape Improvement in Urban Quiet Areas, *Proceedings of Institute of Acoustics* 30, 2008.)

toward quietness (then reported in green), while an increase of the loudness should correspond to an increase in the expected annoyance (then reported in red). The colour yellow has been used to indicate those variations that were comparable with measurement uncertainties.

From Figure 18.2 it can be inferred that:

1. Loudness and L_{Aeq} move toward higher values, which should correspond to an increased annoyance (as registered by the unbiased annoyance).
2. Sharpness and fluctuation strength register only very slight changes and are therefore not good indicators for detecting soundscape changes.
3. Roughness slightly decreases in almost all the cases, but its relative change is much lower than the one of slope.

Roughness is then able to detect a change in the soundscape and possibly its direction but might not be sensitive enough for long-term applications (e.g., it might not be able to distinguish and optimize sound compositions, for instance). It has, however, the advantage that, instead of slope, it can be calculated in real time and it can then be used to adjust the soundscape to various changes of the source.

With all the uncertainties of the case, this study confirms the prominence of indicators based on temporal variations of the sound level in determining perception from the acoustical point of view. More research should therefore be conducted in this direction, necessarily complemented by studies that explore the level of engagement of the individual with the environment and the restorative power of particular sounds within the soundscape ("soundmarks"[11]).

MAPPING PERCEPTION ACROSS EUROPE: A FEW CASE STUDIES

There are beautiful cities around the world, full of art, history, and tradition, that do not sound as nice as they look. However, we are nowadays experiencing a revolutionary change of paradigm: decision makers across Europe agree in stating that it is no longer important how loud a city sounds, but how nicely it does sound.

In this context, it is extremely important to move from mapping noise exposure, complying with the prescriptions of the END, to a real and probably more useful map of annoyance.[45] An exposure noise map would indeed not take into consideration the individual sensitivity to noise or the significance given to particular sounds, all notions that are instead crucial while assessing the reaction of people to noise. Different experiences have been tried across Europe, using more or less calibrated set of indicators and some common trends can be extracted.

Generally, perception mapping has followed two routes: in one (technical) a model of perception (or a set of indicators) is built and then used to predict annoyance; while in the other (aesthetical) soundscapes are mapped as they are, sometimes with georeferenced recordings, leaving the judgment to the end listener.

One example of the technical type comes from Sheffield (UK), where Kang and colleagues proposed and tested soundscape mapping software on one of the most popular squares of the city, called the Peace Gardens, and built below street level.[46] The Peace Gardens' focal point is a large variable water fountain (Goodwin Fountain) with 89 individual jets, which is popular with children. There are plenty of seating areas (grass and benches), which offer to adults an opportunity to relax, and a busy bus road is located a few steps above the gardens. Water cascades guide visitors from the street to the fountain. Questionnaires, designed during a previous study conducted in 14 open public spaces across Europe,[29] were carried out in the Peace Gardens to gain true results of people's soundscape assessments. These are then used to calibrate the model and later test its success in predicting soundscape assessments. The predicted sound level evaluation and acoustic comfort rating of the Peace Gardens' soundscape derived from Kang and coworkers correlated well with the original participants' results ($R = 0.63$ and 0.79). More important, Yu et al.[46] modelled the predicted sound level and acoustic comfort evaluation of different age groups at numerous points throughout the site, thereby creating a perception map of the situation and identifying potentially different responses that may arise due to its various acoustical features.

In the city of Pisa (Italy), the existing noise map has been converted into an annoyance map,[45] first using the classical correlations between L_{den} and %HA (percentage of highly annoyed)[22] applied to modelled values and later using measurements of slope.[39] It was found that the first map could not represent the complaints reported by the population, whereas the second indicator was successful (i.e., within 80% accuracy).

One example of aesthetic mapping can be found in the work by Adams et al.,[47] who conducted soundwalks with 34 residents of Clerkenwell, London, UK, during a summer and a winter. The practice of soundwalks[11] has been adapted and utilised by researchers as a method for investigating the perception and understanding of soundscapes. In the Adams et al. study, participants chose a 10-minute route through their local environment, listening to the soundscape, before being interviewed about their experiences. This allowed the researchers to build a perceptual map of the investigated areas and to notice how individual sounds sources, commonly considered to be noisy and disliked (e.g., road traffic), were often still accepted due to other factors (e.g., feelings of being in control, prior experiences, tolerances, and adaptive behaviours to noise).

The practice of soundwalks, where the users become instrumental to the assessment of the soundscape, has also been used by other authors[44] and has recently had many estimators.[48] One possible evolution, on the tracks of other citizen science activities, allows the end user to build a map of sounds and noises of their favourite areas using a mobile phone.[30] This consideration opens the way to a different type of map, where different sound and noises are recorded and georeferenced for preservation[48] or artistic[50] purposes.

Is it possible then to compare the two approaches? Technical or aesthetical: Is one better than the other? Using a database of sounds in the lab and a set of listening interviews, Davies developed a two-parameter space based on the concepts of "calmness" and "vibrancy," which he used for mapping urban squares.[30] The same parameters' set was then used to map a square in Edinburgh during EURONOISE 2009, while simultaneously Memoli was acquiring slope in a grid of positions. It was found that slope could predict the value of "calmness" (as assessed by questionnaires after each soundwalk) but not the differences in "vibrancy."

Still, the exercise (funded by the NoiseFutures network[51]) allowed testing a distributed set of sensors developed by the National Physical Laboratory and based on MEMS (microelectromechanical systems) microphones. In a context where it is not possible to multiply L_{den} for a fixed coefficient, local measurements reacquire a principal role in weighting the environment around us. They will prove to be crucial when action plans are involved.

PERCEPTION-ORIENTED ACTION PLANS: CHALLENGES AND PERSPECTIVES

We have mentioned how the European Noise Directive[5] prescribes cities to protect their quiet areas and, in the meanwhile, leaves up to the respective authorities what a quiet area is or what it should be to be considered quiet. Mostly the definition so far has been driven by the result of physical parameters, for example, DEFRA suggested to consider an area as quiet if the energetic level in it is on average less than 40 dB.[52] This is for sure a non-noisy place. But how do the citizens appraise these quiet areas? What if it is not silence that they want there? What if they want their own soundscape?

Once it is accepted that a higher annoyance is not always proportional to higher noise levels, new actions needs to be designed, targeted to the sound perception by (groups of) individuals and to the opportunity of reinventing new soundscapes. The effectiveness of these actions seems gravely hindered by the unavoidable subjectivity of pleasant sounds (see section "Soundscape-Based Action Plans"). There is not a single ideal soundscape for all the citizens of a town and, obviously, for all the towns in the world.

From the point of view of the user, however, there are no cultural differences and the individual sensitivity does not count. The sounds that are good for a place are those that allow the people who experience or live there to do what they are supposed to do or they want to do. Conversely, daily activities like conversation, sleep, or recreation are often impeded in many noise-polluted public areas. More important, perhaps, there are sounds that are perceived (by the users) as characteristic of an area and therefore pleasant. Those sounds need to be identified, preserved, and enhanced. The users of the area therefore become the "new experts"[53] and at the same time the target of the action plan and the judges of its ultimate effectiveness.

With this in mind, a good strategy for remodelling the soundscape in a selected location starts with an analysis of the existing sounds of a site, either using the indicators in "The Quest for New Indicators" section or directly asking inhabitants/users to identify the wanted and unwanted sources. The next step is then to shape the acoustic environment in line with the intended uses of the space, integrating the pleasant sounds from local activities. This implies creating zones with different acoustic environments, from lively to quiet, from natural to more urban, which will only be possible if the first-stage analysis of the existing soundscape is performed.

The classical idea of action plans based on removing the noise and regaining the silence is superseded by the creation of acoustically restorative environments, promoting excellent sound quality. A solution that is almost never explored. The principles of traditional noise control and soundscape planning then appear different, or almost opposite, as shown in Table 18.1.

While unwanted sounds need to be removed as waste, a free path has to be opened for sounds that are pleasant and always preferable. The secret of

Table 18.1 Comparison of Noise Control and Soundscape Approaches

Noise Control Approach	Soundscape Approach
Sound as waste	Sound as a resource
Sound creates discomfort	Sound can be pleasant
Analysis of the human responses related only to the level of sound	Low levels are not the objective
Measures by integrating across all sound sources	Differentiation between the sound sources
Actions for reducing the sound level	Action for enhancing the "wanted sounds" and masking the "unwanted sounds"

soundscape managing is simply this: if every place has high quality and bad quality sounds, every place, in theory, has to be adjusted in order to have the good sounds accentuated and the unwanted ones very well masked. If some particular sounds fit well in some environment and they are pleasant for hearing, there is no point in throwing them away.

Case studies

The city of Stockholm in Sweden has pioneered soundscape design as an important tool for achieving a high-quality urban planning. In fact, three permanent sound installations have been recently mounted in central Stockholm, with the purpose of demonstrating how the soundscape in a noise polluted city square can be improved by means of dedicated artefacts of acoustic design.[54] One of these artefacts adds rhythmic sounds (diffused by a loudspeaker) to the background noise produced by the fountain in the central square of Mariatorget. On the same lines, the project "Play Stockholm" shows the musical character of different parts of the city (e.g., the change of the guard, the busy street of Drottninggatan, the trains in central station), thus fighting their noisy nature.

In the town of Antwerp (Belgium), an attempt of redesign of a partially abandoned area, constituted of gasworks, a park, and a transformer station, close to the railway and to a major road, had among its main tasks that of attracting people from the neighbourhood.[54] What the inhabitants wanted was a sort of park that might be contiguous to their dwellings. In planning the reconversion of this area, the correction of the soundscape was taken deeply in consideration, working both on masking of unpleasant sounds and on the distraction of visitor's attention. Proposed solutions include water games and greenery in a purposefully arranged manner (which are more classical), but also loudspeakers or artistic sonic installations (less typical).

In Florence (Italy), different experiments on sonic perception have been going on in the past 2 years inside a beautiful Italian-style garden in a

Figure 18.3 The sonic garden of the Bisarno Castle, Florence, Italy. (From www. soundexperiencedesign.com. With permission.)

private villa.[54] The private nature of the park identifies the end users with the owners of the villa, but the artificial soundscapes under test there also impact on the guests of a touristic residence structurally connected to the garden. The major source of unwanted sounds (i.e., noise) is here a busy highway slip road, located just outside one of the surrounding walls of the garden. The local researchers are testing an intelligent audio system permanently installed in the garden, which broadcasts sound compositions through artistic sound emitters (see Figure 18.3), thus masking the background noise in selected areas of the garden. The loudspeakers have been carefully designed to be aesthetically pleasant, to stimulate curiosity, while blending in the natural landscape. It this case, the software is capable of choosing in real time—with a reaction time of 200 msec—the proper soundtrack from a metacompositive database in order to match the same sound figures of noisy events occurring on the road. Therefore, the software does not reduce the sonic energy in the garden; on the contrary, it adds a small amount of it, reducing the annoyance at the cost of a slightly increased sound level. Surveys among the users of the garden, using standardised protocols, have confirmed a preliminary rise of pleasantness and acoustic comfort when the sound emissions are switched on.[55]

In Berlin (Germany), the Nauener Platz project[54] promotes a new acoustical understanding and interpretation of public places based on noise reduction and audio islands playing the sound selected by the people for that area.[56] This project underlines how the expectations of the users can be an active motor for the design of urban spaces and not just a passive subject of urban planning.

Figure 18.4 The re-soundscaped square: Piazza della Vittoria in Florence, Italy. (From Luzzi S., Soundscapes in the Participatory Design of Florentine Quiet Areas, Proceedings of International Congress of Sound and Vibrations 2010 (ICSV 17), Cairo. With permission.)

In Florence again (Italy), the Strategic Action Plan provided by the END directive has among its choices the redesign of some highly noise-polluted squares, which is the issue of many citizens' complaints. One of these, Piazza della Vittoria, has been the object of a requalification process based on the judgement of the people living and going there.[57] A preliminary campaign of surveys has been organised, specifically targeting the everyday end users of the square: students, residents, mothers and children groups (due to a school nearby), and owners of local activities. The results, collected by the environmental engineering company VIE EN.RO.SE. S.r.l., produced a square divided in four functional subareas, where each category of users can carry on the desired activities. Each sector becomes a "little square in a square" (see Figure 18.4) and it has its own character from both a functional and an acoustic point of view. In the "square of the students," for instance, circular benches installed around a hypothetical "fire sound" encourage socialization and meeting. In the "square of the children," kids and teenagers can spontaneously occupy the space, playing football into a makeshift football pitch excavated below the grass. The "square of sound" is dedicated to rest and it is wisely located next to a nursing home. The "square of games" is where the younger children can play, under the eye of their mothers, under the solicitations of sonic installations mimicking little musical instruments. This small example of requalification reinforces the need of action plans that can be referred mainly to people's perception of sound and to their different concept of quietness.

In Zadar (Croatia), in the middle of the Croatian coast, the outermost pier has been in recent years adapted for accepting modern multistorey cruising ships.[54] This enables the cruiser passengers to literally disembark downtown in the very heart of Zadar. A noise abatement action plan has been quite originally created to mask the noise through a change of sound-scape. A coastal promenade with an attractive staircase made of traditional stonework in the closest vicinity of the cruiser quay was implanted. A group of architects and musicians created in it an instrument made of an organlike series of pipes, which is able to transform the sound of the sea into a melody played by the waves. The Zadar cruiser pier, including the Sea Organ, was opened by a ceremony on 15 April 2005 and it is quickly becoming a worldwide tourist attraction.

Rotterdam and its harbour (Netherlands), regularly experience the dilemma of competing ambitions from industry, infrastructure, and recreation needs.[58] Its citizens are therefore—highly and constantly—exposed to various environmental factors, such as low air quality and noise, which have negative impacts. Quality-of-life-oriented actions are being devised by the local authorities, in which the protection against unwanted sounds is a key element as important as the prevention of negative health impacts. Hence technical and traditional noise abatement measures (e.g., noise barriers or noise absorbing road surfaces) are complemented with other positive approaches, based on the soundscape concept. In particular, three parks—nominated by the citizens as candidate quiet areas in Rotterdam— have been characterised (by careful analysis of sound records) in terms of various typical and nontypical sounds, during day, evening, and night; and during working days and weekends. Simultaneously, the citizens visiting the area will be asked to identify typical characteristics of the acoustic environment (e.g., keynote sounds, signals, and soundmarks) and what they like and dislike. The large amount of data will then be weighted in terms of external environmental factors (meteorology, visual impact) and psychological ones (expectations of the park visitors, their reasons for visiting the park, and their emotional state), for a complete characterisation of the area as a whole, in terms of perceived tranquillity and well-being (positive), or annoyance (negative).

CONCLUSION

The adoption of the European Noise Directive by member states has opened new frontiers to the research on cost-effective action plans for the preservation of urban quiet areas. In these sites, numerical indicators are needed to characterise the quality of quietness in order to judge the effects of the actions taken, to design effective measures, and to encounter the desires of the citizens.

The experiences carried out so far have shown that there is great potential in artificial soundscape design, this being a powerful and unconventional tool for restoring acoustically polluted areas, sometimes condemned to a total lack of human presence or to their insisting complaints.

In these cases, a predictive and quantitative indicator of pure acoustical nature can effectively help the designer of new environments with a technical and quantitative indication of a direction to follow, but it cannot be the only one. In a few words, a good choice of different indicators allows for designing the perception of quietness using artificial soundscapes is more suitable. Classical psychoacoustical indicators present some limits in detecting the change of soundscape. Even positive changes in the perception (assessed by questionnaires) can be classified as negative if only L_{Aeq} is taken as a measure of the change. It is worth highlighting the importance of this result for action plans based on the noise level only (like the ones that the END seems to promote, suggesting L_{den} as main indicator for annoyance). Reducing the energy alone might go in the wrong direction.

Further studies are needed to evaluate the weight of visual factors and to test the robustness of the different indicators. The assessment of perception should also be investigated for people experiencing the installations over longer periods of time (e.g., in a residential environment).

Whereas researchers struggle in this direction, end users appear to be one step forward. Within the limits of existing assessment techniques (featuring both indicators and questionnaires), in fact, mapping exercises and innovative action plans are being tested all across Europe, promising striking developments in the short term. Like the Italian Renaissance has seen the human intellect gain an active role in the development of cities, after centuries of maintaining a passive role, we are seeing to a Renaissance of acoustic perception in the planning of our cities. Exciting times lay ahead.

REFERENCES

1. EU Directorate-General for Regional Policy, Perception survey on quality of life in 75 European cities, http://ec.europa.eu/public_opinion/index_en.htm.
2. European Network of Noise and Health (ENNAH), www.ennah.eu.
3. European Commission, Green Paper on Future Noise Policy, COM(96) 540 (1996).
4. Berglund B., Lindvall T., Schwela, D.H. (editors), Guidelines for Community Noise, World Health Organisation, (1999).
5. Directive 2002/49/EC of the European Parliament and of the Council of 25 June 2002 relating to the assessment and management of environmental noise, http://eur-lex.europa.eu.

6. WG-AEN, Good practice guide on strategic noise mapping (version 2), August (2007).
7. IMAGINE project, Final report, http://www.imagine-project.org/.
8. Licitra G., Memoli G., Limits and advantages of Good Practice Guide to Noise Mapping, Proceedings of Acoustics '08, Paris (2008).
9. De Vos P., Strategies for noise action plans, Proceedings of Acoustics '08, Paris (2008).
10. Järviluoma H., Kytö M., Truax B., Uimonen H., Vikman N., Schafer M., Acoustic environments in change and five village soundscapes, Tampereen ammattikorkeakoulu—Tampere (2009).
11. Schafer M., *The soundscape: Our sonic environment and the tuning of the world* (2nd ed.), Destiny Books, Rochester, VT (1994).
12. Positive Soundscapes Project, www.positivesoundscapes.org.
13. Soundscape Support to Health program, www.soundscape.nu.
14. WG-AEN and WG-HSEA, Quiet areas in agglomerations, Interim Position Paper (2004).
15. Semidor C., Venot-Gbedji F., Fountains as a natural component of urban soundscape, Proceedings of Acoustics '08, Paris (2008).
16. Soundscape of European cities and landscapes, European Commission COST (2008–2012), http://soundscape-cost.org/.
17. Sounder City, The Mayor's Ambient Noise Strategy (2004), http://www.london.gov.uk/mayor/strategies/noise/docs/noise_strategy_all.pdf.
18. "London Plan": Spatial development strategy (2009), http://www.london.gov.uk/thelondonplan/.
19. Eurocities network, http://www.eurocities.eu/main.php.
20. Payne S.R., Davies W.J., Adams M.D., Research into the practical and policy applications of soundscape concepts and techniques in urban areas, final report for DEFRA contract (NANR 200), www.defra.gov.uk/environment/quality/noise/research/.
21. Brown A.L., Kang J., Gjestland T., Towards standardization in soundscape preference assessment, *Applied Acoustics* 72, 387–392 (2011).
22. Miedema H.M.E, Oudshoorn C.G., Annoyance from transportation noise: Relationships with exposure metrics DNL and DENL and their confidence intervals, *Environmental Health Perspectives* 109, 409–416 (2001).
23. City of Westminster, Issues and options for the Westminster noise strategy (2009), www.westminster.gov.uk.
24. Department for Environment, Food and Rural Affairs (DEFRA), Neighbourhood noise policies and practice for local authorities—A management guide, http://www.defra.gov.uk/environment/quality/noise/neighbourhood/.
25. Licitra G., Nolli M., The response of population to urban noise: The results of a socio-acoustic survey performed in a residential area of Pisa, Proceedings of Inter-noise 2006, Honolulu (2006).
26. Pheasant R.J., Fisher M.N., Watts G.R., Whitaker D.J., Horoshenkov K.J., The importance of auditory-visual interaction in the construction of "tranquil space," *Journal of Environmental Psychology* 30, 501–509 (2010).
27. Brambilla G., Di Gabriele M., Maffei L., Verardi P., Can urban squares be recognised by means of their soundscape? Proceedings of Acoustics'08, Paris (2008).
28. Wilson S., Westminster open spaces noise study (2008).

29. Kang J., *Urban sound environment*, Taylor & Francis, Abingdon, UK (2007).
30. Applied Soundscapes Symposium, Manchester, 17 September (2009).
31. Zwicker E., Fastl H., *Psychoacoustics: Facts and models*, 2nd ed., Springer-Verlag, Berlin (1999).
32. Genuit K., Fiebig A., Psychoacoustics and its benefit for the soundscape approach, *Acta Acustica United with Acustica* 92, 1–7 (2006).
33. Rychtarikova M., Vermeir G., Domeckac M., The application of the soundscape approach in the evaluation of the urban public spaces, Proceedings of Acoustics'08, Paris (2008).
34. Fiebig A., Guidati S., Goehrke A., Psychoacoustic evaluation of traffic noise, Proceedings of NAG-DAGA 2009, Rotterdam (2009).
35. Genuit K., Objective evaluation of acoustic quality based on a relative approach, Proceedings of INTERNOISE 1996, Liverpool (1996).
36. DIN 45631/A1—Calculation of loudness level and loudness from the sound spectrum—Zwicker method—Amendment 1: Calculation of the loudness of time variant sound (2008).
37. Sottek R., Genuit K., Models of signal processing in human hearing, *International Journal of Electronics and Communications* 59, 157–165 (2005).
38. De Coensel B., Botteldooren D., De Muer T., 1/f noise in rural and urban soundscapes, *Acta Acustica United with Acustica* 89, 287–295 (2003).
39. Licitra G., Memoli G., Noise indicators and hierarchical clustering in soundscapes, Proceedings of INTERNOISE 2005, Rio de Janeiro (2005).
40. Licitra G., Memoli G., Botteldooren D., De Coensel B., Traffic noise and perceived soundscapes: A case study, Proceedings of Forum Acusticum 2005, Budapest (2005).
41. Ricker D.W., *Echo signal processing*, Kluwer, Boston (2003).
42. Memoli G., Bloomfield A., Dixon M., Soundscape characterisation in selected areas of central London, Proceedings of Acoustics'08, Paris (2008).
43. Memoli G., Garcia I., Aspuru I., Soundscape characterization in the city of Bilbao, Proceedings of NAG-DAGA 2009, Rotterdam (2009).
44. Memoli G., Licitra G., Cerchiai M., Nolli M., Palazzuoli D., Measuring soundscape improvement in urban quiet areas, *Proceedings of Institute of Acoustics* 30 (2008).
45. Memoli G., Licitra G., Cerchiai M., Perspectives for the strategical mapping of soundscapes, Proceedings of Acoustics '08, Paris (2008).
46. Yu L., Kang J., Harrison R., Mapping soundscape evaluation in urban open spaces with artificial neural networks and ordinal logistic regression, Proceedings of the 19th International Congress of Acoustics, Madrid (2007).
47. Adams M., Cox T., Moore G., Croxford B., Refaee M., Sharples S., Sustainable soundscapes: Noise policy and the urban experience, *Urban Studies* 43 (13) 2385–2398 (2006).
48. Semidor C., Listening to a city with the soundwalk method, *Acta Acustica United with Acustica* 92, 959–964 (2006).
49. British Library, Archival Sound Recordings project, http://sounds.bl.uk/maps/Soundscapes.html.
50. London Wireless Soundscape Project, http://www.londonsoundscape.net/index.htm.
51. Noise Futures network, www.noisefutures.org.

52. Sysmonds Group, Definition, Identification and preservation of urban and rural quiet Areas, Final report under Service Contract ENV, C 1/SER/2002/0104R of the European Union, East Grinstead, UK (2003).
53. Schulte-Fortkamp B., Volz R., Jacob A., Using the soundscape approach to develop a public space in Berlin—Perception and evaluation, Proceedings of Acoustics '08, Paris (2008).
54. Designing Soundscape for Sustainable Urban Development, Stockholm (2010), http://www.soundscape-conference.eu/.
55. Licitra G., Cobianchi M. Brusci L., Artificial soundscape approach to noise pollution in urban areas, Proceedings of INTERNOISE 2010, Lisboa (2010).
56. Schulte-Fortkamp B., The daily rhythm of the soundscape "Nauener Platz" in Berlin, *Journal of the Acoustical Society of America* 127, 1774–1774 (2010).
57. Luzzi S., Soundscapes in the participatory design of Florentine quiet areas, Proceedings of International Congress of Sound and Vibrations 2010 (ICSV 17), Cairo (2010).
58. NoMEPorts project, http://nomeports.ecoports.com.

Index

Printed and bound by CPI Group (UK) Ltd, Croydon, CR0 4YY

01/11/2024

01782621-0012